SYMMETRY IN PHYSIC

VOLUME 1: PRINCIPLES AND SIMPLE APPLICATIONS

SYMMETRY IN PHYSICS

VOLUME 1:

PRINCIPLES AND SIMPLE APPLICATIONS

J. P. ELLIOTT and P. G. DAWBER

School of Mathematical and Physical Sciences
University of Sussex, Brighton

M
MACMILLAN

First published 1979
First published in paperback 1984
Reprinted 1985, 1986, 1987, 1989, 1990

Published by
THE MACMILLAN PRESS LTD
Houndmills, Basingstoke, Hampshire RG21 2XS
and London
Companies and representatives
throughout the world

Printed in Hong Kong

British Library Cataloguing in Publication Data
Elliott, J P
Symmetry in physics.
Vol. 1: Principles and simple applications
1. Symmetry (Physics)
I. Title II. Dawber, P G
530 QC793.3.S9
ISBN 0–333–26426–6 hc
ISBN 0–333–38270–6 pb
ISBN 0–333–11820–0 Set of 2 vols hc
ISBN 0–333–38272–2 Set of 2 vols pb

Contents of Volume 1

Contents of Volume 2

Preface to Volume 1

One cannot study any physical system for very long before finding regularities or symmetries which demand explanation and, even though the system may be complex, one expects that the regularities will have a simple explanation. This basic optimism, which pervades not only physics but science in general, is justified in the case of symmetries because there is a theory of symmetry which has application in almost all branches of physics and especially in quantum physics. The object of our book is to describe the theory of symmetry and to study its applications in a wide variety of physical systems.

The book has grown out of several lecture courses which we have given at the University of Sussex during the past ten years. One was a general introductory course on symmetry given to third-year undergraduates, one a postgraduate course on symmetry in solid-state physics and one a postgraduate course on symmetry in atomic, nuclear and elementary-particle physics. As a result, the book may be used by students in any of these categories. We regard chapters 1–5 (inclusive) as a minimum selection for any student wishing to study symmetry, although those students who have taken an undergraduate course on linear algebra will find that much of chapter 3 is familiar and may be read quite rapidly. The remaining chapters 6–12 in volume 1 cover a wide range of applications which is quite sufficient for an undergraduate course. One could even be selective within the first volume by omitting chapters 10–12 on nuclear and elementary particle physics or alternatively by omitting chapters 6 and 9 on the point groups. We would expect the second volume to be used for serious study at the postgraduate level and for occasional reference by the more inquisitive undergraduate.

The first chapter of volume 1 introduces the concept of symmetry with some very simple examples and lists the general consequences. We then leave physics aside for three chapters while preparing the mathematical tools to be used later. The most important of these are group theory and linear algebra which are described in chapters 2 and 3. The fourth chapter brings together these two ideas in a study of group representations and it is this aspect of group theory which is most used in the theory of symmetry. We return to physics in chapter 5 with a brief summary of the basic ideas of quantum mechanics and a general description of the effects of symmetry in quantum systems. The remainder of the book is concerned with applications to different physical systems and the study in greater detail of the relevant groups. We cover a broad range of applications from molecular vibrations to elementary particles and in each case we aim to introduce sufficient background description to enable the reader who has no prior knowledge of that particular physical system to appreciate the role being played by symmetry. Each application is reasonably self-contained and the more sophisticated systems are left until the later chapters. The vibration of molecules is the first phenomenon studied in detail, in chapter 6, and here we are able to illustrate the results of symmetry in classical mechanics before going over to the quantised theory. Chapters 7 and 8 describe the symmetry with respect to rotations with applications to the structure of atoms. It is here that we meet for the first time a continuous group, with an infinite number of elements, or symmetry operations, and the general properties of such groups are described. Chapter 9 describes in some detail the 'point groups', which contain only a finite number of rotations, and uses them to study the influence of a crystal field on atomic states. In chapters 10, 11 and 12 we move on to the more abstract symmetries encountered in nuclear and elementary particle physics but make use of the same general theory that was used for the more concrete applications in earlier chapters. We introduce the groups of unitary transformations in two, three, four and six dimensions and use them to describe the observed symmetry between neutrons and protons and the regularities amongst some of the recently discovered short-lived elementary particles. The ideas of 'strangeness' and 'quarks' are explained.

Volume 2 begins with a further application of the use of 'point groups'—to the motion of electrons in a molecule—and then, in chapter 14, moves away from symmetries with a fixed point to study discrete translations and their applications to crystal structure. The theory of relativity is of profound importance in the philosophy of physics and, when speeds become comparable with that of light, it has practical importance. For all the systems discussed in volume 1 we are able to ignore relativity because the speeds of the particles involved are sufficiently small. Chapter 15 describes the symmetry in four-dimensional space–time which is the origin of relativity theory and discusses its consequences, especially in relation to the classification of elementary particles. The concepts of momentum, energy, mass and spin are interpreted in terms of symmetry using the Lorentz and Poincaré groups and a natural place is found in the theory for particles, like the photon, with zero mass. Chapter 16 is concerned with fields, in contrast to the earlier chapters which dealt with particles or systems of particles. We first describe classical fields, such as the electromagnetic field, using four-dimensional space–time. This is followed by a brief account of the theory of relativistic quantum fields which provides a framework for the creation and annihilation of particles and the existence of antiparticles. Chapters 17 and 18 contain details of two general groups, the 'symmetric' group of all permutations of n objects and the 'unitary' group in N dimensions, and an intimate relation between these two groups is discussed. Particular cases of these two groups have been met earlier. Chapter 19 describes some unexpected symmetries in two familiar potentials, the Coulomb and the harmonic oscillator potentials, and a number of small, unconnected, but interesting topics are collected into the last chapter.

The text includes worked examples and a selection of problems with solutions. A bibliography of references for further reading is given at the end of each chapter for those who wish either to follow the physical applications into more detail or to study some of the mathematical questions to a greater depth.

To aid the reader we have followed the standard convention of using italic type for algebraic symbols such as x, y and z, whereas operators are distinguished by the use of roman type. An operator or matrix will be written T but its matrix elements T_{ij}, which are numbers, will be in italic type. In addition, bold face type will be used for vectors and in chapters 15 and 16 of volume 2 we meet four-vectors $\hat{e}$ which are printed with a circumflex.

Brighton, Sussex, 1979

J. P. E.
P. G. D.

1

Introduction

1.1 The place of symmetry in physics

According to the *Concise Oxford Dictionary*, symmetry is defined as '(Beauty resulting from) right proportion between the parts of the body or any whole, balance, congruity, harmony, keeping'. Although there is much complex detail in physics there is also much beauty and simplicity and it is the symmetry in physical laws and physical systems which is largely responsible for this. Consequently, symmetry plays an important role in physics and one which is increasing in importance with modern developments. It is the purpose of this book to explain in general terms why the existence of symmetry leads to a variety of physical simplicities in both classical and quantum mechanics. To illustrate the general results we shall refer to simple properties of molecules, crystals, atoms, nuclei and elementary particles. Although these physical systems are so obviously different from one another, nevertheless the same theory of symmetry may be applied to them all. The study of symmetry, therefore, helps to unify physics by emphasising the similarity between different fields.

It is true that symmetry plays a part in both classical and quantum physics, but it is in the latter that most interest lies. There are several reasons for this. The first is that there is a much greater scope for symmetry to exist in the microscopic domain since, for example, one electron is identical with any other electron and one atom of carbon (say) is identical with any other. The second reason is that at the microscopic level one must use quantum mechanics which is inherently more complicated than classical mechanics and so provides more scope for simplification through symmetry arguments. For example, a particle is described by a wave *function* rather than a single position. One further reason is that the structure of atomic and subatomic systems is now one of the exciting frontiers of science and the ideas of symmetry are helping to create order out of apparent chaos.

Throughout physics one uses mathematics as the tool with which to investigate the consequences of some assumed theory or model. For example, in the motion of a particle of mass M in one dimension x under some force $f(x)$ the physical law (Newtonian theory) tells us that $f(x) = M(\mathrm{d}^2x/\mathrm{d}t^2)$. To find the position $x(t)$, as a function of time, given $f(x)$, we must solve this differential equation, putting in the initial values of x and $\mathrm{d}x/\mathrm{d}t$. Thus, in Newtonian mechanics, the differential and integral calculus is the appropriate tool. In studying the symmetry of physical systems we are asking about their behaviour under transformations. For example, if a particle moves in one dimension under the influence of a potential $V(x)$, that potential may have reflection symmetry in the origin, i.e. $V(-x) = V(x)$. In this case the potential is said to be invariant (unchanged) under the transformation which replaces x by $-x$. In another example, that of a particle moving in three dimensions, the potential may have spherical symmetry, which means that, in spherical polar coordinates, the potential is independent of angle and may be written $V(r)$. In this case the potential is invariant under any of the transformations which rotate through any angle about any axis through the origin—an infinite number of transformations!

To investigate the physical consequences of the symmetry of a system we must, therefore, learn something about transformations and in particular about the set (collection) of transformations which leave some function, like the potential, invariant. The theory of such sets of transformations is called 'group theory' by mathematicians and this is the appropriate tool for the physicist to use in studying symmetry.

It is fascinating to draw an analogy between the use of calculus in classical mechanics and the use of group theory in quantum mechanics. Historically the discovery of Newton's laws and the invention of the calculus occurred at about the same time in the seventeenth century. Although the ideas of group theory were introduced into mathematics as early as 1810 it was not until the 1920s that the theory of group representations, which is crucial to the study of symmetry, was developed. This was the very time when physicists were formulating the quantum theory. In fact the significance of symmetry in quantum mechanics was realised very early in the classic works of E. Wigner, in 1931, H. Weyl, in 1928, and Van-der-Waerden, in 1932.

There have always been those who have argued that it is unnecessary to use group theory in quantum mechanics. In a sense this is true, since group theory itself is built from elementary algebraic steps. However, the investment of effort in learning to use the sophisticated tool which is group theory is soon rewarded by handsome dividends of simplification and unification in the study of complex quantum mechanical systems. After all, one could argue that the calculus is not necessary in classical mechanics. For example, geometrical arguments could be used to show that the inverse square law of gravitational attraction leads to elliptical orbits. In fact, Newton originally used such a method but in modern times we understand this result through the solution of a differential equation. Looking ahead, it is exciting to speculate that further major advances in mathematics and physics may go hand in hand in the future.

1.2 Examples of the consequences of symmetry

To whet the appetite we now list a number of physical systems which possess symmetry and we point out some features of their behaviour which are direct consequences of the symmetry. Simpler examples are given first. Although in some cases we are able to relate the behaviour to the symmetry without developing new methods this is, of course, not always possible. It is the purpose of this book to describe generally the consequences of symmetry and it will not be until much later in the book that we shall be in a position to understand and to predict the behaviour of systems with intricate symmetries.

1.2.1 One particle in one dimension (classical)

A particle of mass M, moving in one dimension under the influence of a potential $V(x)$, will have its motion governed by the equation

$$M\ddot{x} = -\mathrm{d}V/\mathrm{d}x \tag{1.1}$$

Suppose now that $V(x)$ is a constant, independent of x; in other words that it is invariant under translation. Then clearly $M\ddot{x} = 0$ and, integrating, gives $M\dot{x} = C$, showing the conservation (constancy) of linear momentum $M\dot{x}$.

1.2.2 One particle in two dimensions (classical)

In two dimensions the motion of the particle is governed by the two equations

$$M\ddot{x} = -\partial V/\partial x \quad \text{and} \quad M\ddot{y} = -\partial V/\partial y \tag{1.2}$$

Suppose now that $V(x, y)$ is invariant with respect to rotation about the origin; in other words that $V(x, y)$ is independent of the polar angle θ if expressed in terms of the polar coordinates r, θ rather than the cartesian x and y. In this case $\partial V/\partial\theta = 0$. However,

$$\frac{\partial V}{\partial\theta} = \frac{\partial x}{\partial\theta}\frac{\partial V}{\partial x} + \frac{\partial y}{\partial\theta}\frac{\partial V}{\partial y} = -y\frac{\partial V}{\partial x} + x\frac{\partial V}{\partial y}$$

and using equation (1.2)

$$\frac{\partial V}{\partial \theta} = M(y\ddot{x} - x\ddot{y}) = M\frac{\mathrm{d}}{\mathrm{d}t}(y\dot{x} - x\dot{y})$$

so that the invariance $\partial V/\partial \theta = 0$ implies the constancy of the quantity $M(y\dot{x} - x\dot{y})$ which is the moment of momentum (or angular momentum) about an axis through the origin and perpendicular to the plane.

If the particle were free to move in three dimensions in a potential which was invariant with respect to rotations about *any* axis then this argument shows that any component of the angular momentum is constant. In other words, for a spherically symmetric potential, both the magnitude and the direction of the angular momentum are conserved.

1.2.3 Two particles connected by springs (classical)

Two particles of equal mass M are connected to each other and to fixed supports by equal collinear springs with spring constant λ. Let the natural length of the springs be a and the supports a distance $3a$ apart. Denote the displacements of the two particles from their equilibrium positions by x_1 and x_2. Although the general displacement, illustrated in figure 1.1, has no

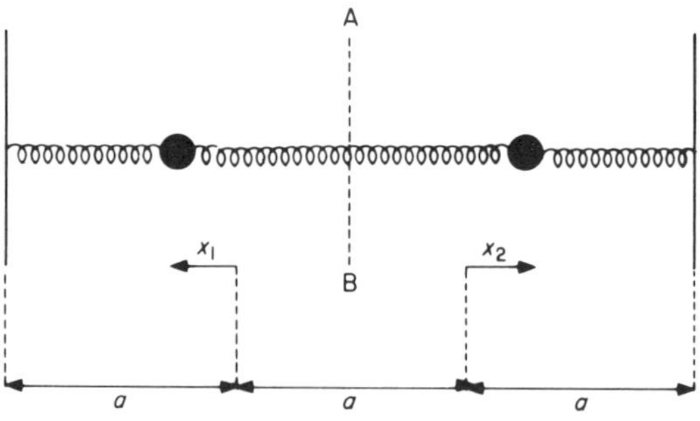

Figure 1.1

symmetry it is intuitively clear that, in some sense, the system has reflection symmetry about the centre. In fact, both the kinetic and potential energies

$$T = \tfrac{1}{2}M(\dot{x}_1^2 + \dot{x}_2^2) \quad \text{and} \quad V = \tfrac{1}{2}\lambda\{x_1^2 + x_2^2 + (x_1 + x_2)^2\}$$

are invariant with respect to the interchange of x_1 and x_2, which is the transformation of coordinates x_1 and x_2 produced by a reflection in the line AB.

The consequences of symmetry are not very dramatic in this case, but the generalisation to the vibration of atoms about their equilibrium positions in a molecule is of considerable importance. It is therefore worth while to solve this simple problem completely. The coupled equations of motion are

$$M\ddot{x}_1 = -\lambda x_1 - \lambda(x_1 + x_2)$$

$$M\ddot{x}_2 = -\lambda x_2 - \lambda(x_1 + x_2)$$

which suggest the definition of new coordinates $Q_1 = x_1 + x_2$ and $Q_2 = x_1 - x_2$. Then by adding and subtracting the two coupled equations we find

$$M\ddot{Q}_1 = -3\lambda Q_1 \quad \text{and} \quad M\ddot{Q}_2 = -\lambda Q_2$$

which shows that the new coordinates Q_1 and Q_2 perform simple harmonic vibrations $Q_i \propto \cos \omega_i t$ with frequencies $\omega_1 = (3\lambda/M)^{\frac{1}{2}}$ and $\omega_2 = (\lambda/M)^{\frac{1}{2}}$, respectively. The original coordinates x_1 and x_2 will be superpositions of the two pure oscillations since $x_1 = \frac{1}{2}(Q_1 + Q_2)$ and $x_2 = \frac{1}{2}(Q_1 - Q_2)$. The new coordinates Q_1, Q_2 and the two frequencies are usually referred to as 'normal' coordinates and 'normal' frequencies.

Notice that the new coordinates Q_1 and Q_2 are even and odd, respectively, under the symmetry reflection. If we denote this reflection by σ, then $\sigma Q_1 = Q_1$ and $\sigma Q_2 = -Q_2$. In other words the new coordinates transform more simply under the reflection than do the original coordinates x_1 and x_2, and furthermore the equations of motion are uncoupled when written in terms of the new coordinates.

In some complicated problems it is not always possible to decouple all equations of motion from one another by such a simple step, but nevertheless by choosing new coordinates to transform 'simply' under the symmetry operations one is able to achieve considerable reduction in the extent of the coupling. Again, although it is clear in this example that Q_1 and Q_2 transform in the simplest possible way, in more complicated symmetries we shall have to define and examine this concept of 'simplicity'. This will be done in chapter 4. For example in the NH_3 molecule which, with four atoms, has altogether twelve degrees of freedom, the use of symmetry considerations can reduce the number of coupled equations of motion for small vibration from twelve to two. A general account of the use of symmetry arguments in the theory of normal modes will be given in chapter 6 and applied to molecular vibrations.

1.2.4 One particle in three dimensions using quantum mechanics—spherical symmetry and degeneracies

In quantum mechanics the energy levels for a particle moving in a spherical potential exhibit degeneracies or, in other words, there is more than one independent wave function with the same energy. The hydrogen atom with the electron moving in the electrostatic field of the proton is such a system. In simple terms the reason for this degeneracy is that in the absence of any preferred direction in space the energy can clearly not depend on the direction of the angular momentum vector. The degeneracy expresses this freedom in the framework of quantum mechanics. If the spherical symmetry is broken, for example by turning on a magnetic field, then this degeneracy is destroyed and a 'multiplet' of several close-lying energy levels is produced, as in the Zeeman splitting.

The spherical symmetry may also be broken by immersing the atom in an external field which, although not spherically symmetric, has nevertheless some symmetry under specific rotations. In this case the degeneracies are only partially destroyed. Such a phenomenon is observed for an atom in a crystal field and will be analysed in detail in chapter 9.

The degeneracies resulting from spherical symmetry have a simple physical explanation in terms of the orientation of the angular momentum vector, but the occurrence of degeneracies is a characteristic result of any symmetry. This is quickly demonstrated by considering the Schrodinger equation $\mathrm{H}\psi = E\psi$, where H is the Hamiltonian operator, E the energy and ψ the wave function. Now introduce a coordinate transformation which transforms ψ into ψ' and H into H$'$ so that, transforming both sides of the Schrodinger equation, we have $\mathrm{H}'\psi' = E\psi'$. Here the energy, being a number, is unchanged by the transformation. Suppose now that H is invariant under the transformation so that $\mathrm{H}' = \mathrm{H}$. Then $\mathrm{H}\psi' = E\psi'$, which tells us that the wave function ψ' is an eigenfunction of the original Hamiltonian with the same energy as ψ. Hence, unless it happens that ψ' is the same function as ψ, apart from a multiplicative constant, there will be at least a two-fold degeneracy at energy E. In chapter 5 we shall see how the structure of the degenerate multiplets follows from the symmetry alone and is independent of the further details of the Hamiltonian.

1.2.5 One particle in one dimension using quantum mechanics—parity and selection rules

Consider a particle moving in one dimension subject to an even potential $V(x) = V(-x)$ in quantum mechanics. This very simple symmetry leads to the result that the eigenfunctions are either even or odd functions of x. The proof of this result is elementary, for if $\mathrm{H}\psi(x) = E\psi(x)$ we have also $\mathrm{H}\psi(-x) = E\psi(-x)$, as in the example in subsection 1.2.4. Hence, $\psi(x)$ and $\psi(-x)$ both have the same energy so that, assuming the absence of degeneracy, these two functions are physically identical. Thus $\psi(-x) = c\psi(x)$, where c is a constant. But this implies that $\psi(x) = c\psi(-x)$ also so that $\psi(x) = c^2\psi(x)$ which gives $c^2 = 1$ and $c = \pm 1$. Hence, $\psi(x)$ is either even or odd, but cannot be a mixture. This even or odd property of the function is referred to as parity. If there is degeneracy we may not conclude that $\psi(-x) = c\psi(x)$. In this case, however, the linear combinations $\{\psi(x) \pm \psi(-x)\}/\sqrt{2}$, which are even and odd, respectively, may be taken as the pair of degenerate eigenfunctions. (In one dimension it is possible to show that there can be no degeneracy but the above argument may be immediately extended to three dimensions where degeneracy may occur.)

One further consequence of symmetry may be illustrated in this example, namely the concept of a 'selection rule' in quantum mechanics. The transition probability for decay from some initial state ψ_i to a final state ψ_f is proportional to the square of the integral

$$I = \int_{-\infty}^{\infty} \psi_f^*(x) g(x) \psi_i(x) \mathrm{d}x$$

where the form of $g(x)$ depends on the particular decay process. Now the integral clearly vanishes if the integrand is odd. Hence if $g(x)$ is an even function of x then the integral is non-zero only if ψ_i and ψ_f are both even or both odd, i.e. the initial and final states have the same 'parity'. On the other hand, if $g(x)$ is odd then the initial and final states must have opposite parity. When I is zero, the probability of the transition taking place is zero—hence the phrase 'selection rule'.

1.2.6 The search for symmetry—elementary particle physics

In describing the theory of symmetry in physics it is natural to explore the consequences of symmetry, but one must always remember that in practice it is the consequences which one observes in the laboratory and one must then ask what is the symmetry which would give rise to them. As with the development of most physical theories it is a two-way process. The experiments suggest possible symmetries and one then explores in detail the consequences of such symmetries, makes predictions and carries out further experiments to test them.

A splendid example of this is provided in elementary particle physics today. The neutron and proton have almost the same mass and so do the π^+, π^- and π^0 mesons. In fact, all the elementary particles occur in such multiplets. Because of the relativistic connection between mass and energy, the existence of mass multiplets suggests a symmetry in the fundamental Hamiltonian. Since the different members of the multiplet have different charges, the symmetry operators will here involve transformations in the 'charge coordinates' of the particles. The development of this idea leads to the concept of 'isospin' which is of great importance both in the structure of nuclei and in the behaviour of elementary particles and will be described in chapter 10.

In recent years more and more elementary particles have been produced in succeeding generations of particle accelerators as their energy has been increased. It seems that several of the charge multiplets, referred to above, have similar masses and the existence of larger multiplets has been postulated. This, in turn, suggests the existence of a symmetry greater than that described by isospin. It is referred to as SU_3 symmetry, a title which will be explained in chapter 11. Unfortunately there is considerable splitting of the SU_3-multiplets, suggesting the existence of appreciable symmetry-breaking terms in the Hamiltonian, just as the Zeeman splitting in an atom suggests the breaking of the spherical symmetry by a magnetic field.

1.3 Summary

The examples given above were taken from a wide variety of physical systems but it is the same general theory of symmetry in each case. It is perhaps worth summarising the most important general consequences of symmetry in a system governed by quantum mechanics. They are: (1) conservation laws; (2) degeneracies in energy; (3) the energy eigenfunctions transform 'simply' under the symmetry operations and may be assigned a symmetry 'label' independent of the details of the Hamiltonian; (4) selection rules; (5) relations between matrix elements of observables. However, it must always be remembered that if the symmetry is too simple some of these consequences will not occur.

All of these consequences have featured at one point or another in the foregoing examples except (5). We cannot hope to be very convincing about consequence (5) at this early stage, but it is seen in the Zeeman splitting of atomic levels. The magnitude of the energy shift of different states of any given multiplet is proportional to their angular momentum about the field direction and this is true irrespective of the detailed structure of the wave functions. Thus we have an example in which the matrix elements are simply related, although the absolute value of any one of them would generally be difficult to calculate.

Another example of consequence (5) would be the mass formulae for nuclei and elementary particles. Here, the small mass differences between members of an isospin multiplet are simply related. More generally, the mass differences between members of an SU_3-multiplet are related by a more complicated formula because of the more complicated symmetry. However, the principle is the same as that governing the more familiar Zeeman splitting.

All of the examples mentioned in this introduction will be described in detail later. First, however, we must make ourselves familiar with the appropriate mathematical tools and this will be done in the next three chapters.

2

Groups and Group Properties

The notion of a group is introduced in this chapter and the most important general properties of group elements are deduced. Illustrations are given from a few very simple groups. The detailed properties of specific groups of importance in physical applications will be left until later chapters.

2.1 Definition of a group

Although we shall soon come to some very concrete examples it is worth beginning with the abstract definition of a group which is incredibly simple and yet leads to so many important consequences.

A set $\mathscr{G}$ of elements G_1, G_2, G_3, . . . is said to form a group if a law of 'multiplication' of the elements is defined which satisfies certain conditions. The result of multiplying two elements G_a and G_b is, naturally, called the 'product' and is written G_aG_b. The conditions to be satisfied are the following:

(1) The product G_aG_b of any two elements is itself an element in the set, i.e.

$$G_aG_b = G_d \text{ for some } G_d \text{ in } \mathscr{G} \qquad (2.1)$$

(2) In multiplying three elements G_a, G_b and G_c together, it does not matter which product is made first. In other words

$$G_a(G_bG_c) = (G_aG_b)G_c \qquad (2.2)$$

where the product inside the brackets is carried out first. This implies that the use of such brackets is unnecessary and we may simply write $G_aG_bG_c$ for the triple product.

(3) One element of the set, usually denoted by E and called the identity, must have the properties

$$EG_a = G_a \quad \text{and} \quad G_aE = G_a \tag{2.3}$$

for any G_a in the set $\mathscr{G}$.

(4) To each element G_a in the set there corresponds another element in the set, denoted by G_a^{-1} and called the 'inverse' of G_a, which has the properties

$$G_aG_a^{-1} = E \quad \text{and} \quad G_a^{-1}G_a = E \tag{2.4}$$

In general it is *not* permissible to change the order of multiplication of group elements; in other words G_aG_b is not in general the same element as G_bG_a. A group for which $G_aG_b = G_bG_a$ for all elements G_a and G_b is very much the exception in physical problems and is called an 'Abelian' group. Its elements are said to 'commute'.

Notice that the inverse of a product G_aG_b is given by

$$(G_aG_b)^{-1} = G_b^{-1}G_a^{-1} \tag{2.5}$$

This is deduced directly from the defining equation

$$(G_aG_b)^{-1}G_aG_b = E$$

by multiplying on the right first by G_b^{-1} and then by G_a^{-1}. A convenient method of recording the multiplication $G_aG_b = G_d$ of elements of a particular group $\mathscr{G}$ is to build the multiplication table in which the rows and columns are labelled by the group elements and the result G_d of the multiplication G_aG_b is entered at the intersection of the row G_a and the column G_b. The definitions of a group imply that every group element must appear once and once only in each row and in each column.

We have deliberately not specified the number of elements in the group. In fact the number may be finite, in which case it is denoted by g and called the 'order' of the group, or it may be infinite. The group is correspondingly called a finite or an infinite group. We shall consider both in this book since both are of importance in physics. For many of the general properties of groups described here and in chapter 4 it will not be necessary to specify whether the group is finite or infinite. In some cases, however, the proofs must be formulated separately for the two possibilities. The finite groups are relatively easy to discuss and for the infinite groups we restrict our attention to 'continuous' groups. This is sufficiently general for the physical problems under discussion and means that the group elements, instead of being distinguished by a discrete label a or b, are labelled by a set of continuous parameters. Thus, by a small change in a parameter, we may pass continuously from one group element to another.

2.2 Examples of groups

The simplest examples of group elements are ordinary numbers with ordinary multiplication and our first two examples are of this kind.

(1) The two numbers 1 and -1 form a group. The identity is clearly 1. The inverse of the identity is always the identity and the inverse of -1 is itself. These properties are contained in the group multiplication table 2.1.

Table 2.1

$G_a \backslash G_b$	1	-1
1	1	-1
-1	-1	1

(2) A slightly larger group of the same kind is the set of numbers 1, -1, i and $-i$ which have the multiplication table 2.2.

Table 2.2

$G_a \backslash G_b$	1	-1	i	$-i$
1	1	-1	i	$-i$
-1	-1	1	$-i$	i
i	i	$-i$	-1	1
$-i$	$-i$	i	1	-1

Both of these groups have the 'cyclic' property which means that all the group elements may be formed by taking powers of a single element. In example (2) the four elements are given by i^k, with $k = 0, 1, 2$ and 3. Further powers will clearly reproduce the same cycle. Since 'ordinary' multiplication is used, these two groups must be Abelian, but a little thought will show that every cyclic group must be Abelian.

The behaviour of physical systems under rotations is of considerable importance in the study of symmetry and various sets of rotations form groups. We now move on to some examples of these. The law of multiplication in every case is that if the rotation R_1 carries a system from position A to position B and if R_2 carries it from B to C, then the product R_2R_1 carries it from A to C. Here we shall find our first example of a non-Abelian group, since although $R_2R_1 = R_1R_2$ when the rotations are about the same axis, in general $R_2R_1 \neq R_1R_2$. Rotations do not in general commute.

To illustrate the non-commutative property of rotations let R_1 be a rotation through an angle $\pi/2$ about the z-axis and R_2 a rotation through π about the

y-axis. (A rotation through a positive angle about a directed axis is defined to be in the sense of a right-handed screw, i.e. clockwise when looking out along the axis.) By following carefully the motion of each of the three axes under the successive operations it is seen that R_2R_1 is the same as a rotation through an angle π about the axis (1, 1, 0), while R_1R_2 is a rotation through π about the axis $(-1, 1, 0)$. But let us begin with simpler examples!

(3) Let E be the identity (a rotation through an angle zero) and let R denote a rotation through an angle π about the z-axis. Then the set E, R forms a group with multiplication table 2.3. (This group is conventionally called C_2).

Table 2.3

$G_a \backslash G_b$	E	R
E	E	R
R	R	E

(4) Let I be the inversion, defined to change the direction of any vector. Clearly $I^2 = E$, the identity, so that the set E, I forms a group, called S_2.

(5) The set E, R_1, R_2 forms a group if R_1 and R_2 denote, respectively, rotations through angles $2\pi/3$ and $4\pi/3$ about the z-axis. The multiplication table 2.4 is readily deduced. (This group is called C_3.)

Table 2.4

$G_a \backslash G_b$	E	R_1	R_2
E	E	R_1	R_2
R_1	R_1	R_2	E
R_2	R_2	E	R_1

(6) The set E, R_1, R_2, R_3, R_4, R_5 forms a group where R_1 and R_2 are rotations through angles $2\pi/3$ and $4\pi/3$ about the z-axis and the remaining elements R_3, R_4 and R_5 are rotations through angles π about the three axes in the xy-plane shown in figure 2.1. (This group is called D_3.) Geometrically this is the group of rotations of an equilateral triangle which carry it into positions indistinguishable from the initial position. One assumes, in this use of the word 'indistinguishable', that there are no marks or imperfections on the triangle to destroy its symmetry! Such motions of a geometrical figure are called 'proper covering operations', the word 'improper' being used if reflections are also allowed.

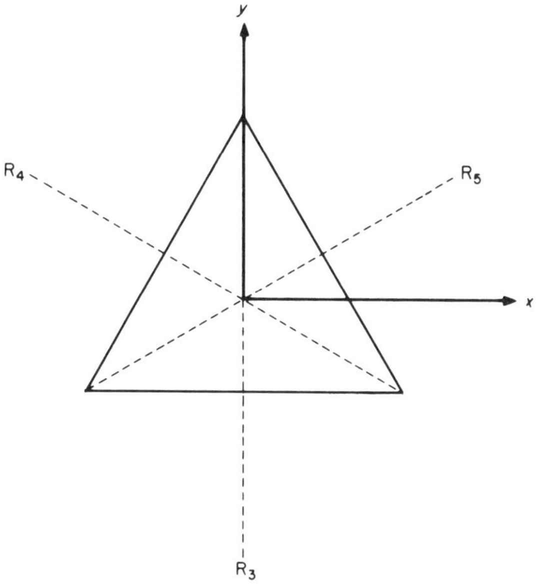

Figure 2.1

To visualise the rotations it is nevertheless helpful to imagine the vertices carrying labels 1, 2 and 3 and to follow the movement of these labels. By this method one may build up the multiplication table 2.5. For example

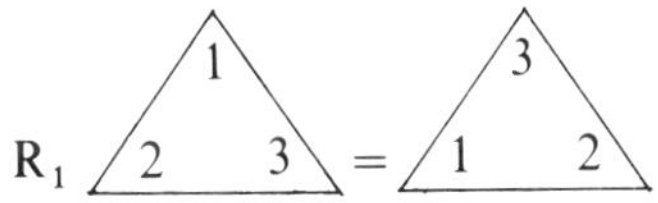

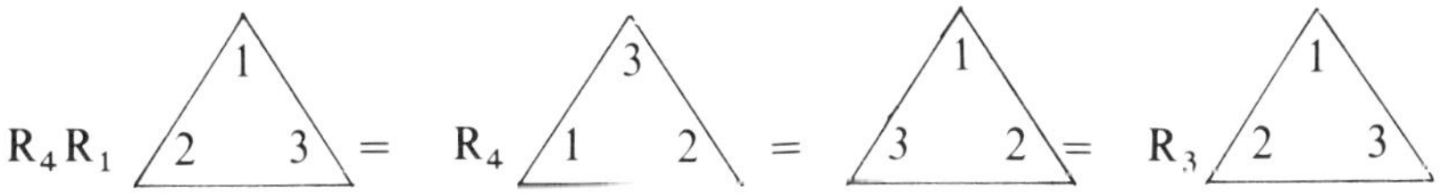

showing that $R_4R_1 = R_3$.

In examples such as this, one may verify that the set forms a group by using the multiplication table. The fact that every entry in the table is one of the set shows that condition (1) of section 2.1 is satisfied. The fact that the identity occurs once, and once only, in every row and column verifies condition (4), while condition (3) is obviously met. The condition (2) is satisfied for all rotations. Alternatively, the very definition of the set to be *all* proper covering operations of the equilateral triangle ensures that they obey the definitions of a

Table 2.5

G_a \ G_b	E	R_1	R_2	R_3	R_4	R_5
E	E	R_1	R_2	R_3	R_4	R_5
R_1	R_1	R_2	E	R_4	R_5	R_3
R_2	R_2	E	R_1	R_5	R_3	R_4
R_3	R_3	R_5	R_4	E	R_2	R_1
R_4	R_4	R_3	R_5	R_1	E	R_2
R_5	R_5	R_4	R_3	R_2	R_1	E

group. The product of two covering operations is clearly another covering operation, as is the inverse.

(7) The triangle of the previous example is also unchanged by a reflection in the plane of the triangle. It is usual to denote this operation by σ_h, imagining that the plane of the triangle is horizontal and using the suffix to refer to the horizontal plane. The inclusion of this new element generates further new elements by taking the products $R_1\sigma_h$, $R_2\sigma_h$, . . ., $R_5\sigma_h$. Geometrically, one sees that the product $R_3\sigma_h$ is just a reflection in the vertical plane containing the R_3 axis and similarly for $R_4\sigma_h$ and $R_5\sigma_h$. It is soon verified that the set of 12 elements

$$E, R_1, R_2, R_3, R_4, R_5, \sigma_h, R_1\sigma_h, R_2\sigma_h, \sigma_3, \sigma_4, \sigma_5$$

where $\sigma_3 = R_3\sigma_h$, etc., form a group which contains D_3 and is usually denoted by D_{3h}. The multiplication table is readily deduced from the table for D_3 and the property $\sigma^2 = E$ for a reflection. Notice that although the group D_{3h} contains improper elements (elements which are not rotations) it does not contain the inversion I—the triangle is clearly not unchanged by the inversion. Elements of the kind $R_1\sigma_h$, involving a rotation combined with a reflection in the plane perpendicular to the rotation axis, are called mirror rotations.

(8) The set of *all* rotations about a single axis form a continuous group, called $\mathscr{R}_2$. The group elements are denoted by $R(a)$, where a is the angle of rotation and $0 \le a < 2\pi$. The multiplication table would here be infinite, but in fact we may write down a general formula for the multiplication of any two elements. Geometrically we see that

$$R(a)R(b) = R(a+b) \tag{2.6}$$

with

$$R(a+2\pi) = R(a)$$

Hence the elements commute and the inverse is given by

$$R^{-1}(a) = R(2\pi - a) \tag{2.7}$$

(9) Extending to three dimensions, the set of all rotations about axes through a fixed point will again form a group. It requires three parameters to specify the rotation. One convenient parameterisation is to give two polar

angles, which fix the axis of rotation, and the angle of rotation about this axis. The rotation $R_k(a)$ would then be used, where the suffix k defines a unit vector along the axis of rotation. Alternatively, we may give the three Euler angles which determine the position of the body relative to its initial position. This group, called $\mathscr{R}_3$, is the set of proper covering operations of the sphere and we shall return to it in chapter 7.

(10) The set of all permutations P of n objects forms a group which is usually called the 'symmetric group' and is denoted by $\mathscr{S}_n$. The product of two permutations is defined as that permutation which produces the final arrangement directly from the initial. It is convenient to use a notation

$$\mathrm{P} = \begin{pmatrix} 1 & 2 & 3 & \ldots & n \\ p_1 & p_2 & p_3 & \ldots & p_n \end{pmatrix} \tag{2.8}$$

to denote a permutation in which the object labelled i is replaced by the object labelled p_i. The numbers $p_1 p_2 p_3 \ldots p_n$ are hence a rearrangement of the numbers $1\ 2\ 3 \ldots n$. There are therefore $n!$ elements, since p_1 may be chosen in n ways, p_2 may be chosen in $(n-1)$ ways, etc. Although it is usual to do so, there is no need to write the numbers in the first row of P in natural order. Thus, for example, the two symbols

$$\begin{pmatrix} 1 & 2 & 3 \\ 2 & 3 & 1 \end{pmatrix} \quad \text{and} \quad \begin{pmatrix} 3 & 1 & 2 \\ 1 & 2 & 3 \end{pmatrix}$$

represent identical permutations and one may immediately write down the inverse of the element P as

$$\mathrm{P}^{-1} = \begin{pmatrix} p_1 & p_2 & p_3 & \ldots & p_n \\ 1 & 2 & 3 & \ldots & n \end{pmatrix}$$

As an example we give the multiplication table for $n = 3$. Using the abbreviations

$$\mathrm{E} = \begin{pmatrix} 1 & 2 & 3 \\ 1 & 2 & 3 \end{pmatrix}, \quad \mathrm{P}_1 = \begin{pmatrix} 1 & 2 & 3 \\ 2 & 1 & 3 \end{pmatrix}, \quad \mathrm{P}_2 = \begin{pmatrix} 1 & 2 & 3 \\ 1 & 3 & 2 \end{pmatrix}, \quad \mathrm{P}_3 = \begin{pmatrix} 1 & 2 & 3 \\ 3 & 2 & 1 \end{pmatrix},$$

$$\mathrm{P}_4 = \begin{pmatrix} 1 & 2 & 3 \\ 2 & 3 & 1 \end{pmatrix}, \quad \mathrm{P}_5 = \begin{pmatrix} 1 & 2 & 3 \\ 3 & 1 & 2 \end{pmatrix}$$

We have, for example

$$\mathrm{P}_1 \mathrm{P}_2 = \begin{pmatrix} 1 & 2 & 3 \\ 2 & 1 & 3 \end{pmatrix} \begin{pmatrix} 1 & 2 & 3 \\ 1 & 3 & 2 \end{pmatrix}$$

$$= \begin{pmatrix} 1 & 3 & 2 \\ 2 & 3 & 1 \end{pmatrix} \begin{pmatrix} 1 & 2 & 3 \\ 1 & 3 & 2 \end{pmatrix} = \begin{pmatrix} 1 & 2 & 3 \\ 2 & 3 & 1 \end{pmatrix} = \mathrm{P}_4$$

In deducing this result the columns of P_1 were rearranged so that the top row of P_1 was identical with the bottom row of P_2. This enables us to use the general result

$$\begin{pmatrix} p_1\, p_2\, p_3 \ldots p_n \\ q_1\, q_2\, q_3 \ldots q_n \end{pmatrix} \begin{pmatrix} 1 & 2 & 3 & \ldots n \\ p_1\, p_2\, p_3 \ldots p_n \end{pmatrix} = \begin{pmatrix} 1 & 2 & 3 & \ldots n \\ q_1\, q_2\, q_3 \ldots q_n \end{pmatrix} \tag{2.9}$$

which is really the definition of the product of two permutations. In this way we find the multiplication table 2.6.

Table 2.6

G_a \ G_b	E	P_1	P_2	P_3	P_4	P_5
E	E	P_1	P_2	P_3	P_4	P_5
P_1	P_1	E	P_4	P_5	P_2	P_3
P_2	P_2	P_5	E	P_4	P_3	P_1
P_3	P_3	P_4	P_5	E	P_1	P_2
P_4	P_4	P_3	P_1	P_2	P_5	E
P_5	P_5	P_2	P_3	P_1	E	P_4

2.3 Isomorphism

Because of the very abstract definition of a group it sometimes happens that two groups whose elements are defined in very different ways may nevertheless be so closely related that they may, for algebraic purposes, be regarded as the same group. Precisely, we say that two groups $\mathscr{G}$ and $\mathscr{H}$ are isomorphic if a one-to-one correspondence $G_a \leftrightarrow H_a$ may be set up between the elements G_a of the group $\mathscr{G}$ and the elements H_a of $\mathscr{H}$, in such a way that if $G_aG_b = G_d$ then $H_aH_b = H_d$. The word homomorphism is used for such a relationship if the one-to-one property is absent. Two isomorphic groups therefore have the same group multiplication table with possible re-ordering of the group elements. Since most of the group properties of interest in the study of symmetry arise from the algebra of the group multiplication table, it follows that one can avoid repetition and draw useful analogies by recognising isomorphisms between groups.

In the examples described in section 2.2 there are already several cases of isomorphism: (a) the correspondence $1 \leftrightarrow E, -1 \leftrightarrow R$ shows the isomorphism between examples (1) and (3); (b) the correspondence $R_1 \leftrightarrow P_5$, $R_2 \leftrightarrow P_4$, $R_3 \leftrightarrow P_1$, $R_4 \leftrightarrow P_2$, $R_5 \leftrightarrow P_3$ shows the isomorphism between examples (6) and (10).

2.4 Subgroups

Given a set of g elements forming a group $\mathscr{G}$ it is often possible to select a smaller number of these elements which satisfy the group definitions among themselves. They are said to form a subgroup of $\mathscr{G}$. It is clear that the identity must be included in the selection, but the crucial feature is that the product of any two selected elements must also lie in the selection. There are a number of examples of subgroups within the groups given in section 2.2. (i) In example (2) the elements 1 and -1 clearly form a subgroup. (ii) The group $\mathscr{R}_2$ is a subgroup of the group $\mathscr{R}_3$. In fact there is an infinity of $\mathscr{R}_2$ subgroups in $\mathscr{R}_3$ obtained by selecting the axis for the rotations of $\mathscr{R}_2$. (iii) All finite rotation groups, and in particular that of example (6), are subgroups of $\mathscr{R}_3$.

The group D_3 of example (6) itself contains several subgroups, as may be seen from the multiplication table. They are E, R_1 and R_2, the cyclic group C_3 of example (5), and three groups E, R_3; E, R_4; and E, R_5; all isomorphic with the group of example (3). The permutation group of example (10) has the same subgroup structure as D_3 because of the isomorphism.

There are a number of rather beautiful theorems concerning subgroups, but the only one which we will mention here is that the order of a subgroup must be a divisor of the order of the group. It is called Lagrange's theorem (see problem 2.4).

From a physical viewpoint the subgroup is of some importance in perturbation theory. A system may be dominated by the symmetry with respect to a group $\mathscr{G}$. Often, however, there is a weak perturbation present which lacks the full symmetry but nevertheless has the lesser symmetry described by a subgroup of $\mathscr{G}$.

2.5 The direct product group

Suppose a group $\mathscr{K}$ contains two subgroups $\mathscr{G}$ and $\mathscr{H}$ whose elements commute so that $G_a H_b = H_b G_a$, where G_a is any element of $\mathscr{G}$ and H_b any element of $\mathscr{H}$. If, furthermore, every element of $\mathscr{K}$ may be written uniquely as a product $G_a H_b$ then $\mathscr{K}$ is called the 'direct product' of $\mathscr{G}$ and $\mathscr{H}$ and is written $\mathscr{K} = \mathscr{G} \times \mathscr{H}$. This definition implies that the identity is the only element common to both $\mathscr{G}$ and $\mathscr{H}$.

An example of a direct product group is provided by combining the groups C_2 and S_2 of examples (3) and (4) of section 2.2. The inversion, which simply changes the sign of all vectors, clearly commutes with all rotations since if $\mathrm{R}\boldsymbol{r} = \boldsymbol{r}'$ then $\mathrm{IR}\boldsymbol{r} = \mathrm{I}\boldsymbol{r}' = -\boldsymbol{r}'$ and $\mathrm{RI}\boldsymbol{r} = -\mathrm{R}\boldsymbol{r} = -\boldsymbol{r}'$. The set of elements E, R, I, RI form the product group $C_2 \times S_2$ which is usually denoted by C_{2h}. Notice that the element RI is in fact a reflection in the xy-plane and is usually denoted by σ. This may be seen by following the motion of the x-, y- and z-axes. The inversion changes the direction of all three axes while the rotation R changes the direction of the x- and y-axes. The product RI therefore changes the direction of the z-axis alone, which is a reflection.

As a second example the group D_{3h} introduced in example (7) of section 2.2 is the product of D_3 with the group of two elements E and σ_h. This latter group is usually denoted by S_1 and we have $D_{3h} = D_3 \times S_1$

More generally, the group $\mathscr{R}_3$ of all rotations in three dimensions may be combined with inversions to form what is known as the 'full orthogonal' group O_3. It has elements $\mathrm{R}_k(a)$ and $\mathrm{IR}_k(a)$.

The advantage of being able to recognise a group as a direct product is that its properties may be deduced from those of the separate groups. This fact will be appreciated later in the book, but for the moment we simply remark that the multiplication table can be immediately written down from the separate tables since

$$(\mathrm{G}_a\mathrm{H}_b)(\mathrm{G}_c\mathrm{H}_d) = (\mathrm{G}_a\mathrm{G}_c)(\mathrm{H}_b\mathrm{H}_d)$$

For the example C_{2h} this gives table 2.7.

Table 2.7

$\mathrm{G}_a \backslash \mathrm{G}_b$	E	R	I	σ
E	E	R	I	σ
R	R	E	σ	I
I	I	σ	E	R
σ	σ	I	R	E

2.6 Conjugate elements and classes

For all but the simplest groups the number of group elements soon becomes large. Already the multiplication tables in examples (5) and (8) of section 2.2 are becoming unwieldy. Fortunately, there is a way of simplifying the study of the structure of groups by collecting together a 'class' of elements with similar properties.

A group element G_a is said to be 'conjugate' to an element G_b if there is an element G_n in the group such that

$$\mathrm{G}_a = \mathrm{G}_n\mathrm{G}_b\mathrm{G}_n^{-1} \tag{2.10}$$

If G_b and G_c are both conjugate to G_a then it quickly follows that G_b and G_c are conjugate to each other for, given

$$\mathrm{G}_a = \mathrm{G}_n\mathrm{G}_b\mathrm{G}_n^{-1} \quad \text{and} \quad \mathrm{G}_a = \mathrm{G}_m\mathrm{G}_c\mathrm{G}_m^{-1}$$

we have

$$\mathrm{G}_b = \mathrm{G}_n^{-1}\mathrm{G}_a\mathrm{G}_n = \mathrm{G}_n^{-1}\mathrm{G}_m\mathrm{G}_c\mathrm{G}_m^{-1}\mathrm{G}_n = (\mathrm{G}_n^{-1}\mathrm{G}_m)\mathrm{G}_c(\mathrm{G}_n^{-1}\mathrm{G}_m)^{-1}$$

Thus one is led to the concept of a 'class' of elements which are all conjugate to each other. Furthermore, no element may belong to more than one class. For suppose G belonged to two classes: it would then be conjugate to all elements in either class, implying that elements in one class would be conjugate also to those in the other. This would unite the two classes into a single class. In consequence, every group may be broken up into separate classes which we denote by the symbols $\mathscr{C}_p$.

In an Abelian group, every group element is in a class by itself since the relation (2.10) implies that $G_a = G_b$, using the commuting property of the elements. For the same reason the identity is always in a class by itself.

2.7 Examples of classes

2.7.1 The rotation group $\mathscr{R}_3$

To find the elements which are in the same class as some chosen rotation $R_{\boldsymbol{k}}(a)$ it is necessary to construct the rotations

$$RR_{\boldsymbol{k}}(a)R^{-1}$$

for arbitrary R. In fact, this triple product has a very simple interpretation. We shall show that it is a rotation through the same angle a about an axis $\boldsymbol{k}'$, where $\boldsymbol{k}'$ is related to the original axis $\boldsymbol{k}$ by

$$\boldsymbol{k}' = R\boldsymbol{k} \tag{2.11}$$

In other words

$$RR_{\boldsymbol{k}}(a)R^{-1} = R_{\boldsymbol{k}'}(a) \tag{2.12}$$

This result is almost obvious if one imagines R as carrying all vectors from old to new (primed) positions. The right-hand side of equation (2.12) is a rotation through angle a about the new $\boldsymbol{k}'$-axis. The left-hand side of the equation first rotates back to the old positions then rotates about the old $\boldsymbol{k}$-axis and then rotates back to the new scheme. For a more careful proof consider

$$[RR_{\boldsymbol{k}}(a)R^{-1}]\boldsymbol{k}' = RR_{\boldsymbol{k}}(a)\boldsymbol{k} = R\boldsymbol{k} = \boldsymbol{k}' \tag{2.13}$$

using (2.11) and the fact that a rotation leaves the vector along the axis of rotation unmoved. But equation (2.13) shows that the left-hand side of (2.12) leaves $\boldsymbol{k}'$ unmoved. It must therefore be a rotation about $\boldsymbol{k}'$. To show, finally, that the rotation angle a is the same follows from the construction of two orthogonal unit vectors $\boldsymbol{e}_1$ and $\boldsymbol{e}_2$ in the plane perpendicular to $\boldsymbol{k}$. From figure 2.2,

$$R_{\boldsymbol{k}}(a)\boldsymbol{e}_1 = \cos a\, \boldsymbol{e}_1 + \sin a\, \boldsymbol{e}_2 \tag{2.14}$$

The rotation R carries $\boldsymbol{k}$ into $\boldsymbol{k}'$ according to equation (2.11) and it will carry $\boldsymbol{e}_1$ and $\boldsymbol{e}_2$ into two new orthogonal unit vectors $\boldsymbol{e}'_1 = R\boldsymbol{e}_1$ and $\boldsymbol{e}'_2 = R\boldsymbol{e}_2$ which lie

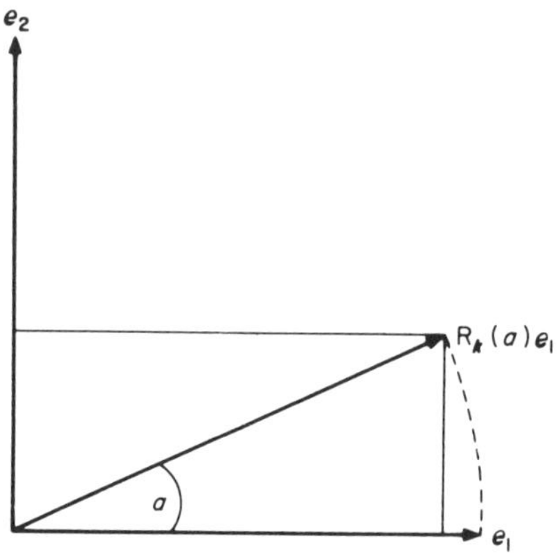

Figure 2.2

in the plane perpendicular to $\boldsymbol{k}'$. Thus, like equation (2.14), we have

$$R_{k'}(a)\boldsymbol{e}'_1 = \cos a\,\boldsymbol{e}'_1 + \sin a\,\boldsymbol{e}'_2 \tag{2.15}$$

but, on the other hand,

$$\begin{aligned}[RR_k(a)R^{-1}]\boldsymbol{e}'_1 &= RR_k(a)\boldsymbol{e}_1 = R(\cos a\,\boldsymbol{e}_1 + \sin a\,\boldsymbol{e}_2)\\ &= \cos a\,\boldsymbol{e}'_1 + \sin a\,\boldsymbol{e}'_2\end{aligned}$$

which, by comparison with equation (2.15), proves equation (2.12).

The class structure of the group $\mathscr{R}_3$ is therefore extremely simple. It is always possible to find a rotation R which takes $\boldsymbol{k}$ into any direction $\boldsymbol{k}'$. Hence, any two rotations through the same angle are in the same class, no matter what their axes. (We are, of course, restricting ourselves always to rotations whose axes pass through a fixed origin.) At the same time equation (2.12) shows that two rotations through different angles cannot be in the same class, since R is arbitrary.

2.7.2 The finite group of rotations D_3 (see example (6) of section 2.2).

On restricting to this subgroup of $\mathscr{R}_3$ we can make equation (2.12) our starting point in the search for the classes. It shows that for two elements of D_3 to be in the same class it is necessary that their rotation angles must be the same, but this is not sufficient. In addition, the rotation R which carries $\boldsymbol{k}$ into $\boldsymbol{k}'$ must be one of the elements of the group D_3. In this way, there are found to be three classes,

$$\begin{aligned}\mathscr{C}_1 &= E\\ \mathscr{C}_2 &= \{R_1, R_2\}\\ \mathscr{C}_3 &= \{R_3, R_4, R_5\}\end{aligned} \tag{2.16}$$

As always, the identity is in a class by itself. The rotation angle of R_1 and R_2

is $2\pi/3$ and that of R_3, R_4 and R_5 is π. (R_2 may be regarded either as a rotation through $4\pi/3$ about the positive z-axis or through $2\pi/3$ about the negative z-axis.) There must, therefore, be at least two classes besides $\mathscr{C}_1$. To show that R_1 and R_2 are in the same class we must find a rotation which carries the axis of R_1 into that of R_2. But this is simply an inversion of the z-axis which is achieved by R_3, R_4 or R_5. In the same way, the rotations R_1 and R_2 carry the axes of R_3, R_4 and R_5 into each other. In detail

$$R_2 = R_3 R_1 R_3^{-1}, \qquad R_3 = R_1 R_4 R_1^{-1} = R_2 R_5 R_2^{-1}$$

2.7.3 The symmetric group $\mathscr{S}_3$ (see example (10) of section 2.2)

Because of the isomorphism the class structure of $\mathscr{S}_3$ is the same as that of D_3, namely

$$\begin{aligned}
\mathscr{C}_1 &= E \\
\mathscr{C}_2 &= \left\{ \begin{pmatrix} 1 & 2 & 3 \\ 2 & 3 & 1 \end{pmatrix}, \begin{pmatrix} 1 & 2 & 3 \\ 3 & 1 & 2 \end{pmatrix} \right\} \\
\mathscr{C}_3 &= \left\{ \begin{pmatrix} 1 & 2 & 3 \\ 2 & 1 & 3 \end{pmatrix}, \begin{pmatrix} 1 & 2 & 3 \\ 3 & 2 & 1 \end{pmatrix}, \begin{pmatrix} 1 & 2 & 3 \\ 1 & 3 & 2 \end{pmatrix} \right\}
\end{aligned} \tag{2.17}$$

Notice here that the 'common' property of the elements in $\mathscr{C}_3$ is that they each leave one of the three objects unmoved, whereas the two elements in $\mathscr{C}_2$ move all three. This is in fact the clue to the class structure of $\mathscr{S}_n$ for general n, but we leave the details of this general case until chapter 17 of volume 2.

2.8 The class structure of product groups

The classes of a direct product group $\mathscr{G} \times \mathscr{H}$ are easily found from the classes of $\mathscr{G}$ and $\mathscr{H}$, as follows. Suppose that $G_a H_b$ and $G_c H_d$ are in the same class. This implies that there is some element $G_e H_f$ such that

$$G_e H_f G_a H_b (G_e H_f)^{-1} = G_c H_d$$

i.e.
$$(G_e G_a G_e^{-1})(H_f H_b H_f^{-1}) = G_c H_d$$

so that

$$G_e G_a G_e^{-1} = G_c \quad \text{and} \quad H_f H_b H_f^{-1} = H_d$$

showing that G_a and G_c are in the same class of $\mathscr{G}$ and that H_b and H_d are in the same class of $\mathscr{H}$. Hence a class of $\mathscr{G} \times \mathscr{H}$ will contain all the products $G_a H_b$ where G_a runs through a class of $\mathscr{G}$ and H_b runs through a class of $\mathscr{H}$. There will be one class of $\mathscr{G} \times \mathscr{H}$ to every pair of classes, one from $\mathscr{G}$ and one from $\mathscr{H}$.

For example, the full orthogonal group O_3 now has two classes associated with each rotation angle a. In one, there are the proper rotations $R_k(a)$ and in the other the improper rotations $IR_k(a)$. These two classes correspond to the two classes E and I of the inversion group. As another example, the group $D_{3h} = D_3 \times S_1$ contains six classes corresponding to combinations of the three classes of D_3 and the two classes E and σ_h of S_1.

2.9 The group rearrangement theorem

In this small section we prove a simple property of a group, called the group rearrangement theorem, which we shall use frequently in later sections. It says that if G_a is any fixed element of a group $\mathscr{G}$ and G_b is allowed to run over all elements of $\mathscr{G}$ then the product element $G_c = G_b G_a$ also runs over all elements of $\mathscr{G}$, each element appearing once only. To prove this result we first note that, given any G_c, the choice $G_b = G_c G_a^{-1}$ will ensure that $G_c = G_b G_a$. (The element G_a^{-1} must exist because of the group postulates.) Secondly, we argue that an element G_c cannot be produced by two different elements G_b and G'_b for this would imply that

$$G_c = G_b G_a = G'_b G_a$$

and hence, multiplying through by G_a^{-1}, one is led to $G_b = G'_b$. This completes the proof.

Bibliography

For a more mathematical, yet elementary, treatment of groups and related structures see

Green, J. A. (1965). *Sets and Groups* (Routledge & Kegan Paul, London)

Problems

2.1 Show that, in a group multiplication table, each element appears once and once only in each row and in each column.

2.2 Show that the elements E, P_4 and P_5 of the group $\mathscr{S}_3$, in the notation of section 2.2(10), form a subgroup which is isomorphic with C_3.

2.3 The group D_4 is the group of proper covering operations (rotations) of a plane square—a generalisation of section 2.2(6). Construct the group multiplication table for D_4 and divide the elements into classes.

2.4 Let $\mathscr{S} \equiv \{S_1, S_2, \ldots, S_n\}$ be a subgroup of a group $\mathscr{G}$ of order g. For any element G_i of $\mathscr{G}$ the set of elements $\mathscr{S}G_i \equiv \{S_1G_i, S_2G_i, \ldots, S_nG_i\}$ are said to form a right coset of $\mathscr{S}$ (a similar definition holds for left cosets $G_i\mathscr{S}$). If G_i is one of the elements of $\mathscr{S}$ show that the coset is identically equal to $\mathscr{S}$. If G_i is not in $\mathscr{S}$ show that the coset contains no element in common with $\mathscr{S}$. Hence show that any two right cosets of $\mathscr{S}$ are either identical or have no elements in common. Finally deduce that the order n of a subgroup must be an integral divisor of the order g of $\mathscr{G}$ and that, given $\mathscr{S}$, a set of g/n elements G_i of $\mathscr{G}$ may be found such that any element of $\mathscr{G}$ may be written in the form $S_p G_i$. (This is known as Lagrange's theorem.)

2.5 Find the classes of the group $\mathscr{S}_3$ of all permutations of three objects. Do this directly rather than relying on the isomorphism referred to in subsection 2.7.3.

2.6 Generate a group by multiplying the following elements: (a) a rotation R through an angle π about the z-axis; (b) a reflection in the xy-plane. Show that it has four elements, construct the multiplication table, find the classes and write the group as a direct product.

2.7 Show that the group D_6 generated by six-fold rotations about the z-axis and two-fold rotations about the x-axis is the direct product $D_3 \times C_2$, where the principal axis of both D_3 and C_2 is the z-axis. Divide the elements into classes.

3

Linear Algebra and Vector Spaces

In applying group theory to the study of symmetry, one is not so much interested in the structure of the groups themselves, but rather in the transformations or changes which are produced in various 'things' by the group elements. The things will generally be either the coordinates of the constituents of some physical system or, in quantum mechanics, the wave functions of those coordinates. In either case it is possible to describe the things as vectors in some space. It is the purpose of this chapter to describe in general terms, and therefore rather abstractly, the properties of vectors and their transformations. The reader will undoubtedly have met most of the concepts in connection with vectors in ordinary three-dimensional space and probably also in a standard first course on linear algebra and matrices. To many, therefore, this chapter will be revision, but it should not be treated too lightly since it provides the foundations, framework and notation for the following chapter in which all the crucial group theoretical properties are deduced. It should perhaps be pointed out that there will be no use of group theory in the present chapter. The study of the transformations induced in a vector space by the elements of a group is the subject of chapter 4.

3.1 Linear vector space

The concept of a vector is a very general one. We may think of a simple vector in three-dimensional space which specifies the position of a single particle with respect to some origin. With N particles we should need $3N$ coordinates which may be regarded as the components of a vector in $3N$-dimensional space. In classical mechanics this $3N$-dimensional vector space is the appropriate framework in which to work. However, in quantum mechanics the appropriate framework is a set of (wave) functions of the coordinates. The concept of the vector space may, conveniently, be extended to such a set of functions, although it means that the dimensionality of the space becomes infinite. (In practice one is often concerned with finite-dimensional subspaces of such an infinite-dimensional space of functions.)

A Fourier series provides a simple example of a space of functions. Expanding an arbitrary function in a Fourier sine series

$$f(x) = \sum_{n=1}^{\infty} c_n \sin nx$$

may be considered as the decomposition of a 'vector' $f(x)$ into the infinite set of 'orthogonal' basis vectors sin nx. The coefficients c_n may be regarded as the components of $f(x)$ and a suitable definition of scalar product enables us to justify the use of the word 'orthogonal'. This extension of one's simple ideas of a vector first to a finite number of dimensions greater than three and then to an infinite number is a classic example of the use of abstraction in mathematics. One may discuss the properties of an abstract vector space without saying whether it has three or $3N$ or an infinite number of dimensions. After these introductory remarks, let us set up the formal definitions of a linear vector space.

A set $r_1, r_2, \ldots$ is said to form a 'linear vector space L' if the sum of any two members produces another in the set and if multiplication by a complex number c also produces another in the set. If c is restricted to be real, the space is called real. There is clearly an infinite number of members in any such space and they are called vectors.

A set of vectors $\boldsymbol{r}_1, \boldsymbol{r}_2, \ldots, \boldsymbol{r}_p$ is said to be 'linearly independent' if the members are not related by an equation

$$\sum_{k=1}^{p} c_k \boldsymbol{r}_k = 0 \tag{3.1}$$

for any values of the coefficients c_k other than $c_k = 0$.

The 'dimension' of L is defined as the greatest number of vectors in L which form a linearly independent set. We denote the dimension by the symbol s. (N.B. The dimension is not the number of vectors in L.) In an s-dimensional vector space L, any set of s linearly independent vectors are said to form a 'basis'. Let us denote one such set by $\boldsymbol{e}_i$ where $i = 1, 2, \ldots, s$. The importance of a basis is that for any vector $\boldsymbol{r}$ we may write

$$\boldsymbol{r} = \sum_{i=1}^{s} r_i \boldsymbol{e}_i \tag{3.2}$$

The proof is immediate for, if there was a vector $\boldsymbol{r}$ for which (3.2) was not true this would imply that the set of $(s+1)$ vectors $\boldsymbol{r}$, $\boldsymbol{e}_i$ were linearly independent. But this is impossible since it would imply that L had dimension $(s+1)$ contradicting the definition of L as having dimension s. (This is an example of a 'reductio ad absurdum' type of argument which we shall often find useful.)

In the physical space of three dimensions it is usual to take three mutually perpendicular vectors as the basis. Clearly this choice of perpendicular (or orthogonal) vectors makes life simpler in three dimensions and suggests that one should try to extend the concept of orthogonality to the abstract vector space. This may readily be done through a general definition of a 'scalar product' (sometimes called an 'inner product'). There is considerable freedom in the definition of a scalar product but in order to be called a scalar product it must have the following properties:

(1) It is a complex number associated with each ordered pair of vectors $\boldsymbol{r}_1$, $\boldsymbol{r}_2$ and is denoted by $(\boldsymbol{r}_1, \boldsymbol{r}_2)$.

(2) $(\boldsymbol{r}_2, \boldsymbol{r}_1) = (\boldsymbol{r}_1, \boldsymbol{r}_2)^*$ (3.3a)

where the asterisk denotes complex conjugate.

(3) $(\boldsymbol{r}_1, c\boldsymbol{r}_2) = c(\boldsymbol{r}_1, \boldsymbol{r}_2)$ (3.3b)

(4) $(\boldsymbol{r}_1 + \boldsymbol{r}_2, \boldsymbol{r}_3) = (\boldsymbol{r}_1, \boldsymbol{r}_3) + (\boldsymbol{r}_2, \boldsymbol{r}_3)$ (3.3c)

(5) $(\boldsymbol{r}, \boldsymbol{r}) \geq 0$, the equality holding only for the trivial vector $\boldsymbol{r} = 0$.

The positive square root $(\boldsymbol{r}, \boldsymbol{r})^{\frac{1}{2}}$ is called the 'norm' (or length) of the vector $\boldsymbol{r}$. If a vector $\boldsymbol{r}$ has $(\boldsymbol{r}, \boldsymbol{r})^{\frac{1}{2}} = 1$ it is said to be 'normalised'. Two vectors $\boldsymbol{r}_1$ and $\boldsymbol{r}_2$ are said to be 'orthogonal' if their scalar product vanishes, $(\boldsymbol{r}_1, \boldsymbol{r}_2) = 0$. A set of vectors which are each normalised and mutually orthogonal is called an 'orthonormal' set. It is always possible to choose an orthonormal set as a basis for a vector space and hereafter we shall always assume that the basis $\boldsymbol{e}_i$ has been so chosen, i.e.

$$(\boldsymbol{e}_i, \boldsymbol{e}_j) = \delta_{ij} \tag{3.4}$$

where δ_{ij} is the so-called Kronecker delta which is defined by

$$\delta_{ij} = 0 \quad \text{for} \quad i \neq j$$
$$\delta_{ij} = 1 \quad \text{for} \quad i = j.$$

It may not be obvious that we are always free to choose an orthonormal basis, but this is readily shown. Suppose we have a basis $\grave{\boldsymbol{e}}_i$ which is not orthonormal then, by a chain process starting with $\grave{\boldsymbol{e}}_1$, we may construct a basis $\boldsymbol{e}_i$ of mutually orthogonal vectors as follows:

(1) Divide $\tilde{e}_1$ by its norm, to form a normalised vector $e_1 = \tilde{e}_1/(\tilde{e}_1, \tilde{e}_1)^{\frac{1}{2}}$.

(2) Construct $\tilde{e}_2 - (e_1, \tilde{e}_2)e_1$ and normalise to form e_2. Then clearly $(e_1, e_2) \propto (e_1, \tilde{e}_2) - (e_1, \tilde{e}_2)(e_1, e_1) = 0$.

(3) Construct $\tilde{e}_3 - (e_1, \tilde{e}_3)e_1 - (e_2, \tilde{e}_3)e_2$ and normalise to form e_3. Immediately $(e_1, e_3) = (e_2, e_3) = 0$.

This process continues, resulting in the orthonormal set e_i with $i = 1, 2, \ldots, s$. The new set e_i is linearly independent because they are orthogonal. For suppose there is a relation.

$$\sum_{i=1}^{s} c_i e_i = 0 \tag{3.5}$$

then by taking the scalar product of some e_j with both sides of equation (3.5) we have $c_j = 0$. Hence no non-trivial relation (3.5) exists and the set is linearly independent. To complete the argument it is necessary to comment that none of the e_i can vanish identically for this would imply a linear dependence among the original set $\tilde{e}_i$ and they were assumed to form a basis and therefore to be linearly independent. This method of forming an orthonormal set is called the Schmidt process.

With an orthonormal basis, the coefficient r_i of equation (3.2) is sometimes called the ith 'component' (or coordinate) of r and is related simply to the scalar product, namely

$$r_i = (e_i, r) \tag{3.6}$$

This follows on taking the scalar product of some chosen basis vector e_j with both sides of (3.2) and using the orthonormality (3.4). In terms of components, the scalar product of two arbitrary vectors is given by

$$(r_1, r_2) = \sum_{i=1}^{s} \sum_{j=1}^{s} r_{1i}^{*} r_{2j}(e_i, e_j) = \sum_{i=1}^{s} r_{1i}^{*} r_{2i} \tag{3.7}$$

Given a vector space L with some orthonormal basis e_i with $i = 1, 2, \ldots, s$ it is possible to form a 'subspace' L_1 by choosing a subset of the e_i with (say) $i = 1, 2, \ldots, s_1 < s$ as the new basis. The remaining basis vectors form another subspace L_2, called the 'orthogonal complement' to L_1. We often write $L = L_1 + L_2$. Notice that this does not say that all vectors of L lie either in L_1 or L_2 but that any vector in L may be expressed as the sum of a vector in L_1 and an orthogonal vector in L_2.

3.2 Examples of linear vector spaces

3.2.1 Displacements in three dimensions

The possible displacements of a particle from the origin in ordinary three-dimensional space provide the most natural example of a real linear vector space. Clearly the sum of any two displacements produces another displacement. The usual basis would be the three unit vectors $\boldsymbol{e}_x, \boldsymbol{e}_y, \boldsymbol{e}_z$ along the x-, y- and z-axes, respectively, and the scalar product would be defined as the product of the lengths of the two vectors and the cosine of the angle between them. Notice that this definition is independent of the basis and the reader may verify geometrically that it possesses the necessary properties (3.3).

3.2.2 Displacement of a set of N particles in three dimensions

In this generalisation of the previous example the most natural basis would be the set of $3N$ unit vectors. $\boldsymbol{e}_{tx}, \boldsymbol{e}_{ty}, \boldsymbol{e}_{tz}$ where $t = 1, 2, \ldots, N$ and $\boldsymbol{e}_{tx}$ represents a displacement of particle t by one unit in the x-direction with all other particles being unmoved. The scalar product might be defined as the sum over particles of the ordinary scalar product defined above in subsection 3.2.1. An arbitrary vector

$$\boldsymbol{r} = \sum_{t=1}^{N} \sum_{i=x,y,z} r_{ti}\boldsymbol{e}_{ti} \tag{3.8}$$

would then represent a displacement of all particles in which particle t is moved a distance r_{tx} in the x-direction, etc.

3.2.3 Function spaces

Consider next the set of all continuous functions $\psi(x)$ with x in the interval $a \leq x \leq b$ and with boundary conditions $\psi(a) = \psi(b) = 0$. It is soon verified that this set forms a vector space according to the definitions in section 3.1, but it is by no means obvious how we should define a scalar product. In fact, the definition

$$(\psi', \psi) = \int_a^b \psi'^*(x)\psi(x)\mathrm{d}x \tag{3.9}$$

may be seen to satisfy the formal requirements (3.3). Two functions ψ' and ψ are then orthogonal if the integral (3.9) vanishes and ψ is normalised if $\int_a^b |\psi(x)|^2\mathrm{d}x = 1$. If, for simplicity, we take $a = 0$, $b = \pi$ then the set

$$\psi_n(x) = (2/\pi)^{\frac{1}{2}} \sin nx$$

with n any integer, forms an orthonormal basis of infinite dimension. Although it is quickly shown that

$$(\psi_m, \psi_n) = \frac{2}{\pi}\int_0^{\pi} \sin mx \sin nx \, \mathrm{d}x = \delta_{n,m} \tag{3.10}$$

a careful discussion of completeness is necessary to justify the expansion

$$\psi(x) = \sum_{n=1}^{\infty} c_n \psi_n(x) \tag{3.11}$$

for any function in the set and hence to show that the functions $\psi_n(x)$ form a basis. The coefficients c_n in the expansion (3.11) are exactly analogous to the coefficients r_i in (3.2), and corresponding to (3.6) we have

$$c_n = (\psi_n, \psi) = (2/\pi)^{\frac{1}{2}} \int_0^{\pi} \sin nx\ \psi(x)\mathrm{d}x \tag{3.12}$$

The pair of equations (3.11) and (3.12) are recognised as the Fourier sine series and the formula for its coefficients.

3.2.4 Function space with finite dimension

As an example, consider the six-dimensional space L of functions $\psi(\boldsymbol{r}) = c_1x^2 + c_2y^2 + c_3z^2 + c_4yz + c_5zx + c_6xy$ of the coordinates x, y, z of a particle where the choice of the six complex parameters c_i determines the function $\psi(\boldsymbol{r})$. The scalar product of two functions $\psi(\boldsymbol{r})$ and $\psi'(\boldsymbol{r})$ in L will be defined by the integral

$$(\psi', \psi) = \iiint_V \psi'^*(\boldsymbol{r})\psi(\boldsymbol{r})\mathrm{d}V \tag{3.13}$$

where V is the unit sphere. One must be careful to distinguish such scalar products from the scalar product of vectors in the ordinary three-dimensional space of $\boldsymbol{r}$. To help in this distinction we use the notation $\boldsymbol{r}'.\boldsymbol{r}$, with a 'dot', for the scalar product in ordinary three-dimensional space rather than the general notation (ψ', ψ) with a comma. We may choose the six functions $\psi_1 = x^2$, $\psi_2 = y^2$, $\psi_3 = z^2$, $\psi_4 = yz$, $\psi_5 = zx$, $\psi_6 = xy$ as a basis in L and they may be written in terms of $\boldsymbol{r}$ by

$$\psi_1 = (\boldsymbol{e}_x.\boldsymbol{r})^2,\ \psi_4 = (\boldsymbol{e}_y.\boldsymbol{r})(\boldsymbol{e}_z.\boldsymbol{r}),\ \text{etc.} \tag{3.14}$$

This simple basis is not orthonormal but the linear independence of its members is clear.

3.2.5 Wave functions

In quantum mechanics the properties of a system of N particles in a particular state are described by a wave function $\psi(\boldsymbol{r}) \equiv \psi(\boldsymbol{r}_1\boldsymbol{r}_2, \ldots, \boldsymbol{r}_N)$ of the coordinates $\boldsymbol{r}_i$ of the particles. Depending on the system, the wave functions satisfy specified boundary conditions and the set of wave functions describing all possible states of the system forms a vector space. The scalar product is defined as

$$(\psi', \psi) = \int \psi'^*\psi\mathrm{d}V$$

where the integral runs over all values of all coordinates and the volume element $\mathrm{d}V = \mathrm{d}\boldsymbol{r}_1 \mathrm{d}\boldsymbol{r}_2, \ldots, \mathrm{d}\boldsymbol{r}_N$ is the product of the volume elements for each particle. The functions ψ are assumed to have finite norm (ψ, ψ). If the system is governed by a Hermitian Hamiltonian H then the eigenfunctions ψ_i', which satisfy the Schrodinger equation $\mathrm{H}\psi_i = E_i\psi_i$ and describe the stationary states ψ_i of the system, may be shown to be orthonormal in the sense that $(\psi_i, \psi_j) = \int \psi_i^* \psi_j \,\mathrm{d}V = \delta_{ij}$. Consequently they form a basis for the vector space. If there is degeneracy at some energy E with a set of s linearly independent orthonormal eigenfunctions ψ_i at that energy then the set ψ_i with $i = 1, 2, \ldots, s$ forms the basis for an s-dimensional subspace, any vector of which is an eigenfunction of H with energy E.

3.3 Linear operators

Having prepared the framework of the vector space L we now move on to the most crucial concept of the book, that of a transformation (or mapping) which carries each vector of L into another vector of the same space. Suppose that an arbitrary vector $\boldsymbol{r}$ is carried into $\boldsymbol{r}'$, then we define the 'operator' T by the equation

$$\mathrm{T}\boldsymbol{r} = \boldsymbol{r}' \tag{3.15}$$

Furthermore, T is called a 'linear operator' if it satisfies the relations

$$\mathrm{T}(\boldsymbol{r}_1 + \boldsymbol{r}_2) = \mathrm{T}\boldsymbol{r}_1 + \mathrm{T}\boldsymbol{r}_2 \tag{3.16a}$$

$$\mathrm{T}c\boldsymbol{r} = c\mathrm{T}\boldsymbol{r} \tag{3.16b}$$

for any vectors $\boldsymbol{r}_1$, $\boldsymbol{r}_2$ and $\boldsymbol{r}$, where c is any complex number.

Bearing in mind that even a finite-dimensional vector space contains an infinite numbers of vectors, it would seem, at first sight, that the definition (3.15) of T would require an infinite number of parameters. However, the restriction (3.16) to linear operators greatly simplifies the problem and implies that if we specify the transformation of the basis vectors we have specified the transformation of the general vector. From a practical point of view all operators of interest in the study of symmetry in this book will be linear operators, with one small exception in subsection 15.7.4 of volume 2.

Consider then the transformation of some basis vector $\boldsymbol{e}_i$ belonging to a vector space of finite dimension s and denote the new vector by $\boldsymbol{e}_i'$ so that

$$\mathrm{T}\boldsymbol{e}_i = \boldsymbol{e}_i' \tag{3.17}$$

But, since the set $\boldsymbol{e}_1, \boldsymbol{e}_2, \ldots, \boldsymbol{e}_s$ forms a basis, it must be possible to expand

$$\boldsymbol{e}_i' = \sum_{j=1}^{s} T_{ji}\boldsymbol{e}_j$$

Thus with (3.17) we have

$$\mathrm{T}\boldsymbol{e}_i = \sum_{j=1}^{s} T_{ji}\boldsymbol{e}_j$$

but, since the coefficients T_j will clearly depend on the label i of the original vector, it is advisable to add this label to the coefficient T_j writing T_{ji}, so that

$$\boldsymbol{e}'_i = \mathrm{T}\boldsymbol{e}_i = \sum_{j=1}^{s} T_{ji}\boldsymbol{e}_j \tag{3.18}$$

The transformation of the entire basis is therefore given by the set of s^2 coefficients T_{ji}. From these coefficients, the transformation $\mathrm{T}\boldsymbol{r} = \boldsymbol{r}'$ of an arbitrary vector $\boldsymbol{r} = \sum_{i=1}^{s} r_i \boldsymbol{e}_i$ may be deduced,

$$\begin{aligned}\boldsymbol{r}' = \mathrm{T}\boldsymbol{r} &= \sum_{i=1}^{s} r_i T\boldsymbol{e}_i = \sum_{i=1}^{s} r_i \sum_{j=1}^{s} T_{ji}\boldsymbol{e}_j \\ &= \sum_{j=1}^{s} \left\{ \sum_{i=1}^{s} T_{ji} r_i \right\} \boldsymbol{e}_j \end{aligned} \tag{3.19}$$

If we introduce the components r'_j of $\boldsymbol{r}'$,

$$\boldsymbol{r}' = \sum_{j=1}^{s} r'_j \boldsymbol{e}_j$$

then a comparison with equation (3.19) shows that they are related to the components r_i of $\boldsymbol{r}$ by the equation

$$r'_j = \sum_{i=1}^{s} T_{ji} r_i \tag{3.20}$$

It is important to realise that the transformation coefficients T_{ji} come into both equations (3.18) and (3.20), but that the two equations say different things. The first relates the transformed basis vectors to the original set, while the second relates the components of the transformed arbitrary vector to its original components, all components being with respect to the original basis. Notice, however, that whereas (3.18) contains a sum over the first suffix j of T_{ji}, the second equation (3.20) contains a sum over the second suffix i.

The set of coefficients T_{ji}, in fact, form a matrix T where T_{ji} is the matrix element in the jth row and ith column. If the basis vector $\boldsymbol{e}_i$ is represented by a column matrix with 0 in every row except the ith, which contains a 1, then

$$\mathrm{T}\boldsymbol{e}_i = \begin{pmatrix} T_{11} & T_{12} & \cdots & T_{1s} \\ T_{21} & & & \cdot \\ \cdot & & & \cdot \\ \cdot & & & \cdot \\ T_{s1} & \cdot & \cdots & T_{ss} \end{pmatrix} \begin{pmatrix} 0 \\ \cdot \\ 1 \\ \cdot \\ 0 \end{pmatrix} = \begin{pmatrix} T_{1i} \\ T_{2i} \\ \cdot \\ \cdot \\ T_{si} \end{pmatrix}$$

$$= T_{1i}\begin{pmatrix}1\\0\\ \cdot\\ \cdot\\0\end{pmatrix} + T_{2i}\begin{pmatrix}0\\1\\0\\ \cdot\\0\end{pmatrix} + \ldots T_{si}\begin{pmatrix}0\\ \cdot\\ \cdot\\ \cdot\\1\end{pmatrix}$$

$$= \sum_{j=1}^{s} T_{ji}\boldsymbol{e}_j$$

which is just another way of expressing equation (3.18).

Similarly, the general vector $\boldsymbol{r}$ is represented by

$$\boldsymbol{r} = \begin{pmatrix}r_1\\r_2\\ \cdot\\ \cdot\\r_s\end{pmatrix} \text{ with } \mathrm{T}\boldsymbol{r} = \begin{pmatrix}T_{11} & T_{12} & \ldots & T_{1s}\\ T_{21} & & & \cdot\\ \cdot & & & \cdot\\ \cdot & & & \cdot\\ T_{s1} & \cdot & \ldots & T_{ss}\end{pmatrix}\begin{pmatrix}r_1\\r_2\\ \cdot\\ \cdot\\r_s\end{pmatrix} = \begin{pmatrix}\sum_{i=1}^{s} T_{1i}r_i\\ \cdot\\ \cdot\\ \cdot\\ \sum_{i=1}^{s} T_{si}r_i\end{pmatrix}$$

Another interpretation may be given to the matrix elements T_{ji} if the basis $\boldsymbol{e}_i$ is orthonormal. By constructing the scalar product of $\boldsymbol{e}_j$ with $\boldsymbol{e}'_i$

$$(\boldsymbol{e}_j, \boldsymbol{e}'_i) = (\boldsymbol{e}_j, \mathrm{T}\boldsymbol{e}_i) = \sum_k T_{ki}(\boldsymbol{e}_j, \boldsymbol{e}_k) = \sum_k T_{ki}\,\delta_{jk} = T_{ji} \qquad (3.21)$$

Thus, the matrix element T_{ji} is simply the scalar product $(\boldsymbol{e}_j, \mathrm{T}\boldsymbol{e}_i)$ which has the operator T 'sandwiched' between the two basis vectors $\boldsymbol{e}_j$ and $\boldsymbol{e}_i$. In quantum mechanics one often uses the expression 'matrix element of an operator' to mean just such a scalar product.

3.4 The multiplication, inverse and transformation of operators

The product TS of two operators T and S in a vector space L is defined as the result first of operating with S and then with T.

If

$$\mathrm{S}\boldsymbol{e}_i = \sum_j S_{ji}\boldsymbol{e}_j \quad \text{and} \quad \mathrm{T}\boldsymbol{e}_j = \sum_k T_{kj}\boldsymbol{e}_k$$

then

$$\mathrm{TS}\boldsymbol{e}_i = \sum_j S_{ji}\,\mathrm{T}\boldsymbol{e}_j = \sum_j\sum_k S_{ji}\,T_{kj}\boldsymbol{e}_k = \sum_k\left\{\sum_j T_{kj}S_{ji}\right\}\boldsymbol{e}_k$$

so that the matrix elements of the product TS are given by

$$(\mathrm{TS})_{ki} = \sum_j T_{kj} S_{ji} \tag{3.22}$$

showing that the matrix of the product operator TS is the conventional matrix product (3.22) of the matrices of T and S, in that order. Note that operators, like matrices, do not generally 'commute', i.e. the order of multiplication affects the result. The difference TS − ST between the two ways of multiplying the operators is called the 'commutator' of T and S. It is denoted by [T, S] and is itself an operator. Notice also our convention that the operators always operate on vectors to the right. Consequently, in a product TS one imagines S operating first, followed by T.

Suppose we are given an operator T in a vector space L, and as usual we write $\mathrm{T}\boldsymbol{r} = \boldsymbol{r}'$ and call $\boldsymbol{r}'$ the transformed vector. In general, it is then possible to define the 'inverse' operator T^{-1} by the relation

$$\boldsymbol{r} = \mathrm{T}^{-1}\boldsymbol{r}' \tag{3.23}$$

The inverse T^{-1} has the obvious properties $\mathrm{T}^{-1}\mathrm{T} = \mathrm{T}\mathrm{T}^{-1} = \mathrm{E}$, where E is the identity operator which leaves all vectors unchanged and, in any orthonormal basis, is given by the diagonal unit matrix. The matrix of T^{-1} will be the conventional inverse matrix of T. It is, of course, assumed that the inverse exists and this in turn puts a constraint on T, namely that it is a one-to-one transformation or, in terms of the matrix of T, that its determinant shall not vanish. In the problems of interest in this book the transformation will almost always be one-to-one. Notice that the inverse of a product TS is given by $(\mathrm{TS})^{-1} = \mathrm{S}^{-1}\mathrm{T}^{-1}$, in which the order of multiplication of the inverses is reversed. This is a familiar result from matrix theory and is quickly proved. By definition, the inverse satisfies $(\mathrm{TS})^{-1}\mathrm{TS} = \mathrm{E}$ and multiplying both sides on the right by S^{-1} gives $(\mathrm{TS})^{-1}\mathrm{T} = \mathrm{S}^{-1}$ and again multiplying on the right by T^{-1} gives the required result $(\mathrm{TS})^{-1} = \mathrm{S}^{-1}\mathrm{T}^{-1}$. In practical terms this result is obvious from a study of the operations of putting on one's socks and shoes. In dressing, the socks are put on first followed by the shoes, but in the inverse operation of undressing it is the shoes which must come off first, followed by the socks!

Suppose the operator T transforms $\boldsymbol{r}$ to $\boldsymbol{r}'$ and also transforms a vector $\tilde{\boldsymbol{r}}$ to $\tilde{\boldsymbol{r}}'$, i.e. $\mathrm{T}\boldsymbol{r} = \boldsymbol{r}'$ and $\mathrm{T}\tilde{\boldsymbol{r}} = \tilde{\boldsymbol{r}}'$. Then, if there is a second operator S which transforms $\boldsymbol{r}$ to $\tilde{\boldsymbol{r}}$, i.e. $\mathrm{S}\boldsymbol{r} = \tilde{\boldsymbol{r}}$, we may wish to know the operator S′ which transforms $\boldsymbol{r}'$ to $\tilde{\boldsymbol{r}}'$, i.e. $\mathrm{S}'\boldsymbol{r}' = \tilde{\boldsymbol{r}}'$. If we call the vectors $\boldsymbol{r}'$ and $\tilde{\boldsymbol{r}}'$ the transformed vectors (transformed by T) then the operator S′ is called the 'transformed operator'. To express S′ in terms of the original operators we note that $\tilde{\boldsymbol{r}}' = \mathrm{T}\tilde{\boldsymbol{r}} = \mathrm{TS}\boldsymbol{r} = \mathrm{TST}^{-1}\boldsymbol{r}'$ so that

$$\mathrm{S}' = \mathrm{TST}^{-1} \tag{3.24}$$

The concept of a transformed operator will become more clear when examples are given in section 3. 8 but figure 3.1 may help to visualise the process. Here, a

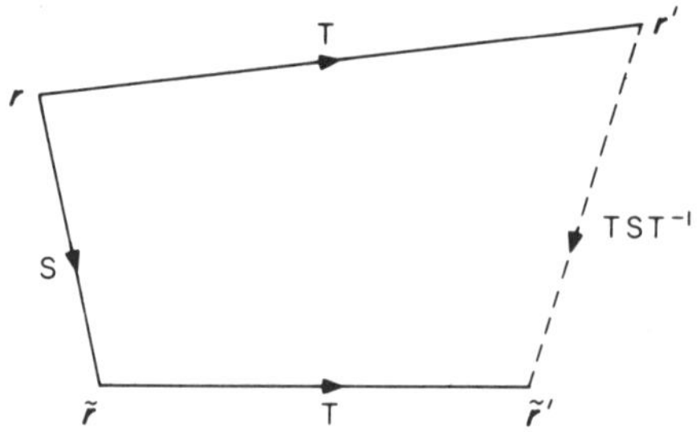

Figure 3.1

vector is represented by a point and an operator by a line with an arrow. No significance is to be attached to distance or angle. Thus the equation $\mathrm{S}\boldsymbol{r} = \tilde{\boldsymbol{r}}$ is represented by the line labelled S from $\boldsymbol{r}$ to $\tilde{\boldsymbol{r}}$. To construct the required transformed operator from $\boldsymbol{r}'$ to $\tilde{\boldsymbol{r}}'$, which is drawn as a dotted line, we may pass round the other three sides of the quadrilateral. This entails first going against the T-arrow, representing the inverse, T^{-1}, to be followed by S and T, giving the previous result TST^{-1}.

More generally, the operator T may be taken to transform from a space L to another space L' so that $\boldsymbol{r}'$, $\tilde{\boldsymbol{r}}'$ and the operator S′ lie in L'.

In the spirit of the earlier example of putting on socks and shoes we might here regard T as the operation of putting on a shoe and S as the operation of mending a hole in the shoe. If the shoe is being worn, then the mending operation is more complex and could be denoted by TST^{-1}, meaning first take off the shoe (T^{-1}), then mend it (S) and finally put it back on (T).

3.5 The adjoint of an operator—unitary and Hermitian operators

The *adjoint* of T is denoted by $\mathrm{T}^\dagger$ and is defined to satisfy the equation

$$(\boldsymbol{r}, \mathrm{T}^\dagger \boldsymbol{s}) = (\mathrm{T}\boldsymbol{r}, \boldsymbol{s}) \tag{3.25}$$

for all vectors $\boldsymbol{r}$ and $\boldsymbol{s}$ in L. By choosing in particular $\boldsymbol{r} = \boldsymbol{e}_i$ and $\boldsymbol{s} = \boldsymbol{e}_j$ and using equation (3.20) it follows that, in an orthonormal basis, the matrices of $\mathrm{T}^\dagger$ and T are related by

$$(\mathrm{T}^\dagger)_{ij} = (T_{ji})^* \tag{3.26}$$

The most direct definition of a 'unitary operator' is to say that T is unitary if it satisfies the equation

$$\mathrm{T}^\dagger\mathrm{T} = \mathrm{E} \tag{3.27}$$

or in other words

$$\mathrm{T}^\dagger = \mathrm{T}^{-1} \tag{3.28}$$

The significance of such a restriction on T is that a transformation by T preserves the scalar product in the sense that, if $\boldsymbol{r}'$ and $\boldsymbol{s}'$ are defined by $\mathrm{T}\boldsymbol{r} = \boldsymbol{r}'$ and $\mathrm{T}\boldsymbol{s} = \boldsymbol{s}'$ then

$$(\boldsymbol{r}', \boldsymbol{s}') = (\mathrm{T}\boldsymbol{r}, \mathrm{T}\boldsymbol{s}) = (\boldsymbol{r}, \mathrm{T}^{\dagger}\mathrm{T}\boldsymbol{s}) = (\boldsymbol{r}, \boldsymbol{s}) \qquad (3.29)$$

In terms of the basis, this means that if the $\boldsymbol{e}_i$ form an orthonormal basis and the $\boldsymbol{e}'_i$ are defined by $\mathrm{T}\boldsymbol{e}_i = \boldsymbol{e}'_i$, then from (3.29) the $\boldsymbol{e}'_i$ form a new orthonormal basis. A unitary operator therefore transforms from one orthonormal basis to another.

The matrix elements of a unitary matrix are clearly related through equation (3.28). In detail this gives

$$\sum_{j=1}^{s} T_{ji}^{*} T_{jk} = \delta_{ik} \qquad (3.30a)$$

and

$$\sum_{j=1}^{s} T_{ij} T_{kj}^{*} = \delta_{ik} \qquad (3.30b)$$

although (3.30b) is implied by (3.30a). If the matrix T is real, then $\mathrm{T}^{\dagger}$ is the transpose and the asterisk may be dropped from these two equations. A real unitary matrix is called an 'orthogonal' matrix.

The word 'Hermitian' (or self-adjoint) is used to describe an operator which is equal to its adjoint. Thus an operator H is Hermitian if

$$\mathrm{H}^{\dagger} = \mathrm{H} \qquad (3.31)$$

which means that

$$(\mathrm{H}\boldsymbol{r}, \boldsymbol{s}) = (\boldsymbol{r}, \mathrm{H}\boldsymbol{s}) \qquad (3.32)$$

for all $\boldsymbol{r}$ and $\boldsymbol{s}$.

Finally, notice that the concepts of unitary and Hermitian operators have meaning only in relation to a well-defined vector space and scalar product.

3.6 The eigenvalue problem

Given an operator T in a vector space L we may ask if there exist vectors $\boldsymbol{r}$ in L which have the very special property that the transformed vector $\mathrm{T}\boldsymbol{r}$ is just a constant multiplied by $\boldsymbol{r}$. In other words, given T can we find $\boldsymbol{r}$ such that

$$\mathrm{T}\boldsymbol{r} = \lambda \boldsymbol{r} \qquad (3.33)$$

for some λ? To solve this problem it is best to express $\boldsymbol{r}$ in some basis $\boldsymbol{e}_i$ and then, using equation (3.19), we have

$$\sum_{j=1}^{s} T_{ij} r_j = \lambda r_i$$

i.e.

$$\sum_{j=1}^{s} (T_{ij} - \lambda\delta_{ij}) r_j = 0 \qquad (i = 1, 2, \ldots, s) \qquad (3.34)$$

where the r_j are the components of $\boldsymbol{r}$. The condition for consistency of the set

(3.34) of s linear homogeneous equations in s unknowns is that the determinant of the coefficients must vanish. This gives an sth order polynomial equation for λ with s roots λ, called eigenvalues of T. To each root λ_i there is a corresponding solution $\boldsymbol{r} = \xi^i$ called an eigenvector of T. Because (3.33) is homogeneous we are still free to normalise the ξ^i. Furthermore, if T is Hermitian it follows that the eigenvectors are orthogonal, or can be made so. Thus the eigenvectors of a Hermitian operator may be taken to provide an orthonormal basis for the space L. The orthogonality proof is straightforward, for if

$$\mathrm{T}\,\xi^i = \lambda_i\,\xi^i$$

and

$$\mathrm{T}\,\xi^j = \lambda_j\,\xi^j$$

then

$$(\xi^j, \mathrm{T}\,\xi^i) - (\xi^i, \mathrm{T}\,\xi^j)^* = \lambda_i(\xi^j, \xi^i) - \lambda_j^*(\xi^i, \xi^j)^*$$

so that

$$(\xi^j, \mathrm{T}\,\xi^i) - (\mathrm{T}\,\xi^j, \xi^i) = (\lambda_i - \lambda_j^*)(\xi^j, \xi^i)$$

and hence

$$(\lambda_i - \lambda_j^*)(\xi^j, \xi^i) = (\xi^j, \mathrm{T}\,\xi^i) - (\xi^j, \mathrm{T}\,\xi^i) = 0 \tag{3.35}$$

using the property (3.32) for a Hermitian operator T. In this result we first put $i = j$ and hence deduce that $\lambda_i - \lambda_i^* = 0$. Thus the eigenvalues of T are real. Then for $i \neq j$ and assuming $\lambda_i \neq \lambda_j$ we have the orthogonality

$$(\xi^j, \xi^i) = 0$$

and by normalising the ξ^i we have, for the eigenvectors of a Hermitian operator,

$$(\xi^i, \xi^j) = \delta_{ij} \tag{3.36}$$

This result can be ensured even when $\lambda_i = \lambda_j$ since, although the proof of orthogonality given above would then fail, the Schmidt process of section 3.1 may be used to form an orthonormal basis within the set of degenerate eigenvectors, i.e. those with the same eigenvalue. The important point here is that any linear combination of degenerate eigenvectors is itself an eigenvector.

3.7 Induced transformation of functions

In subsections 3.2.3 and 3.2.4 it was shown how a set of functions with a suitably defined scalar product formed a vector space. One important way to construct an operator in function space is through the transformation of functions 'induced' by a transformation of the variables on which the function depends. Let us denote the coordinates of some system by a vector $\boldsymbol{r} = (r_1, r_2, \ldots, r_s)$ in s dimensions and consider some group of transformations G of the system and thus of $\boldsymbol{r}$. We now consider the transformations induced in a vector space of functions $\psi(\boldsymbol{r})$ by these transformations G of $\boldsymbol{r}$. We denote this induced transformation by T(G) and define it by the equation

$$T(G)\psi(\boldsymbol{r}) = \psi(G^{-1}\boldsymbol{r}) \tag{3.37}$$

We introduce the notation $\psi'(\boldsymbol{r})$ for the transformed function $T(G)\psi(\boldsymbol{r})$. It is important to realise that there are two vector spaces involved: the space of the coordinate vector $\boldsymbol{r}$, in which G is defined, and the function space of the $\psi(\boldsymbol{r})$, in which the induced transformation T(G) is defined. In terms of the components of $\boldsymbol{r}$ in some basis $\boldsymbol{e}_i$, the transformation (3.37) may be written

$$T(G)\psi(r_1, r_2, \ldots, r_s) = \psi(\bar{r}_1, \bar{r}_2, \ldots, \bar{r}_s) \tag{3.38}$$

where $\bar{r}_i$ is the ith component of $\bar{\boldsymbol{r}} = G^{-1}\boldsymbol{r}$. If the matrix of G is known in the basis $\boldsymbol{e}_i$ then from equation (3.20)

$$\bar{r}_j = \sum_i (G^{-1})_{ji} r_i = \sum_i G^*_{ij} r_i \tag{3.39}$$

for unitary G. The numbers $\bar{r}_j$ can also be interpreted as the components of $\boldsymbol{r}$ in the transformed basis $\boldsymbol{e}'_i = G\boldsymbol{e}_i$, since

$$(\boldsymbol{e}'_i, \boldsymbol{r}) = (G\boldsymbol{e}_i, \boldsymbol{r}) = (\boldsymbol{e}_i, G^{-1}\boldsymbol{r}) = \bar{r}_i \tag{3.40}$$

To find $\psi'(\boldsymbol{r})$ given $\psi(\boldsymbol{r})$ and the matrix of G one simply makes the substitution (3.39) into (3.38).

It may seem perverse to use the inverse G^{-1} in the definition (3.37), since we could equally well have used G. However, the convention of using G^{-1} leads to important simplifications later and has some physical significance. For, suppose $\psi(\boldsymbol{r})$ represents the temperature of a body at the point $\boldsymbol{r}$ in three dimensions and G denotes a rotation of the body about the origin. Then the definition (3.37) ensures that the new function $\psi'(\boldsymbol{r})$ represents the temperature

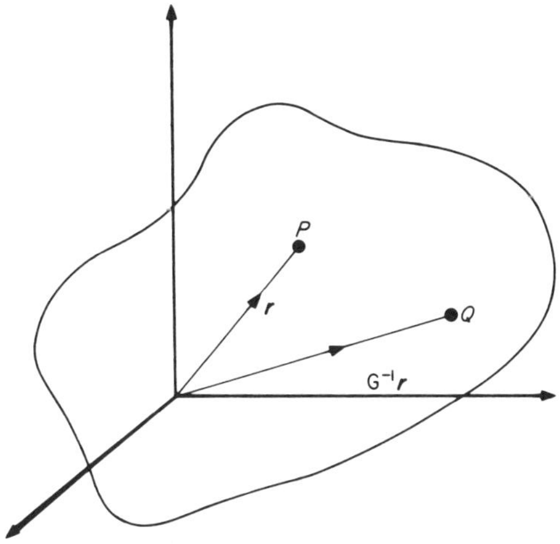

Figure 3.2

at $\boldsymbol{r}$ after the rotation. This is illustrated in figure 3.2, where Q denotes the point $G^{-1}\boldsymbol{r}$. Thus, after rotation G the point Q is carried to P, since $G(G^{-1}\boldsymbol{r}) = \boldsymbol{r}$. Thus, after the rotation the temperature at P denoted by $\psi'(\boldsymbol{r})$ is what it was at Q before the rotation, namely $\psi(G^{-1}\boldsymbol{r})$. We therefore have $\psi'(\boldsymbol{r}) = \psi(G^{-1}\boldsymbol{r})$ and the transformation from $\psi(\boldsymbol{r})$ to $\psi'(\boldsymbol{r})$ defines the operator $T(G)$ in function space. In this book we shall be more often interested in wave functions $\psi(\boldsymbol{r})$ than in temperatures, but the argument is the same and the temperature of a body is somewhat easier to visualise than the value of a wave function at a point.

Mathematically, the reason for choosing the inverse G^{-1} in equation (3.37) becomes clear when two successive operations are considered. Suppose that $T(G_2)$ given as in (3.37) by $T(G_2)\psi(\boldsymbol{r}) = \psi(G_2^{-1}\boldsymbol{r}) = \psi'(\boldsymbol{r})$ defines ψ'. Then we have

$$\begin{aligned} T(G_1)T(G_2)\psi(\boldsymbol{r}) &= T(G_1)\psi(G_2^{-1}\boldsymbol{r}) = T(G_1)\psi'(\boldsymbol{r}) = \psi'(G_1^{-1}\boldsymbol{r}) \\ &= \psi(G_2^{-1}G_1^{-1}\boldsymbol{r}) = \psi((G_1 G_2)^{-1}\boldsymbol{r}) \end{aligned} \tag{3.41}$$

showing that the product $G_1 G_2$ of transformations on the vectors $\boldsymbol{r}$ induces a product $T(G_1)T(G_2)$ of operators in function space, in the same order. The use of G rather than G^{-1} in the definition (3.37) would have led to the association of $T(G_2)T(G_1)$ with $G_1 G_2$ involving a confusing change of order.

An important special case of the transformation (3.37) of functions is provided by the components themselves which may be regarded as functions of $\boldsymbol{r}$ by writing $r_i = (\boldsymbol{e}_i, \boldsymbol{r})$. Equation (3.37) transforms r_i into

$$T(G)r_i = (\boldsymbol{e}_i, G^{-1}\boldsymbol{r}) = {}'\bar{r}_i \tag{3.42}$$

the ith component of $G^{-1}\boldsymbol{r}$.

In the discussion above we have considered only a scalar function of position, with $\psi(\boldsymbol{r})$ associating a number with each $\boldsymbol{r}$. The generalisation to functions with several components, which associate several numbers with each $\boldsymbol{r}$, will be left to chapter 8.

3.8 Examples of linear operators

3.8.1 Rotation of vectors in the xy-plane

Consider a rotation of all vectors in the xy-plane about the z-axis through an angle a. This operation, which is denoted by $R(a)$, is a linear operator in the two-dimensional space of the xy-plane. It clearly satisfies the defining properties (3.16). The matrix is quickly found using equations (3.21) and (2.14):

$$R_{ji}(a) = \boldsymbol{e}_j . R(a)\boldsymbol{e}_i = \boldsymbol{e}_j . \boldsymbol{e}'_i$$

giving $R_{11} = \cos a$, $R_{12} = -\sin a$, $R_{21} = \sin a$, $R_{22} = \cos a$

so that
$$R = \begin{pmatrix} \cos a & -\sin a \\ \sin a & \cos a \end{pmatrix}$$

A rotation preserves distance and angle so that it clearly preserves the scalar product. The matrix is therefore an orthogonal one satisfying equation (3.30).

3.8.2 Permutations

A permutation may often be regarded as a linear operator. Suppose the permutation P_{12} exchanges the vectors $\boldsymbol{e}_1$ and $\boldsymbol{e}_2$ in a two-dimensional space. Then P_{12} has the matrix

$$\begin{pmatrix} 0 & 1 \\ 1 & 0 \end{pmatrix}$$

and satisfies the definitions (3.16). The multiplication of operators is illustrated by considering the product $(P_{12})^2 = E$ which may be checked from the matrix product

$$\begin{pmatrix} 0 & 1 \\ 1 & 0 \end{pmatrix}\begin{pmatrix} 0 & 1 \\ 1 & 0 \end{pmatrix} = \begin{pmatrix} 1 & 0 \\ 0 & 1 \end{pmatrix} = E$$

The eigenvalue problem can also be illustrated in this example. From the matrix, the eigenvalues λ of P_{12} must satisfy

$$\begin{vmatrix} -\lambda & 1 \\ 1 & -\lambda \end{vmatrix} = 0$$

giving $\lambda = \pm 1$ with eigenvectors $(\frac{1}{2})^{\frac{1}{2}}(\boldsymbol{e}_1 \pm \boldsymbol{e}_2)$, respectively. From a geometrical point of view P_{12} is equivalent to a reflection in the 45° diagonal. Clearly, P_{12} leaves the vector $(\boldsymbol{e}_1 + \boldsymbol{e}_2)$ unchanged, while $(\boldsymbol{e}_1 - \boldsymbol{e}_2)$ has its direction changed, corresponding to eigenvalue -1.

3.8.3 Multiplication by a function in function space

The vector space of the example in subsection 3.2.5 consists of all functions which satisfy specified boundary conditions and have finite norm. Multiplication of some such function $\psi(\boldsymbol{r})$ by a continuous function $S(\boldsymbol{r})$ will produce another function $\phi(\boldsymbol{r})$ in the space, given by

$$\phi(\boldsymbol{r}) = S(\boldsymbol{r})\psi(\boldsymbol{r}) \tag{3.43}$$

In this sense, multiplication by $S(\boldsymbol{r})$ may be regarded as a linear operator in the space. Some care must be taken with this idea since, in the example in subsection 3.2.4 of a finite-dimensional space of quadratic functions, multiplication by any function $S(\boldsymbol{r})$ (except a constant) will not give a quadratic and hence is *not* an operator in the space defined.

Following a transformation (3.37) in the function space, the transformed operator (3.24) is given by $S'(\boldsymbol{r}) = S(G^{-1}\boldsymbol{r}) = S(\bar{\boldsymbol{r}})$ in the notation of (3.38).

3.8.4 Differentiation in function space

Given a function $\psi(\boldsymbol{r}) \equiv \psi(x, y, z)$, a new function $\phi(x, y, z) = (\partial/\partial x)$ $\psi(x, y, z)$ may be constructed by differentiation. In this sense, $\partial/\partial x$ is a linear operator in function space provided of course that the function $\phi(x, y, z)$ lies in the given space. In the example in subsection 3.2.3, although $\partial/\partial x$ is not an operator in the space of quadratic functions, it is clear that $y(\partial/\partial x)$ is a linear operator in the space. Notice that the operators defined by multiplication by functions, as in subsection 3.8.3, will commute with each other, since clearly $S_1(\boldsymbol{r})S_2(\boldsymbol{r})\psi(\boldsymbol{r}) = S_2(\boldsymbol{r})S_1(\boldsymbol{r})\psi(\boldsymbol{r})$. However, this is no longer true generally when differentiation is involved since, for example, if we take $\psi(\boldsymbol{r}) = xy$, $S(\boldsymbol{r}) = x$ we have

$$\left.\begin{aligned} &S(\boldsymbol{r})\psi(\boldsymbol{r}) = x^2 y, \text{ giving } \frac{\partial}{\partial x} S(\boldsymbol{r})\psi(\boldsymbol{r}) = 2xy \\ \text{while} \quad &\frac{\partial}{\partial x}\psi(\boldsymbol{r}) = y, \text{ giving } S(\boldsymbol{r})\frac{\partial}{\partial x}\psi(\boldsymbol{r}) = xy \end{aligned}\right\} \tag{3.44}$$

Generally the commutator (see section 3.4) of two operators is non-zero. In this example, the commutator may be deduced explicitly since for any $\psi(\boldsymbol{r})$

$$\begin{aligned} \left[\frac{\partial}{\partial x}, S(\boldsymbol{r})\right]\psi(\boldsymbol{r}) &= \frac{\partial}{\partial x} S(\boldsymbol{r})\psi(\boldsymbol{r}) - S(\boldsymbol{r})\frac{\partial}{\partial x}\psi(\boldsymbol{r}) \\ &= \frac{\partial S(\boldsymbol{r})}{\partial x}\psi(\boldsymbol{r}) \end{aligned}$$

and we may therefore write

$$\left[\frac{\partial}{\partial x}, S(\boldsymbol{r})\right] = \frac{\partial S(\boldsymbol{r})}{\partial x} \tag{3.45}$$

where the result $\partial S/\partial x$ is itself an operator in the sense of the example in subsection 3.8.3, i.e. a function to be multiplied. This result is true for any $S(\boldsymbol{r})$ and, in the particular case above, $\partial S/\partial x = 1$ which agrees with equation (3.44).

3.8.5 Induced transformation of functions

As an example of the general ideas of section 3.7 consider functions $\psi(\boldsymbol{r})$, where $\boldsymbol{r}$ is a vector in two dimensions. Introducing the polar coordinates r, ϕ we have $\psi(\boldsymbol{r}) = \psi(r, \phi)$. Then, according to the definition (3.37), the transformation induced in ψ by a rotation $\mathrm{R}(a)$ is given by

$$\mathrm{T}(\mathrm{R}(a))\psi(\boldsymbol{r}) = \psi(\mathrm{R}^{-1}(a)\boldsymbol{r}) = \psi(r, \phi - a)$$

i.e.

$$\mathrm{T}(\mathrm{R}(a))\psi(r, \phi) = \psi(r, \phi - a)$$

Consider for example, the pair of functions

$$\psi_1(r, \phi) = \cos\phi, \qquad \psi_2(r, \phi) = \sin\phi$$

which give $T(R(a))\psi_1(r,\phi) = \cos(\phi - a) = \cos\phi\cos a + \sin\phi\sin a$
$T(R(a))\psi_2(r,\phi) = \sin(\phi - a) = \sin\phi\cos a - \cos\phi\sin a$

Thus
$$T\psi_1 = \cos a\,\psi_1 + \sin a\,\psi_2$$
$$T\psi_2 = -\sin a\,\psi_1 + \cos a\,\psi_2$$

which illustrates the transformation of ψ_1 and ψ_2 induced by the transformation $R(a)$. One may clearly form more complicated functions than this. Again, if we take

$$\psi(r,\phi) = \exp(im\phi)$$

then
$$T\psi = \exp[im(\phi - a)] = \exp(-ima)\psi$$

showing that this particular function is actually an eigenfunction of T with eigenvalue $\exp(-ima)$. Notice that the functions of interest in this example will normally be periodic functions of ϕ with period 2π, since an increase of 2π in the angle ϕ produces the same vector $\boldsymbol{r}$. Functions of the type $\exp(im\phi)$ will have this property so long as m is an integer, positive or negative. Even with this restriction, there is still an infinite number of eigenvalues $\exp(-ima)$ of the operator T, reflecting the fact that the function space is infinite dimensional.

3.8.6 Further example of induced transformation of functions

As a further example of the transformations induced in a space of functions consider the quadratic functions of subsection 3.2.4 and the rotation R through an angle $2\pi/3$ about the z-axis. For the function $\psi_1 = x^2$ we have, from equation (3.38),

$$T(R_1)\psi_1 = \bar{x}^2$$

where, from equation (3.39) and subsection 3.8.1, $\bar{x} = x\cos 2\pi/3 + y\sin 2\pi/3 = -\frac{1}{2}x + (\frac{3}{4})^{\frac{1}{2}}y$. Thus finally, $T(R_1)\psi_1 = \frac{1}{4}x^2 - (\frac{3}{4})^{\frac{1}{2}}xy + \frac{3}{4}y^2 = \frac{1}{4}\psi_1 - (\frac{3}{4})^{\frac{1}{2}}\psi_6 + \frac{3}{4}\psi_2$, which shows that ψ_1 is transformed into a linear combination of the six functions ψ_i.

3.8.7 Transformed operator

In section 3.4 we defined a transformed operator $S' = TST^{-1}$ which had the same matrix in the transformed basis $\boldsymbol{e}_i' = T\boldsymbol{e}_i$ as had the operator S in the original basis $\boldsymbol{e}_i$. As an example of this, let us take an operator S of the kind discussed in subsection 3.8.3 and a function space and transformation T of the kind also discussed in that subsection. In particular, let us take the operator S as multiplication by a function $\sin 3\phi$ and the vector as $\psi = \cos\phi$ with the transformation $T = T(R(a))$ so that $T\psi = \cos(\phi - a)$. Thus we write

$$\psi' = T\psi = \cos(\phi - a)$$
$$\tilde{\psi} = S\psi = \sin 3\phi\cos\phi$$
$$\tilde{\psi}' = T\tilde{\psi} = \sin 3(\phi - a)\cos(\phi - a)$$

so that the transformed operator S′, which is defined by the relation $S'\psi' = \tilde{\psi}'$, is in this case simply multiplication by the function $\sin 3(\phi - a)$.

Bibliography

For a more detailed account of linear algebra see

Birkhoff, G. and Maclane, S. (1965). *A Survey of Modern Algebra* (Macmillan, London)

Problems

3.1 Show that, for three-dimensional vectors in physical space, the scalar product $(\boldsymbol{r}_1, \boldsymbol{r}_2) = |\boldsymbol{r}_1|\,|\boldsymbol{r}_2| \cos\theta$, where θ is the angle between $\boldsymbol{r}_1$ and $\boldsymbol{r}_2$, satisfies the criteria of section 3.1 for a scalar product.

3.2 If L_1 denotes the one-dimensional subspace defined by the vectors $(a, a, 0)$, show that the orthogonal complement L_2 is defined by the vectors $(b, -b, c)$ and express the vector (1, 2, 3) as a sum of vectors in L_1 and L_2.

3.3 If the scalar product of two functions is defined by $(f, g) = \int_{-1}^{1} f(x)^* g(x)\mathrm{d}x$ show that $f_0(x) = (\frac{1}{2})^{\frac{1}{2}}$ and $f_1(x) = (\frac{3}{2})^{\frac{1}{2}}x$ are orthonormal. Use the Schmidt process to construct a quadratic $f_2(x)$ which is orthonormal to both f_0 and f_1. Continue the process to find a cubic $f_3(x)$.

3.4 Taking the $f_n(x)$ of the previous example as basis vectors and $T = x^2$ as an operator, use equation (3.21) to find the matrix elements T_{01}, T_{11} and T_{02}.

3.5 Show that the operator T in the previous example is Hermitian and discuss the boundary conditions which need to be applied to the functions $f(x)$ at the boundaries ± 1 in order that the operator $i\mathrm{d}/\mathrm{d}x$ should be Hermitian, with the scalar product of question 3.3.

3.6 The transformation R is a rotation through 90° about the z-axis. Find $\mathrm{R}\boldsymbol{e}_i = \boldsymbol{e}_i'$ for the three unit vectors $\boldsymbol{e}_i$ along the x-, y-, and z-axes. Using equations (3.38) and (3.39) or (3.40) find the induced transformations $\mathrm{T(R)}\psi(\boldsymbol{r})$ when $\psi(\boldsymbol{r})$ is (a) x, (b) y, (c) x^2, and (d) xy. Repeat the problem for R a rotation of 45° about the z-axis.

3.7 Starting from the function $f(x, y) = x^2$, use the Schmidt process to construct an orthonormal basis for the space of functions x^2, y^2 and xy of the previous problem when the scalar product is defined to be

$$(f(x, y), g(x, y)) = \int_{-1}^{1} \mathrm{d}x \int_{-1}^{1} \mathrm{d}y\, f(x, y)g(x, y).$$

Hence construct the matrix T(R) for the case when R is a 90° rotation about the z-axis. Diagonalise this matrix and find the eigenvalues and eigenvectors.

3.8 If R is a rotation through 45° about the z-axis and T a rotation through 90° about the x-axis, show geometrically that TRT^{-1} is a rotation through 45° about the negative y-axis. Verify this result by multiplying the matrices of these operators in the basis $\boldsymbol{e}_x$, $\boldsymbol{e}_y$, $\boldsymbol{e}_z$. (This is an example of the general result (2.12).)

3.9 Show that if S is the operator $\partial/\partial r_i$ then the transformed operator TST^{-1} is given by $\partial/\partial \bar{r}_i$ in the notation of equation (3.38).

4

Group Representations

In the previous two chapters we have introduced two mathematical ideas—group theory in chapter 2 and vector spaces in chapter 3. These two ideas are now brought together and we examine the association between group elements and transformations in a vector space. At a number of points in this chapter, from section 4.6 onwards, we need to perform a sum over all elements of a group. This raises no difficulties for a finite group but for a continuous group (see section 2.2) the sum would have to be replaced by an integral and questions of convergence arise. We leave such questions until chapter 7. For the moment we simply remark that for most continuous groups of physical interest, the sum may be replaced by an appropriately defined integral.

4.1 Definition of a group representation

If we can find a set T of linear operators $\mathrm{T}(\mathrm{G}_a)$ (see section 3.3), in a vector space L, which correspond to the elements G_a of a group $\mathscr{G}$ in the sense that

$$\mathrm{T}(\mathrm{G}_a)\mathrm{T}(\mathrm{G}_b) = \mathrm{T}(\mathrm{G}_a\mathrm{G}_b), \qquad \mathrm{T}(\mathrm{E}) = 1 \tag{4.1}$$

then this set of operators is said to form a 'representation' of the group $\mathscr{G}$ in the space L. A representation of a group $\mathscr{G}$ is thus a 'mapping' of the elements G_a on to the operators $\mathrm{T}(\mathrm{G}_a)$ in the vector space L. If the dimension of the space L

is s then the representation is said to be s-dimensional. The space L is called the 'representation space' of T.

The representation $\mathrm{T}(\mathrm{G}_a)$ is said to be 'faithful' if there is a one-to-one relationship between the operations $\mathrm{T}(\mathrm{G}_a)$ and the group elements G_a (an isomorphism). In general, the mapping may be many-to-one, with several group elements being represented by the same operator. For instance there is the extreme example of the 'identity representation' in which all elements are represented by the unit operator 1.

4.2 Matrix representations

In practice, one usually defines an operator by its matrix with respect to some chosen basis. Hence by choosing a basis $\boldsymbol{e}_1, \boldsymbol{e}_2, \ldots, \boldsymbol{e}_s$ in L we can form a matrix for each operator $\mathrm{T}(\mathrm{G}_a)$ as in section 3.3 by the equations

$$\mathrm{T}(\mathrm{G}_a)\boldsymbol{e}_i = \sum_j T_{ji}(\mathrm{G}_a)\boldsymbol{e}_j \tag{4.2}$$

The set of matrices $\mathrm{T}(\mathrm{G}_a)$ with matrix elements $T_{ji}(\mathrm{G}_a)$ form a matrix representation of the group. As one would expect, the matrices $\mathrm{T}(\mathrm{G}_a)$ obey equation (4.1) under ordinary matrix multiplication

$$\mathrm{T}(\mathrm{G}_a)\mathrm{T}(\mathrm{G}_b) = \mathrm{T}(\mathrm{G}_a\mathrm{G}_b) \tag{4.3}$$

This follows directly from (4.1) and (4.2), since

$$\begin{aligned}\mathrm{T}(\mathrm{G}_a)\mathrm{T}(\mathrm{G}_b)\boldsymbol{e}_i &= \sum_j \mathrm{T}(\mathrm{G}_a)\, T_{ji}(\mathrm{G}_b)\boldsymbol{e}_j \\ &= \sum_j \sum_k T_{ji}(\mathrm{G}_b)\, T_{kj}(\mathrm{G}_a)\boldsymbol{e}_k\end{aligned}$$

and

$$\begin{aligned}\mathrm{T}(\mathrm{G}_a)\mathrm{T}(\mathrm{G}_b)\boldsymbol{e}_i &= \mathrm{T}(\mathrm{G}_a\mathrm{G}_b)\boldsymbol{e}_i \\ &= \sum_k T_{ki}(\mathrm{G}_a\mathrm{G}_b)\boldsymbol{e}_k\end{aligned}$$

so that

$$T_{ki}(\mathrm{G}_a\mathrm{G}_b) = \sum_j T_{kj}(\mathrm{G}_a)\, T_{ji}(\mathrm{G}_b)$$

To find the matrix elements in practice, when using an orthonormal basis, it is often more convenient to use the relation

$$T_{ji}(\mathrm{G}_a) = (\boldsymbol{e}_j, \mathrm{T}(\mathrm{G}_a)\boldsymbol{e}_i) \tag{4.4}$$

which follows directly from equation (4.2).

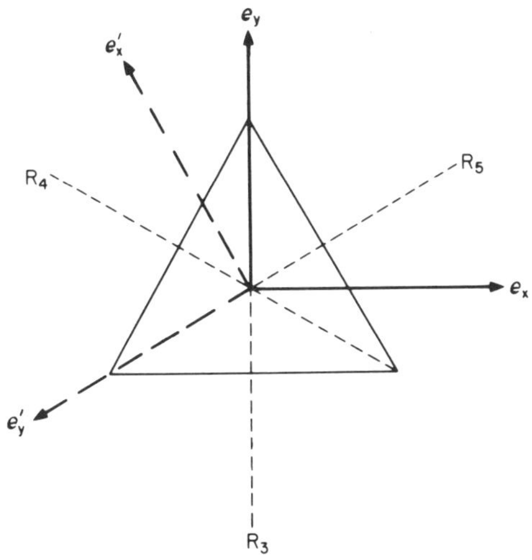

Figure 4.1

4.3 Examples of representations

4.3.1 The group D_3

In order to bring out the physical significance of a representation we will look now at the group D_3 introduced in subsection 2.2(6) and construct a matrix representation for it. We can form a faithful representation immediately by writing down the transformations induced by each operation in the ordinary three-dimensional Cartesian space. Let us choose basis vectors $\boldsymbol{e}_x$ and $\boldsymbol{e}_y$, as in figure 4.1, with $\boldsymbol{e}_z$ pointing up out of the paper. Then, for example, the group element R_1 which rotates through 120° about the z-axis gives rise to the mapping

$$\begin{aligned} \mathrm{T}(\mathrm{R}_1)\boldsymbol{e}_x &= \boldsymbol{e}'_x = -\tfrac{1}{2}\boldsymbol{e}_x + (\tfrac{3}{4})^{\frac{1}{2}}\boldsymbol{e}_y \\ \mathrm{T}(\mathrm{R}_1)\boldsymbol{e}_y &= \boldsymbol{e}'_y = -(\tfrac{3}{4})^{\frac{1}{2}}\boldsymbol{e}_x - \tfrac{1}{2}\boldsymbol{e}_y \\ \mathrm{T}(\mathrm{R}_1)\boldsymbol{e}_z &= \boldsymbol{e}'_z = \boldsymbol{e}_z \end{aligned} \tag{4.5}$$

and hence, using equation (4.4), to the matrix

$$\mathrm{T}(\mathrm{R}_1) = \begin{pmatrix} -\frac{1}{2} & -\sqrt{\frac{3}{4}} & 0 \\ \sqrt{\frac{3}{4}} & -\frac{1}{2} & 0 \\ 0 & 0 & 1 \end{pmatrix}$$

In the same way, for the other group elements

$$\mathrm{T}(\mathrm{R}_2)=\begin{pmatrix} -\frac{1}{2} & \sqrt{\frac{3}{4}} & 0 \\ -\sqrt{\frac{3}{4}} & -\frac{1}{2} & 0 \\ 0 & 0 & 1 \end{pmatrix}, \qquad \mathrm{T}(\mathrm{R}_3)=\begin{pmatrix} -1 & 0 & 0 \\ 0 & 1 & 0 \\ 0 & 0 & -1 \end{pmatrix}$$

$$\mathrm{T}(\mathrm{R}_4)=\begin{pmatrix} \frac{1}{2} & -\sqrt{\frac{3}{4}} & 0 \\ -\sqrt{\frac{3}{4}} & -\frac{1}{2} & 0 \\ 0 & 0 & -1 \end{pmatrix}, \qquad \mathrm{T}(\mathrm{R}_5)=\begin{pmatrix} \frac{1}{2} & \sqrt{\frac{3}{4}} & 0 \\ \sqrt{\frac{3}{4}} & -\frac{1}{2} & 0 \\ 0 & 0 & -1 \end{pmatrix}$$

$$\text{with } \mathrm{T}(\mathrm{E})=\begin{pmatrix} 1 & 0 & 0 \\ 0 & 1 & 0 \\ 0 & 0 & 1 \end{pmatrix}$$

It is readily verified that these matrices possess the same multiplication table as the group elements. Thus for example we have

$$\mathrm{T}(\mathrm{R}_1)\mathrm{T}(\mathrm{R}_4)=\mathrm{T}(\mathrm{R}_5)$$

which is consistent with the entry $\mathrm{R}_1\mathrm{R}_4=\mathrm{R}_5$ in table 2.5.

By taking the one-dimensional space of the vector $\boldsymbol{e}_z$ alone, we may generate a very simple one-dimensional representation which we denote by $\mathrm{T}^{(2)}$,

$$\mathrm{T}^{(2)}(\mathrm{R}_1)=1,\ \mathrm{T}^{(2)}(\mathrm{R}_2)=1,\ \mathrm{T}^{(2)}(\mathrm{R}_3)=-1,\ \mathrm{T}^{(2)}(\mathrm{R}_4)=-1,$$
$$\mathrm{T}^{(2)}(\mathrm{R}_5)=-1,\ \mathrm{T}^{(2)}(\mathrm{E})=1$$

Notice that $\mathrm{T}^{(2)}$ is not the same as the identity representation which we denote by $\mathrm{T}^{(1)}(\mathrm{R}_i)=1$, associating $+1$ with every group element. Notice also that the three numbers -1 are associated with the three elements which belong to the same class $\mathscr{C}_3$ (see section 2.7) of the group D_3. This is an example of a general feature. It will be explained in section 4.9 that representations of elements in the same class have some common properties.

Because of the presence of the zeros in the third row and third column of the matrices $\mathrm{T}(\mathrm{R}_i)$ it is clear that the 2×2 matrices formed by the first two rows and columns are themselves a representation of the group D_3, which we denote by $\mathrm{T}^{(3)}$.

4.3.2 The group $\mathscr{R}_2$

We may use the same space as in the previous example to generate a representation of the infinite group $\mathscr{R}_2$ of rotations about the z-axis. The group elements $\mathrm{R}(a)$ are now labelled by the continuous parameter a in the range $0 \leq a < 2\pi$ and the matrix was given in section 3.8 as

$$\mathrm{T}(a) = \begin{pmatrix} \cos a & -\sin a & 0 \\ \sin a & \cos a & 0 \\ 0 & 0 & 1 \end{pmatrix} \tag{4.6}$$

One immediately verifies that

$$\mathrm{T}(a)\mathrm{T}(b) = \mathrm{T}(a+b) \tag{4.7}$$

for any a and b, consistent with the multiplication rule $\mathrm{R}(a)\,\mathrm{R}(b) = \mathrm{R}(a+b)$ for the group elements.

4.3.3 Function spaces

The first two examples of representations were constructed in the familiar physical space of three dimensions and it may be difficult to visualise at this stage how representations of dimension greater than three may be constructed for groups like D_3 and $\mathscr{R}_2$. To show how this may be achieved we consider the transformations of functions induced by such coordinate rotations as in equation (3.37), to form representations in function space. These will be most important in applications of group theory in quantum mechanics.

Suppose we have a space L of functions $\psi(\boldsymbol{r})$ of some coordinates $\boldsymbol{r}$ which is invariant under a group of coordinate transformations G_a in the sense that if $\psi(\boldsymbol{r})$ belongs to the space L then so also does $\psi(\mathrm{G}_a^{-1}\boldsymbol{r})$ for all G_a in the group. We can then define a representation T in the function space L by the transformations

$$\mathrm{T}(\mathrm{G}_a)\psi(\boldsymbol{r}) = \psi(\mathrm{G}_a^{-1}\boldsymbol{r}) \tag{4.8}$$

of the kind discussed in section 3.7. Again it is easily verified that this definition satisfies the condition (4.1) since, defining $\psi'(\boldsymbol{r}) = \psi(\mathrm{G}_b^{-1}\boldsymbol{r})$, we have

$$\begin{aligned}\mathrm{T}(\mathrm{G}_a)\mathrm{T}(\mathrm{G}_b)\psi(\boldsymbol{r}) &= \mathrm{T}(\mathrm{G}_a)\psi(\mathrm{G}_b^{-1}\boldsymbol{r}) = \mathrm{T}(\mathrm{G}_a)\psi'(\boldsymbol{r}) = \psi'(\mathrm{G}_a^{-1}\boldsymbol{r}) \\ &= \psi(\mathrm{G}_b^{-1}\mathrm{G}_a^{-1}\boldsymbol{r}) = \psi((\mathrm{G}_a\mathrm{G}_b)^{-1}\boldsymbol{r}) = T(\mathrm{G}_a\mathrm{G}_b)\psi(\boldsymbol{r})\end{aligned}$$

(It is vital in this proof to introduce the new function $\psi'(\boldsymbol{r}) = \psi(\mathrm{G}_b^{-1}\boldsymbol{r})$ since, in general, $\mathrm{T}(\mathrm{G}_a)\psi(\mathrm{G}_b^{-1}\boldsymbol{r}) \neq \psi(\mathrm{G}_a^{-1}\mathrm{G}_b^{-1}\boldsymbol{r})$.)

The matrix representation in function space is obtained from the general expression, equation (4.2), by introducing a basis $\psi_i(\boldsymbol{r})$ in function space and expanding

$$\mathrm{T}(\mathrm{G}_a)\psi_i(\boldsymbol{r}) = \psi_i(\mathrm{G}_a^{-1}\boldsymbol{r}) = \psi_i'(\boldsymbol{r}) = \sum_j T_{ji}(\mathrm{G}_a)\psi_j(\boldsymbol{r}) \tag{4.9}$$

The $\psi_i(\boldsymbol{r})$ are particular examples of the abstract $\boldsymbol{e}_i$.

As an example, the six-dimensional space L of quadratic functions introduced in subsection 3.2.4 is clearly invariant under any rotation and in particular under the group D_3. For example, $T(R_1)\psi_1 = \frac{1}{4}\psi_1 - (\frac{3}{4})^{\frac{1}{2}}\psi_6 + \frac{3}{4}\psi_2$. Continuing in this way, we find the matrix

$$T(R_1) = \begin{pmatrix} \frac{1}{4} & \frac{3}{4} & 0 & 0 & 0 & \sqrt{\frac{3}{4}} \\ \frac{3}{4} & \frac{1}{4} & 0 & 0 & 0 & -\sqrt{\frac{3}{4}} \\ 0 & 0 & 1 & 0 & 0 & 0 \\ 0 & 0 & 0 & -\frac{1}{2} & \sqrt{\frac{3}{4}} & 0 \\ 0 & 0 & 0 & -\sqrt{\frac{3}{4}} & -\frac{1}{2} & 0 \\ -\sqrt{\frac{3}{4}} & \sqrt{\frac{3}{4}} & 0 & 0 & 0 & -\frac{1}{2} \end{pmatrix}$$

The other five matrices may be deduced in a similar way and will again be found to satisfy the same multiplication table as the group elements.

It is worth remarking that multiplication of the functions $\psi(\boldsymbol{r})$ by any scalar function $f(r)$, where $r = |\boldsymbol{r}|$, leaves the representation quite unchanged and, provided that $f(r)$ decreases sufficiently rapidly for large r, enables the volume V of the scalar product to be extended to infinity. The function $\psi(\boldsymbol{r})$ could then represent a wave function for a particle moving in a spherically symmetric potential about the origin, like the electron in a hydrogen atom.

4.4 The generation of an invariant subspace

If an operator T is defined in a space L then it may be possible to find some subspace L_1 of L with the property that if $\boldsymbol{r}_1$ is any vector in L_1 then the transformed vector $\boldsymbol{r}'_1 = T\boldsymbol{r}_1$ also lies in the subspace L_1. Such a subspace is called 'invariant' with respect to the operator T. More generally, given a set of operators $T(G_a)$ in L which form a representation of a group $\mathscr{G}$, it may be possible to find a subspace L_1 which is invariant with respect to all $T(G_a)$ as G_a runs through a group $\mathscr{G}$. The subspace L_1 is then said to be invariant with respect to the transformations induced by the group $\mathscr{G}$ and we generally use the term 'invariant subspace' in this latter sense.

In the example of subsection 4.3.1 it is clear from the matrices that the subspace defined by the vectors $\boldsymbol{e}_x$ and $\boldsymbol{e}_y$ is invariant as is its orthogonal complement (see section 3.1), the one-dimensional space defined by $\boldsymbol{e}_z$. The six-dimensional function space of the example in subsection 4.3.3 is itself a subspace of the space of all continuous functions which has an infinite number of dimensions.

The main point of this section is to show how to generate an invariant subspace starting from a single vector of the space. Let $\boldsymbol{r}$ be an arbitrary vector in L and define the set of g vectors $\boldsymbol{r}_a$ by the equation

$$\boldsymbol{r}_a = T(G_a)\boldsymbol{r} \tag{4.10}$$

where G_a runs through the g elements of a group $\mathscr{G}$. It follows immediately that the set of vectors $\boldsymbol{r}_a$ span an invariant subspace of L since for any b,

$$\begin{aligned} T(G_b)\boldsymbol{r}_a &= T(G_b)T(G_a)\boldsymbol{r} = T(G_bG_a)\boldsymbol{r} \\ &= T(G_c)\boldsymbol{r} = \boldsymbol{r}_c \end{aligned} \tag{4.11}$$

where the group element G_c is given by $G_c = G_bG_a$.

If all the vectors $\boldsymbol{r}_a$ are linearly independent, they will form a basis for a g-dimensional representation of the group, since equation (4.11) is a particular form of the general equation (4.2) for a representation. In fact the matrix of $T(G_b)$ would be given by $T_{ji}(G_b) = 1$ if the group elements labelled by those indices b, i and j satisfy the relation $G_bG_i = G_j$. Otherwise, the matrix elements are all zero.

In general, the vectors $\boldsymbol{r}_a$ will not be linearly independent but it will always be possible to construct a number $s \leq g$ of linearly independent basis vectors as linear combinations of the $\boldsymbol{r}_a$. It is usually convenient to make the independent basis vectors orthonormal by the Schmidt process. To illustrate this generation procedure we return to the example in subsection 4.3.3. In particular we generate from the single function $\psi_1(\boldsymbol{r}) = x^2$ using the group D_3,

$$\begin{aligned}
T(E)\psi_1 &= \psi_1 \\
T(R_1)\psi_1 &= \tfrac{1}{4}x^2 + \tfrac{3}{4}y^2 - (\tfrac{3}{4})^{\frac{1}{2}}xy \\
T(R_2)\psi_1 &= \tfrac{1}{4}x^2 + \tfrac{3}{4}y^2 + (\tfrac{3}{4})^{\frac{1}{2}}xy \\
T(R_3)\psi_1 &= \psi_1 \\
T(R_4)\psi_1 &= \tfrac{1}{4}x^2 + \tfrac{3}{4}y^2 - (\tfrac{3}{4})^{\frac{1}{2}}xy \\
T(R_5)\psi_1 &= \tfrac{1}{4}x^2 + \tfrac{3}{4}y^2 + (\tfrac{3}{4})^{\frac{1}{2}}xy
\end{aligned}$$

By inspection it is clear that the generated functions are not linearly independent, that the generated space has only three dimensions and that the three functions x^2, y^2 and xy will provide a basis, though not an orthornormal one. The matrix representation in this basis is then deduced using equation (4.2), giving

$$T(R_1) = \begin{pmatrix} \frac{1}{4} & \frac{3}{4} & \sqrt{\frac{3}{4}} \\ \frac{3}{4} & \frac{1}{4} & -\sqrt{\frac{3}{4}} \\ -\sqrt{\frac{3}{4}} & \sqrt{\frac{3}{4}} & -\frac{1}{2} \end{pmatrix}, \quad T(R_2) = \begin{pmatrix} \frac{1}{4} & \frac{3}{4} & -\sqrt{\frac{3}{4}} \\ \frac{3}{4} & \frac{1}{4} & \sqrt{\frac{3}{4}} \\ \sqrt{\frac{3}{4}} & -\sqrt{\frac{3}{4}} & -\frac{1}{2} \end{pmatrix}, \text{ etc.}$$

The Schmidt procedure for finding an orthonormal basis (section 3.1) involves rather more arithmetic. One first writes $\phi_1 = Ax^2$ with A chosen to normalise ϕ_1 using the scalar product (3.13)

$$\iiint_V \phi_1^* \phi_1 \, dV = 1$$

giving

$$\phi_1 = \left(\frac{35}{4\pi}\right)^{\frac{1}{2}} x^2$$

Then put $\phi_2 = B(y^2 - C\phi_1)$ and find C and B by insisting on the orthogonality

$(\phi_1, \phi_2) = 0$ and the normalisation $(\phi_2, \phi_2) = 1$. This gives

$$\phi_2 = \left(\frac{35}{32\pi}\right)^{\frac{1}{2}} (3y^2 - x^2)$$

It is soon verified that, because it is an odd function in x and y, the function xy is orthogonal to both ϕ_1 and ϕ_2. Normalising, we have

$$\phi_3 = \left(\frac{105}{4\pi}\right)^{\frac{1}{2}} xy$$

4.5 Irreducibility

The examples in section 4.4 show that it is possible to construct matrix representations of ever-increasing size by increasing the complexity of the function space. A study of the possible representations of even a simple group like D_3 would therefore seem a very daunting proposition. But we are saved by the following very remarkable property of group representations. For a finite group all representations may be 'built up' from a finite number of 'distinct' irreducible representations. The group D_3, for example, has only three distinct irreducible representations, two of which are one dimensional and one is two dimensional. We have introduced several new words here which will be carefully defined very soon but let us first look back at the example in subsection 4.3.1 for an illustration. In that example the representation had dimension three. However, a glance at the matrices reveals that the 2×2 matrices obtained by taking only the first two rows and columns form a two-dimensional representation, while the diagonal matrix elements in the third row and column form a one-dimensional representation. This is possible because of the zeros in the coupling positions between the first two rows and columns and the third. In terms of the vector space this means that the two vectors $\boldsymbol{e}_x$ and $\boldsymbol{e}_y$ form an invariant vector space, while the single vector $\boldsymbol{e}_z$ forms a second invariant vector space which is orthogonal to the first. In such a situation we say that the three-dimensional representation has reduced into a 'sum' of a two-dimensional representation and a one-dimensional representation. It is obvious that the one-dimensional representation cannot be further reduced and by trial one could see that the two-dimensional representation also cannot be reduced. By this we mean that it is impossible to find a new basis $\boldsymbol{e}_1 = \alpha\boldsymbol{e}_x + \beta\boldsymbol{e}_y$ and $\boldsymbol{e}_2 = \beta\boldsymbol{e}_x - \alpha\boldsymbol{e}_y$ such that the matrix elements

$$T_{12}(\mathrm{R}_a) = (\boldsymbol{e}_1, \mathrm{T}(\mathrm{R}_a)\boldsymbol{e}_2) \text{ and } T_{21}(\mathrm{R}_a) = (\boldsymbol{e}_2, \mathrm{T}(\mathrm{R}_a)\boldsymbol{e}_1)$$

in the new basis are zero for all elements R_a of the group D_3. Such a representation, which cannot be reduced, is called 'irreducible'.

From a physical point of view the concept of irreducibility is of crucial importance since, as we show in the next chapter, the wave functions describing the stationary states of a symmetrical system with the same energy will in general provide the basis functions for an irreducible representation of the group of symmetry operations.

Let us now define the concept of reduction in more general terms. Let L be a space which is invariant with respect to the transformations $\mathrm{T}(\mathrm{G}_a)$ induced by some group $\mathscr{G}$ of elements G_a. Then if L_1 is a subspace of L which is invariant and if L_2, the orthogonal complement (see section 3.1) of L_1, is also invariant then the representation T is said to reduce. If it is impossible to find such a subspace, then the representation T is said to be 'irreducible'. It is essential in this definition of reducibility that both the subspace L_1 and its orthogonal complement L_2 should be invariant. Once again there is a fortunate simplification since, if the representation operators $\mathrm{T}(\mathrm{G}_a)$ are unitary, the invariance of L_1 implies the invariance of L_2. The proof is immediate. Let us denote the basis vectors of L_1 by $\boldsymbol{e}_i$ and those of L_2 by $\tilde{\boldsymbol{e}}_j$, so that by definition of L_2, $(\boldsymbol{e}_i, \tilde{\boldsymbol{e}}_j) = 0$. From the invariance of L_1 we have that $(\mathrm{T}(\mathrm{G}_a)\boldsymbol{e}_i, \tilde{\boldsymbol{e}}_j) = 0$ for all G_a and from the unitarity of T this is equivalent to $(\boldsymbol{e}_i, \mathrm{T}(\mathrm{G}_a^{-1})\tilde{\boldsymbol{e}}_j) = 0$ which implies also that $(\boldsymbol{e}_i, \mathrm{T}(\mathrm{G}_a)\tilde{\boldsymbol{e}}_j) = 0$ for all G_a. This last equation shows that the vectors $\mathrm{T}(\mathrm{G}_a)\tilde{\boldsymbol{e}}_j$ are orthogonal to the $\boldsymbol{e}_i$ and so must lie in L_2. Hence L_2 is invariant, as required. The restriction to unitary representations is not severe since, as we shall show in section 4.6, almost all of the representations of interest in physical problems are unitary. It is therefore possible to divide any space L into a sum of subspaces L_q (not necessarily only two in number) each of which is invariant and irreducible, although this division is not necessarily unique. We may write $L = L_1 + L_2 + L_3 + \ldots$ where each L_q is irreducible and invariant under the transformations $\mathrm{T}(\mathrm{G}_a)$. Correspondingly, we write the reduction of the representation as

$$\mathrm{T}(\mathrm{G}_a) = \mathrm{T}^{(1)}(\mathrm{G}_a) \dot{+} \mathrm{T}^{(2)}(\mathrm{G}_a) \dot{+} \mathrm{T}^{(3)}(\mathrm{G}_a) \dot{+} \ldots \tag{4.12}$$

where $\mathrm{T}^{(q)}(\mathrm{G}_a)$ is the irreducible representation induced in the space L_q. This equation is to be interpreted as a sum of operators $\mathrm{T}^{(q)}(\mathrm{G}_a)$ which operate in different spaces L_q, and the 'dot' over the plus sign reminds us of this fact.

In terms of matrices, if the ordering of the basis vectors is chosen so that those belonging to L_1 are written first to be followed by those of L_2 and so on, then the matrix will appear in a 'block diagonal' form with zeros elsewhere, as illustrated below. Here, the matrix $\mathrm{T}^{(q)}(\mathrm{G}_a)$ is a square matrix of dimension equal to that of the subspace L_q, while the entries 0 denote rectangular zero matrices,

$$\begin{pmatrix} \mathrm{T}^{(1)} & 0 & 0 & . & . & 0 \\ 0 & \mathrm{T}^{(2)} & 0 & . & . & 0 \\ 0 & 0 & \mathrm{T}^{(3)} & . & . & 0 \\ . & . & . & & & \\ . & . & . & & & \\ 0 & 0 & 0 & & & \end{pmatrix}$$

Each $T^{(q)}(G_a)$ is an irreducible matrix representation of the group $\mathscr{G}$. Given an arbitrary matrix representation T it is, of course, necessary to transform carefully to the new basis appropriate to the subspaces L_q in order to achieve the simple block diagonal form for the matrix. We write the equation

$$T(G_a) = T^{(1)}(G_a) \dot{+} T^{(2)}(G_a) \dot{+} T^{(3)}(G_a) + \ldots \tag{4.13}$$

for the matrix reduction, analogous to equation (4.12) for the operators. The 'dot' over the plus sign again reminds us that this is not the usual matrix addition but signifies that T is composed of the square matrices $T^{(q)}(G_a)$ arranged down the diagonal.

Given an arbitrary vector $\boldsymbol{r}$ we may construct an invariant space L as described in section 4.4 by the operations $T(G_a)\boldsymbol{r}$. If this space is then reduced into irreducible subspaces L_q it follows that the vector $\boldsymbol{r}$ may be written $\boldsymbol{r} = \sum_q \boldsymbol{r}_q$, where $\boldsymbol{r}_q$ lies in L_q. Thus the arbitrary vector $\boldsymbol{r}$ is said to be analysed into irreducible components $\boldsymbol{r}_q$.

4.6 Equivalent representations

The reduction process described in section 4.5 enables us in principle to reduce any representation to its constituent irreducible representations. Henceforth, therefore, we shall largely restrict our attention to the properties of irreducible representations, knowing that the properties of any reducible representation may be deduced from them. Even so, there will still be an infinite number of possible irreducible representations, as may be seen from the example in subsection 4.3.1. The space defined by the two vectors $\boldsymbol{e}_x$ and $\boldsymbol{e}_y$ was irreducible and provided a two-dimensional irreducible representation of D_3. However, a different choice of basis vectors within the space would give rise to a different set of matrices $T(G_a)$. One might hope that such a trivial change of basis would leave the essential properties of the representation unchanged and this is indeed true. We now introduce the concept of the equivalence of representations which puts this idea into precise form.

Let $T(G_a)$ denote a representation of a group $\mathscr{G}$ in a space L. Then if A is a mapping from L on to a space L' with the same dimension, it follows that the set of operators

$$T'(G_a) = AT(G_a)A^{-1} \tag{4.14}$$

which act in L' also form a representation of $\mathscr{G}$. This is readily proved, since

$$\begin{aligned} T'(G_a)T'(G_b) &= AT(G_a)A^{-1}AT(G_b)A^{-1} \\ &= AT(G_a)T(G_b)A^{-1} \\ &= AT(G_aG_b)A^{-1} \\ &= T'(G_aG_b) \end{aligned}$$

which is the defining property (4.10) for a representation. The two representations T′ and T are said to be 'equivalent'. It is vital that the mapping A be the same for all group elements G_a. As a particular case L' and L may be the same space.

If T′ and T are equivalent and if T″ and T′ are equivalent then it follows that T″ and T are also equivalent, so that one has the concept of a class of mutually equivalent representations.

For matrix representations a change of basis produces an equivalent matrix representation. In detail, let $T(G_a)$ be a set of operators with matrices $T_{ji}(G_a)$ in a basis $\boldsymbol{e}_i$. Then if the new basis is given by $\boldsymbol{e}'_i = A\boldsymbol{e}_i$ the matrix of $T(G_a)$ in the new basis is given by the matrix product $A^{-1}T(G_a)A$, since

$$T\boldsymbol{e}'_i = TA\boldsymbol{e}_i = \sum_j (TA)_{ji}\boldsymbol{e}_j = \sum_{jk} (TA)_{ji}(A^{-1})_{kj}\boldsymbol{e}'_k$$
$$= \sum_k (A^{-1}TA)_{ki}\boldsymbol{e}'_k$$

The reason for the interchange of A and A^{-1} as compared with equation (4.14) is that whereas (4.14) described a new operator, the present discussion concerns the same operator $T(G_a)$ but referred to a new basis.

One might expect that the important properties of a representation would be common to any two equivalent representations and indeed this will be found to be the case. As a result, we may restrict our attention to only one representation from each class of equivalent representations. In particular, we need consider only unitary representations because of a result known as Maschke's theorem which states that, for finite groups, every class of equivalent representations contains unitary representations†. The theorem is also true for most infinite groups of interest in physics.

4.6.1 Proof of Maschke's theorem

We need to show that any representation is equivalent to a unitary representation. Given a representation $T(G_a)$ we must find an operator S such that the equivalent representation $T'(G_a) = ST(G_a)S^{-1}$ is unitary. In fact we shall show that the operator $S = \{\sum_b T^\dagger(G_b)T(G_b)\}^{\frac{1}{2}}$ will suffice, although we shall not attempt to explain the inspiration which led to this choice. To prove unitarity we must show that

$$T'(G_a)^\dagger = T'(G_a)^{-1} \tag{4.15}$$

The first step is to write

$$\begin{aligned} T^\dagger(G_a)S^2T(G_a) &= \sum_b T^\dagger(G_a)T^\dagger(G_b)T(G_b)T(G_a) \\ &= \sum_b T^\dagger(G_bG_a)T(G_bG_a) \\ &= \sum_c T^\dagger(G_c)T(G_c) \\ &= S^2 \end{aligned} \tag{4.16}$$

† This is not the form usually attributed to Maschke but it is most convenient for our purpose.

where $G_c = G_b G_a$. We have used the group property that, if G_a is fixed and the element G_b runs through the group, covering each element once and once only then G_c also runs through the group elements once and once only—see section 2.9. Now, post-multiplying both sides of (4.16) by $T^{-1}(G_a)S^{-1}$ and pre-multiplying by S^{-1} we have

$$S^{-1}T^{\dagger}(G_a)S = ST^{-1}(G_a)S^{-1}$$

i.e.

$$(ST(G_a)S^{-1})^{\dagger} = (ST(G_a)S^{-1})^{-1}$$

which is the required unitarity condition (4.15). We have used the fact that the operator S is Hermitian.

4.7 Inequivalent irreducible representations

Two representations T and T′ are said to be 'inequivalent' if there exists no operator A which satisfies equation (4.14) for all G_a of the group. It is also convenient to refer to equivalent irreducible representations as the same representation. (So far as the matrix is concerned this implies that we shall always imagine the basis to be chosen so that the matrices are identical.) In terms of the reduction (4.12) of a reducible representation into its irreducible constituents this means that an irreducible representation may appear several times in the reduction. We therefore write the reduction as

$$T = \dot{\sum_{\alpha}} m_{\alpha} T^{(\alpha)} \tag{4.17}$$

where α runs only over the inequivalent irreducible representations and the integer m_α gives the number of times that the irreducible representation $T^{(\alpha)}$ occurs in the reduction. (The 'dot' over the summation sign has the same significance as in equation (4.13).)

For example, the six-dimensional representation in subsection 4.3.3 contains the identity representation twice in its reduction. The two independent functions (x^2+y^2) and z^2 are both invariant under the group D_3 and each therefore is the basis function for the one-dimensional identity representation $T^{(1)}$ which associates the number 1 with each group element. It can also be shown (see problem 4.9) that the two-dimensional representation $T^{(3)}$ encountered in subsection 4.3.1 also appears twice in this reduction which may therefore be written $T = 2T^{(1)} \dot{+} 2T^{(3)}$.

4.8 Orthogonality properties of irreducible representations

Through the arguments of the previous three sections, the study of group representations has been reduced to the study of the inequivalent irreducible representations which, we now find, possess important 'orthogonality' properties. These properties are central to the mathematical theory of representations, enabling us to manipulate them in a very simple way, and are also the secret of most of the physical phenomena characteristic of symmetry. The derivation of the orthogonality properties flows from two results known as Schur's lemmas, which were deduced by Schur in 1905. The word 'lemma' is used by mathematicians to describe a result intermediate in the derivation of a theorem or property. The two lemmas are first stated below and then used. Their proofs are left until the end of the section.

Schur's first lemma

Let $T(G_a)$ be an irreducible representation of a group $\mathscr{G}$ in a space L and let A be a fixed operator in L. Schur's first lemma states that, if $T(G_a)A = AT(G_a)$ for all G_a in $\mathscr{G}$ then $A = \lambda 1$, where 1 is the identity (or unit) operator. In other words, any fixed operator which commutes with the operators $T(G_a)$ of an irreducible representation for all G_a in $\mathscr{G}$ is a constant multiple of the unit operator.

Schur's second lemma

Let $T^{(1)}(G_a)$ and $T^{(2)}(G_a)$ be two irreducible representations of $\mathscr{G}$ in two spaces L_1 and L_2, respectively, of dimensions s_1 and s_2 and let A be an operator which transforms vectors from L_2 into L_1. Schur's second lemma states that, if $T^{(1)}$ and $T^{(2)}$ are inequivalent and $T^{(1)}(G_a)A = AT^{(2)}(G_a)$ for all G_a in $\mathscr{G}$ then $A = 0$, the null (or zero) operator.

We now use Schur's lemmas to derive the orthogonality relations for matrix representations. Consider two irreducible representations $T^{(\alpha)}(G_a)$ and $T^{(\beta)}(G_a)$ of a group $\mathscr{G}$, where $T^{(\alpha)}(G_a)$ operates in L_α and $T^{(\beta)}(G_a)$ operates in L_β. Let X be any operator which carries vectors of L_β into vectors of L_α. We can now show that the operator A defined by

$$A = \sum_b T^{(\alpha)}(G_b) X T^{(\beta)}(G_b^{-1}) \tag{4.18}$$

has just the properties needed for the operator A in Schur's lemmas, for

$$\begin{aligned}
T^{(\alpha)}(G_a)A &= \sum_b T^{(\alpha)}(G_a)T^{(\alpha)}(G_b)XT^{(\beta)}(G_b^{-1}) \\
&= \sum_b T^{(\alpha)}(G_a G_b)XT^{(\beta)}(G_b^{-1})T^{(\beta)}(G_a^{-1})T^{(\beta)}(G_a) \\
&= \sum_b T^{(\alpha)}(G_a G_b)XT^{(\beta)}((G_a G_b)^{-1})T^{(\beta)}(G_a) \\
&= \sum_c T^{(\alpha)}(G_c)XT^{(\beta)}(G_c^{-1})T^{(\beta)}(G_a) \\
&= AT^{(\beta)}(G_a)
\end{aligned}$$

We have used the group property that if a is fixed and b runs through the group, then c also runs through the group if c is defined by the equation $G_a G_b = G_c$, see section 2.9. We consider the two cases

(1) $T^{(\alpha)}$ and $T^{(\beta)}$ are the same representation, in which case Schur's lemma gives $A = \lambda 1$;

(2) $T^{(\alpha)}$ and $T^{(\beta)}$ are inequivalent, when Schur's lemma gives $A = 0$.

The two cases may be combined into the single equation

$$A = \lambda \delta_{\alpha\beta} 1 \tag{4.19}$$

where it is to be understood that $\delta_{\alpha\beta} = 0$ if the irreducible representations $T^{(\alpha)}$ and $T^{(\beta)}$ are inequivalent and that $\delta_{\alpha\beta} = 1$ if $T^{(\alpha)}$ and $T^{(\beta)}$ are the same representation. The case when $T^{(\alpha)}$ and $T^{(\beta)}$ are equivalent, but not the same, is not covered by the equation but that is a situation which will be of no interest to us.

The content of Schur's lemmas with the choice (4.18) for A may therefore be concentrated into the single equation

$$\sum_{a=1}^{g} \sum_{m=1}^{s_\beta} \sum_{k=1}^{s_\alpha} T_{ik}^{(\alpha)}(G_a) X_{km} T_{mj}^{(\beta)}(G_a^{-1}) = A_{ij} = \lambda \delta_{\alpha\beta} \delta_{ij} \tag{4.20}$$

in which the rectangular matrix X is quite arbitrary but λ will be determined by the choice of X. We make use of this freedom by choosing $X_{km} = \delta_{kp}\delta_{mq}$; in other words we take for X the matrix with zero elements everywhere except for the element in the row p and column q which is taken to be unity. With this choice, two of the summations in equation (4.20) disappear leaving

$$\sum_{a=1}^{g} T_{ip}^{(\alpha)}(G_a) T_{qj}^{(\beta)}(G_a^{-1}) = \lambda \delta_{\alpha\beta} \delta_{ij} \tag{4.21}$$

The value of λ is relevant only when $\alpha = \beta$ and $i = j$, in which case by summing both sides of equation (4.21) over i, we have

$$\sum_{i=1}^{s_\alpha} \sum_{a=1}^{g} T_{ip}^{(\alpha)}(G_a) T_{qi}^{(\alpha)}(G_a^{-1}) = \lambda \sum_{i=1}^{s_\alpha} 1 = \lambda s_\alpha$$

i.e.

$$\sum_{a=1}^{g} T_{qp}^{(\alpha)}(E) = \lambda s_\alpha$$

giving

$$\lambda = g \delta_{pq} / s_\alpha$$

since the identity E is represented by the unit matrix. Putting this value for λ into equation (4.21) we have

$$\sum_{a=1}^{g} T_{ip}^{(\alpha)}(G_a) T_{qj}^{(\beta)}(G_a^{-1}) = g \delta_{\alpha\beta} \delta_{ij} \delta_{pq} / s_\alpha \tag{4.22}$$

If the matrix representation $T^{(\beta)}$ is unitary then there is a further simplification. Since

$$\mathrm{T}^{(\beta)}(\mathrm{G}_a^{-1})\mathrm{T}^{(\beta)}(\mathrm{G}_a) = \mathrm{T}^{(\beta)}(\mathrm{E}) = 1$$

it follows that

$$\mathrm{T}^{(\beta)}(\mathrm{G}_a^{-1}) = (\mathrm{T}^{(\beta)}(\mathrm{G}_a))^{-1}$$

so that if T is unitary then

$$T_{qj}^{(\beta)}(\mathrm{G}_a^{-1}) = T_{jq}^{(\beta)}(\mathrm{G}_a)^*$$

On substitution into equation (4.22), we have finally

$$\sum_{a=1}^{g} T_{ip}^{(\alpha)}(\mathrm{G}_a)\, T_{jq}^{(\beta)}(\mathrm{G}_a)^* = g\,\delta_{\alpha\beta}\delta_{ij}\delta_{pq}/s_\alpha \tag{4.23}$$

This orthogonality relation is extremely powerful. Notice that on the left-hand side the four matrix element indices are quite arbitrary and the sum runs only over the group elements. The relation shows that this sum is zero if α and β are inequivalent irreducible representations. Even if α and β refer to the same representation, the sum is still zero if different matrix elements are used on the left, i.e. if $i \neq j$ or $p \neq q$. The remaining non-zero case may be expressed as

$$\sum_{a=1}^{g} |T_{ip}^{(\alpha)}(\mathrm{G}_a)|^2 = g/s_\alpha$$

for any i and p.

The sum over the group elements which enters the orthogonality is often referred to as a 'group averaging' process. More strictly the group average is obtained by dividing by the total number g of group elements. The use of the word 'orthogonality relation' to describe equation (4.23) implies the vanishing of a scalar product in some vector space. The present use of the word is justified by considering the set of matrix elements $T_{ip}^{(\alpha)}(\mathrm{G}_a)$ for fixed α, i and p as the components, labelled by a, of a vector in a space of dimension g. The scalar product of two vectors in this space is then defined in the usual way, equation (3.7), by summing over components. Equation (4.23) then states the orthogonality of such vectors for different sets of labels α, i and p.

It is important to realise that the orthogonality is valid only for irreducible representations and in fact, as we shall soon see, this leads to a simple algebraic test to determine whether a given representation is irreducible or not.

Before proceeding with the general development we give two applications of the results of this section. First we show that the irreducible representations of an Abelian group are one-dimensional. Let $\mathrm{T}^{(\alpha)}(\mathrm{G}_a)$ be an irreducible representation of an Abelian group $\mathscr{G}$. Then, since the group elements commute, by definition of an Abelian group, we have

$$\mathrm{T}^{(\alpha)}(\mathrm{G}_a)\mathrm{T}^{(\alpha)}(\mathrm{G}_b) = \mathrm{T}^{(\alpha)}(\mathrm{G}_b)\mathrm{T}^{(\alpha)}(\mathrm{G}_a)$$

for all G_b in $\mathscr{G}$ and for each G_a. It therefore follows from Schur's first lemma that $\mathrm{T}^{(\alpha)}(\mathrm{G}_a)$ is a constant times the unit operator, $\mathrm{T}^{(\alpha)}(\mathrm{G}_a) = \lambda_a^{(\alpha)}1$. Thus the representation $\mathrm{T}^{(\alpha)}(\mathrm{G}_a)$ is diagonal for all G_a and is therefore reducible (a contradiction!) unless it is one-dimensional. Hence all irreducible representations of Abelian groups are one-dimensional.

As a second application we illustrate the orthogonality for irreducible representations of the group D_3. In subsection 4.3.1 we found a one-dimensional representation $T^{(2)}$ and a two-dimensional representation $T^{(3)}$. In addition, every group has an identity representation $T^{(1)}$. These three representations are displayed in table 4.1. They are all irreducible (see problem 4.7) and may be used to illustrate the orthogonality. The dimensions are clearly $s_1 = 1$, $s_2 = 1$, $s_3 = 2$, while $g = 6$. We may then verify, as particular cases of equation (4.23), that

$$\sum_{a=1}^{6} [T_{ip}^{(3)}(G_a)]^2 = 6/2 = 3 \text{ for any } i \text{ and } p$$

$$\sum_{a=1}^{6} T_{ip}^{(3)}(G_a)T^{(2)}(G_a) = 0 \text{ for any } i \text{ and } p$$

$$\sum_{a=1}^{6} T^{(1)}(G_a)T^{(2)}(G_a) = 0$$

$$\sum_{a=1}^{6} [T^{(2)}(G_a)]^2 = 6/1 = 6, \text{ etc.}$$

4.8.1 Proof of Schur's first lemma

Let $\boldsymbol{r}$ be an eigenvector of A in L with eigenvalue λ so that $A\boldsymbol{r} = \lambda\boldsymbol{r}$. If we now apply a transformation $T(G_a)$ to $\boldsymbol{r}$ to produce a new vector $\boldsymbol{r}_a = T(G_a)\boldsymbol{r}$, then $\boldsymbol{r}_a$ is also an eigenvector of A with the same eigenvalue λ, since

$$A\boldsymbol{r}_a = AT(G_a)\boldsymbol{r} = T(G_a)A\boldsymbol{r} = T(G_a)\lambda\boldsymbol{r} = \lambda T(G_a)\boldsymbol{r} = \lambda\boldsymbol{r}_a$$

The set of vectors $\boldsymbol{r}_a$ generated as G_a runs through the group $\mathcal{G}$ must form an invariant subspace since

$$T(G_b)\boldsymbol{r}_a = T(G_b)T(G_a)\boldsymbol{r} = T(G_b G_a)\boldsymbol{r} = \boldsymbol{r}_c$$

where c is defined from the group multiplication table using $G_b G_a = G_c$. But since L is, by definition, irreducible it cannot have an invariant subspace. Thus the space of vectors $\boldsymbol{r}_a$ must be the whole space L. Hence for any vector $\boldsymbol{R} = \sum_a c_a\boldsymbol{r}_a$ in L we have

$$A\boldsymbol{R} = A\sum_a c_a\boldsymbol{r}_a = \sum_a c_a A\boldsymbol{r}_a = \sum_a c_a \lambda\boldsymbol{r}_a = \lambda\boldsymbol{R}$$

but since $\boldsymbol{R}$ is any vector in L we may therefore write $A = \lambda 1$. In matrix form, A would simply be λ multiplied by the unit matrix.

Table 4.1

Group element G_a / Representation	E	R_1	R_2	R_3	R_4	R_5
$T^{(1)}$	1	1	1	1	1	1
$T^{(2)}$	1	1	1	-1	-1	-1
$T^{(3)}$	$\begin{pmatrix} 1 & 0 \\ 0 & 1 \end{pmatrix}$	$\begin{pmatrix} -\frac{1}{2} & -\sqrt{\frac{3}{4}} \\ \sqrt{\frac{3}{4}} & -\frac{1}{2} \end{pmatrix}$	$\begin{pmatrix} -\frac{1}{2} & \sqrt{\frac{3}{4}} \\ -\sqrt{\frac{3}{4}} & -\frac{1}{2} \end{pmatrix}$	$\begin{pmatrix} -1 & 0 \\ 0 & 1 \end{pmatrix}$	$\begin{pmatrix} \frac{1}{2} & -\sqrt{\frac{3}{4}} \\ -\sqrt{\frac{3}{4}} & -\frac{1}{2} \end{pmatrix}$	$\begin{pmatrix} \frac{1}{2} & \sqrt{\frac{3}{4}} \\ \sqrt{\frac{3}{4}} & -\frac{1}{2} \end{pmatrix}$

4.8.2 Proof of Schur's second lemma

Let us first consider the case $s_2 \leq s_1$. Then A will generate from L_2 a subspace L_A of L_1 with some dimension $s_A \leq s_2 \leq s_1$. The subspace L_A is composed of the vectors $A\boldsymbol{r}$, where $\boldsymbol{r}$ is any vector in L_2. It follows immediately that L_A is invariant under $\mathscr{G}$ since

$$T^{(1)}(G_a)A\boldsymbol{r} = AT^{(2)}(G_a)\boldsymbol{r} = A\boldsymbol{r}_a$$

which lies in L_A because $\boldsymbol{r}_a[=T^{(2)}(G_a)\boldsymbol{r}]$ lies in L_2. However, $T^{(1)}$ is by definition irreducible so that L_1 cannot have an invariant subspace. Thus there is a contradiction unless L_A is either a null space ($s_A = 0$) or the whole space L_1 ($s_A = s_1$). In other words we deduce that either (1) $A\boldsymbol{r} = 0$ for all $\boldsymbol{r}$ in L_2 so that $A = 0$, or (2) $s_A = s_1 = s_2$, the second equality following from the inequality $s_A \leq s_2$ and the assumption $s_2 \leq s_1$.

The second of these alternatives is excluded by the assumption that $T^{(1)}$ and $T^{(2)}$ are inequivalent representations. It would imply that L_1 and L_2 have the same dimension so that A has an inverse and hence from the assumption $T^{(1)}(G_a)A = AT^{(2)}(G_a)$ we have $T^{(1)}(G_a) = AT^{(2)}(G_a)A^{-1}$, showing that $T^{(1)}$ and $T^{(2)}$ are equivalent. Thus there is no alternative to the conclusion that $A = 0$.

The proof must be reworded for the remaining case $s_2 > s_1$. Then, necessarily $s_A < s_2$ so that there must be vectors $\boldsymbol{r}$ of L_2 which are mapped on to zero by A, i.e. $A\boldsymbol{r} = 0$. Let us denote the subspace of such vectors in L_2 by L_B. It will have dimension $(s_2 - s_A)$. Then L_B must be invariant since if $\boldsymbol{r}_a = T^{(2)}(G_a)\boldsymbol{r}$ then $A\boldsymbol{r}_a = AT^{(2)}(G_a)\boldsymbol{r} = T^{(1)}(G_a)A\boldsymbol{r} = 0$, showing that $\boldsymbol{r}_a$ also belongs to L_B. This contradicts the irreducibility of $T^{(2)}$ unless $L_B = L_2$, in other words $A\boldsymbol{r} = 0$ for *all* vectors $\boldsymbol{r}$ in L_2. Thus again we conclude that $A = 0$.

4.9 Characters of representations

As mentioned in section 4.6, we can construct an infinite number of equivalent matrix representations from any given representation by a change of basis (or similarity transformation). We want to find some well-defined properties of a representation which are independent of such transformations. In fact there are many invariants which could be constructed since all the eigenvalues of a matrix remain the same under a similarity transformation. However, for most purposes a single characteristic is sufficient and the most useful choice is the sum of all the eigenvalues, called the 'trace' of the matrix, which is also given by the sum of the diagonal elements in any basis. This trace of a matrix representation $T(G_a)$ is denoted by $\chi(G_a)$. The set of numbers $\chi(G_a)$, as the group element G_a runs through the group, is called the 'character' of the representation T and is denoted by χ. We have

$$\chi(G_a) = \sum_{i=1}^{s} T_{ii}(G_a) \tag{4.24}$$

We note immediately that the character is invariant under a similarity transformation since if $T'(G_a) = AT(G_a)A^{-1}$ then[†]

$$\begin{aligned}\chi'(G_a) &= \sum_i T'_{ii}(G_a)\\ &= \sum_{ijk} A_{ij}T_{jk}(G_a)(A^{-1})_{ki}\\ &= \sum_{jk} T_{jk}(G_a)(A^{-1}A)_{kj}\\ &= \sum_j T_{jj}(G_a) = \chi(G_a)\end{aligned}$$

We can also see by a similar argument that all elements in the same class $\mathscr{C}_p$ (see section 2.6) must have the same character which we will denote by χ_p. For, suppose that G_a and G_b are in the same class so that they are related by an equation $G_a = G_m G_b G_m^{-1}$. Then for any representation T,

$$\begin{aligned}\chi(G_a) = \sum_i T_{ii}(G_a) &= \sum_i T_{ii}(G_m G_b G_m^{-1})\\ &= \sum_{ijk} T_{ij}(G_m)T_{jk}(G_b)T_{ki}(G_m^{-1})\\ &= \sum_{jk} T_{jk}(G_b)T_{kj}(G_m^{-1}G_m)\\ &= \sum_j T_{jj}(G_b) = \chi(G_b)\end{aligned}$$

4.10 Orthogonality relation for characters of irreducible representations

For irreducible representations we can use the orthogonality relation (4.23) to deduce a relationship between the characters. By putting $p = i$ and $q = j$ in equation (4.23) and summing both sides over i and j we get

$$\sum_{a=1}^{g} \sum_i T_{ii}^{(\alpha)}(G_a) \sum_j T_{jj}^{(\beta)}(G_a)^* = g\delta_{\alpha\beta}$$

so that

$$\sum_{a=1}^{g} \chi^{(\alpha)}(G_a)\chi^{(\beta)}(G_a)^* = g\delta_{\alpha\beta} \qquad (4.25a)$$

If we collect conjugate elements together into classes $\mathscr{C}_p$ containing a number c_p of elements, this relation may be written

$$\sum_{p=1}^{n} c_p \chi_p^{(\alpha)} \chi_p^{(\beta)*} = g\delta_{\alpha\beta} \qquad (4.25b)$$

[†] The symbol A used in this argument is not to be confused with the particular operator defined in equation (4.18).

with the sum running over the n classes $\mathscr{C}_p$ of $\mathscr{G}$.

In the particular case $\alpha = \beta$ we get the relation

$$\sum_{a=1}^{g} |\chi^{(\alpha)}(G_a)|^2 = \sum_{p=1}^{n} c_p |\chi_p^{(\alpha)}|^2 = g \tag{4.26}$$

Again we refer to the result (4.25) as an 'orthogonality' relation for the characters. In equation (4.25b) the character χ would be regarded as a vector with components $(c_p)^{\frac{1}{2}} \chi_p$ in a vector space of dimension n, the number of classes of $\mathscr{G}$. The irreducible characters form a set of orthogonal vectors in this space. Hence it is clear that the number of inequivalent irreducible representations cannot be greater than the number of classes in the group. We shall see, in section 4.13, that these two numbers are, in fact, equal.

The example in subsection 4.3.1 may be used to illustrate the orthogonality of the characters of irreducible representations. It is convenient to form a table (table 4.2) showing the character $\chi_p^{(\alpha)}$ for each class p and irreducible representation α, although now there are fewer columns, one for each class rather than one for each group element. It is important to remember to include the number c_p of elements in each class when using the character orthogonality relation.

Table 4.2

Class / *Representation*	$\mathscr{C}_1$ (E)	$\mathscr{C}_2$ (R_1, R_2)	$\mathscr{C}_3$ (R_3, R_4, R_5)
$T^{(1)}$	1	1	1
$T^{(2)}$	1	1	-1
$T^{(3)}$	2	-1	0

4.11 Use of group characters in reducing a representation

We have seen in sections 4.5 and 4.7 how, in principle, an arbitrary representation T may be reduced into its irreducible constituents. By summing the diagonal elements of T it follows immediately from equation (4.17) that the character χ of T is related to the irreducible characters by a similar equation. If χ_p denotes the character of T for an element in a class $\mathscr{C}_p$ then

$$\chi_p = \sum_{\alpha} m_\alpha \chi_p^{(\alpha)} \tag{4.27}$$

where m_α is the number of times that each inequivalent irreducible representation $T^{(\alpha)}$ occurs in the reduction of T. It is obviously of interest, given T, to be able to deduce the m_α and this may be done very simply if the irreducible characters $\chi_p^{(\alpha)}$ are known. For, using the orthogonality relation (4.25b) with equation (4.27) we have

$$\frac{1}{g}\sum_p c_p \chi_p^{(\beta)*}\chi_p = \frac{1}{g}\sum_\alpha m_\alpha \sum_p c_p \chi_p^{(\beta)*}\chi_p^{(\alpha)}$$

$$= \frac{1}{g}\sum_\alpha m_\alpha g \delta_{\alpha\beta} = m_\beta \qquad (4.28)$$

This equation takes the scalar product of the character χ with the irreducible character $\chi^{(\beta)}$ and is analogous to equation (3.6) for the 'component' m_β of χ in the direction of the 'vector' $\chi^{(\beta)}$. As illustrations of this process, consider the three-dimensional representation T of subsection 4.3.1. It has character $\chi = (3, 0, -1)$, where the three numbers give the χ_p for the three classes of D_3 taken in the same order as in table 4.2. Writing

$$\mathrm{T} = m_1 \mathrm{T}^{(1)} \dotplus m_2 \mathrm{T}^{(2)} \dotplus m_3 \mathrm{T}^{(3)}$$

in the notation of the table we find, using equation (4.28),

$$m_1 = \tfrac{1}{6}(3+0-3) = 0$$

$$m_2 = \tfrac{1}{6}(3+0+3) = 1$$

$$m_3 = \tfrac{1}{6}(6+0+0) = 1$$

showing that T reduces into a sum of $\mathrm{T}^{(2)}$ and $\mathrm{T}^{(3)}$. This example is almost trivial since the reduction was apparent from the matrices. It does, however, illustrate the method which has considerable safeguards against error, since the numbers m_α must come out as positive integers or zero.

The six-dimensional representation of the example in subsection 4.3.3 provides a more substantial illustration—see problem 4.9.

4.12 A criterion for irreducibility

From the character of a representation we may immediately deduce whether it is irreducible or not. We have seen in section 4.10 that, if χ is irreducible, then

$$\sum_p c_p |\chi_p|^2 = g \qquad (4.29)$$

It also follows that, if equation (4.29) is satisfied, then χ is irreducible, making this condition both necessary and sufficient. The proof of the latter follows from equations (4.25) and (4.27) which lead to

$$\sum_p c_p |\chi_p|^2 = \sum_{\alpha\beta p} c_p m_\alpha m_\beta \chi_p^{(\alpha)} \chi_p^{(\beta)*} = g \sum_\alpha m_\alpha^2$$

Hence, if equation (4.29) is satisfied then $\sum_{\alpha} m_{\alpha}^2 = 1$ and since the m_{α} are integers this implies that all $m_{\alpha} = 0$ except for one which is unity, say $m_{\gamma} = 1$. This shows that $T = T^{(\gamma)}$ which is irreducible. As an illustration of this we can show that the two-dimensional representation $T^{(3)}$ of D_3 is irreducible since, from table 4.2,

$$\sum_{p} c_p |\chi_p|^2 = (4+2+0) = 6 = g$$

4.13 How many inequivalent irreducible representations?—The regular representation

In section 4.10 it was shown that the number of inequivalent irreducible representations of a finite group $\mathscr{G}$ could not exceed the number of classes. In the example of D_3, table 4.2 shows that three inequivalent irreducible representations had been found. But there are only three classes in D_3 which enables us to conclude that $T^{(1)}$, $T^{(2)}$ and $T^{(3)}$ are the only inequivalent irreducible representations of D_3. In general it will be useful to deduce beforehand the precise number of inequivalent irreducible representations and we shall now show, with the help of a rather artificial device, that the number of inequivalent irreducible representations is always equal to the number of classes. The device is to construct a very special representation whose dimension is equal to g, the number of elements in the group. It is called the 'regular' representation and is denoted by $T^{(R)}$.

The matrices $T^{(R)}(G_a)$ for the regular representation are defined by the equations

$$G_a G_b = \sum_{c} T_{cb}^{(R)}(G_a) G_c \tag{4.30}$$

It is easily shown that the matrices $T^{(R)}(G_a)$ do indeed form a representation. By multiplying both sides of equation (4.30) by a further group element G_d,

$$\begin{aligned} G_d G_a G_b &= \sum_{c} T_{cb}^{(R)}(G_a) G_d G_c \\ &= \sum_{c} \sum_{e} T_{cb}^{(R)}(G_a) T_{ec}^{(R)}(G_d) G_e \\ &= \sum_{e} \left[\sum_{c} T_{ec}^{(R)}(G_d) T_{cb}^{(R)}(G_a) \right] G_e \end{aligned}$$

But also

$$G_d G_a G_b = (G_d G_a) G_b = \sum_{e} T_{eb}^{(R)}(G_d G_a) G_e$$

so that, comparing with the previous equation,

$$T^{(R)}(G_dG_a) = T^{(R)}(G_d)T^{(R)}(G_a)$$

which is the defining property (4.1) for a representation. We notice that since the product G_aG_b is itself some group element, only one term appears in the sum on the right-hand side of equation (4.30). Hence for given a and b all the matrix elements $T^{(R)}_{cb}(G_a)$ will be zero, except for one value of c for which the matrix element is equal to 1. Thus each column of the matrix $T^{(R)}(G_a)$ will be zero except for one entry of 1 and this entry will be on the diagonal only when G_a is the identity element E. Hence the character of the regular representation will be zero for all elements except the identity for which it is equal to the dimension g of the representation, i.e.

$$\chi^{(R)}(G_a) = 0 \qquad G_a \neq E \tag{4.31}$$
$$\chi^{(R)}(E) = g$$

Let us now study the reduction of the regular representation

$$T^{(R)}(G_a) = \dot{\sum_{\alpha}} m_\alpha T^{(\alpha)}(G_a)$$

Using equations (4.28) and (4.31)

$$m_\alpha = \frac{1}{g}\sum_a \chi^{(\alpha)}(G_a)^* \chi^{(R)}(G_a) = \frac{1}{g} g\chi^{(\alpha)}(E) = s_\alpha \tag{4.32}$$

where s_α is the dimension of the irreducible representation $T^{(\alpha)}$. Thus the number of times that the irreducible representation $T^{(\alpha)}$ appears in the reduction of the regular representation is equal to its dimension. We notice that this implies that the regular representation must contain all the irreducible representations.

By equating the dimension g of $T^{(R)}$ to the total dimension of its constituents we have the important result

$$g = \sum_\alpha m_\alpha s_\alpha = \sum_\alpha s_\alpha^2 \tag{4.33}$$

Notice that, whereas equation (4.32) described a property of the regular representation, the result (4.33) is a general property of the group, namely that the sum of the squares of the dimensions of all possible inequivalent irreducible representations of the group is equal to the number of group elements. We now use this result to prove that the number of inequivalent irreducible representations of a group is equal to the number of classes of that group.

Proof. In section 4.8 it was shown that the matrix elements $T^{(\alpha)}_{ij}(G_a)$ may be regarded as components of a set of orthogonal vectors $\boldsymbol{T}^{(\alpha)}_{ij}$ in a g-dimensional space with basis vectors $\boldsymbol{e}_a$ labelled by $a = 1, 2, \ldots, g$. The total number of such vectors is found by taking all possible values of i, j and α and is given by the sum $\sum_{\alpha} s_{\alpha}^2$ over all inequivalent irreducible representations. But this sum has just been shown to be equal to g. Hence the number of orthogonal vectors $\boldsymbol{T}_{ij}^{(\alpha)}$ is equal to the dimension of the space and so the $\boldsymbol{T}_{ij}^{(\alpha)}$ must span the space. Thus any vector $\boldsymbol{v}$ in this space can be expanded as a linear combination of the vector $T_{ij}^{(\alpha)}$,

$$\boldsymbol{v} = \sum_{\alpha ij} c(\alpha ij) \boldsymbol{T}_{ij}^{(\alpha)}$$

and for its components v_a in the basis $\boldsymbol{e}_a$,

$$v_a = \sum_{\alpha ij} c(\alpha ij)\, T_{ij}^{(\alpha)}(\mathrm{G}_a) \tag{4.34}$$

Let us now restrict our attention to those vectors $\boldsymbol{v}$ which have the same 'component' in all 'directions' $\boldsymbol{e}_a$ which correspond to elements in the same class of the group $\mathscr{G}$, i.e. $v_a = v_a$ if $\mathrm{G}_c = \mathrm{G}_b^{-1}\mathrm{G}_a\mathrm{G}_b$ for any G_b in $\mathscr{G}$. Hence we may write

$$\begin{aligned}
v_a &= \frac{1}{g}\sum_{b=1}^{g} v_c, \text{ where } \mathrm{G}_c = \mathrm{G}_b^{-1}\mathrm{G}_a\mathrm{G}_b \\
&= \frac{1}{g}\sum_{b}\sum_{\alpha ij} c(\alpha ij)\, T_{ij}^{(\alpha)}(\mathrm{G}_b^{-1}\mathrm{G}_a\mathrm{G}_b), \text{ from equation (4.34)} \\
&= \frac{1}{g}\sum_{b}\sum_{\alpha ij}\sum_{kl} c(\alpha ij)\, T_{ik}^{(\alpha)}(\mathrm{G}_b^{-1})\, T_{kl}^{(\alpha)}(\mathrm{G}_a)\, T_{lj}^{(\alpha)}(\mathrm{G}_b) \\
&= \frac{1}{g}\sum_{\alpha ij}\sum_{kl} c(\alpha ij)\, T_{kl}^{(\alpha)}(\mathrm{G}_a)\delta_{ij}\delta_{kl}g/s_\alpha, \text{ using equation (4.22)} \\
&= \sum_{\alpha}\frac{1}{s_\alpha}\sum_{i} c(\alpha ii)\,\chi^{(\alpha)}(\mathrm{G}_a)
\end{aligned} \tag{4.35}$$

These vectors $\boldsymbol{v}$ form a subspace of dimension n, where n is the number of classes and equation (4.35) shows that the characters, which are a set of orthonormal vectors in this subspace (see section 4.10), span the subspace. Hence there must be exactly n such characters $\chi^{(\alpha)}$ which is the desired result, that the number of inequivalent irreducible representations is equal to n the number of classes of the group $\mathscr{G}$.

4.14 The second orthogonality relation for group characters

The equality between the number of classes and the number of inequivalent irreducible representations means that the character table is square, the columns labelled by the classes and the rows by the inequivalent irreducible representations. The orthogonality relation (4.25) means that any two rows of the table are orthogonal and from this we may deduce that any two columns of the table also satisfy an orthogonality relation.

To deduce this relation, define an $n \times n$ matrix B with matrix elements

$$B_{\alpha p} = \left(\frac{c_p}{g}\right)^{\frac{1}{2}} \chi_p^{(\alpha)}$$

where n is the number of classes and also the number of inequivalent irreducible representations. The orthogonality relation (4.25) leads to the equation

$$\sum_p B_{\beta p} B^*_{\alpha p} = 1 \qquad \text{for any } \alpha \text{ and } \beta$$

or in matrix form $\mathrm{B}\mathrm{B}^\dagger = 1$. Since B is a square matrix, this implies that the determinant of B has modulus unity, that B has an inverse and that $\mathrm{B}^{-1} = \mathrm{B}^\dagger$. Hence we also have $\mathrm{B}^\dagger \mathrm{B} = 1$ which in terms of matrix elements gives

$$\sum_\alpha B^*_{\alpha p} B_{\alpha q} = \delta_{pq}$$

and in terms of the group characters this becomes

$$\sum_\alpha \chi_p^{(\alpha)*} \chi_q^{(\alpha)} = \frac{g}{c_p} \delta_{pq} \tag{4.36}$$

which is an orthogonality relation between columns of the character table.

4.15 Construction of the character table

In the character table (table 4.2) for the group D_3 the entries were deduced by summing the diagonal matrix elements of the representation matrices which had been constructed previously. This is not the simplest way to find a character table. In practice, the character tables for all the groups which are likely to arise have been calculated long ago and tables are given in appendix 1, but it is nevertheless of interest to see how the character tables might be deduced from first principles for the most common finite groups. This can be donc remaıkably simply. In fact, the properties of the characters of irreducible representations which we have already deduced are in many cases sufficient to deduce the characters unambiguously. We list those properties below.

(1) The number of irreducible representations = the number of classes.

(2) The dimension s_α of the irreducible representations must satisfy the equation $\sum_\alpha s_\alpha^2 = g$, which in many cases has a unique solution for the integers s_α. Then, since the character of the identity E is equal to the dimension of the representation, the entries in the first column of the character table are just the integers s_α.

(3) For every group there is always an identity representation which is one-dimensional and has $T(G_a) = 1$ and hence $\chi(G_a) = 1$. This determines one row of the table, usually taken as the first row.

(4) Rows are orthogonal with weighting factors c_p and normalised to g,

$$\sum_p c_p \chi_p^{(\alpha)} \chi_p^{(\beta)*} = g\delta_{\alpha\beta}, \text{ which is equation (4.25)}$$

In particular by choosing β to be the identity representation we have

$$\sum_p c_p \chi_p^{(\alpha)} = 0 \text{ for } \alpha \text{ not the identity representation}$$

(5) Columns are orthogonal and normalised to g/c_p,

$$\sum_\alpha \chi_p^{(\alpha)} \chi_q^{(\alpha)*} = \frac{g}{c_p}\delta_{pq}, \text{ which is equation (4.36)}$$

In particular by choosing $\mathscr{C}_q$ to be the class E we have for all other columns p

$$\sum_\alpha s_\alpha \chi_p^{(\alpha)} = 0$$

It is possible in this way to deduce the character tables of some of the smaller groups. The only information required about the group in using this method is the order g of the group and the number and size of its classes. For larger groups this method fails to provide enough information to deduce the character table. It is then necessary to return to the group multiplication table and deduce further properties of the characters—see appendix 3.3 of volume 2. Alternatively, if a group possesses a normal subgroup this may be used to generate the representations of the group from those of its normal subgroup (as shown in section 20.3 of volume 2). We stress that in most applications to physical problems one need simply look up the character table.

4.16 Orthogonality of basis functions for irreducible representations

In this section and those which follow we shall turn our attention more towards the properties of the basis vectors of the space in which an irreducible representation is generated. These basis vectors may be regarded abstractly as, for example, in section 4.2 and denoted by $\boldsymbol{e}_i^{(\alpha)}$ with $i = 1, 2, \ldots, s_\alpha$ or, more specifically, they may be functions $\psi_i^{(\alpha)}(\boldsymbol{r})$ in some vector space of functions as in subsection 4.3.3. Since in applications to quantum mechanics we shall usually be dealing with functions, we prefer in this and some following sections to use the notation $\psi_i^{(\alpha)}$ rather than $\boldsymbol{e}_i^{(\alpha)}$ for the basis vectors of the representation $T^{(\alpha)}$. For brevity we drop the argument $\boldsymbol{r}$ of the functions.

It was shown in section 4.4 how to generate an invariant function space and it is usual to choose an orthonormal set of functions as a basis for such a space. Suppose now that we have an invariant function space which is also irreducible so that the representation induced in the space by some group operations is an irreducible representation. We may now use the orthogonality properties (4.23) of the irreducible representations to show that basis functions belonging to two inequivalent irreducible representations are orthogonal.

Let the function $\phi_i^{(\alpha)}$ transform according to the ith row of the irreducible representation $\mathrm{T}^{(\alpha)}$, by which we mean that

$$\mathrm{T}(\mathrm{G}_a)\phi_i^{(\alpha)} = \sum_l T_{li}^{(\alpha)}(\mathrm{G}_a)\phi_l^{(\alpha)} \tag{4.37}$$

and let the function $\psi_j^{(\beta)}$ transform according to the jth row of the irreducible representation $\mathrm{T}^{(\beta)}$. Suppose further that the operators $\mathrm{T}(\mathrm{G}_a)$ are unitary with respect to some definition of scalar product which may be used for all the functions considered, so that

$$\begin{aligned}(\phi_i^{(\alpha)}, \psi_j^{(\beta)}) &= (\mathrm{T}(\mathrm{G}_a)\phi_i^{(\alpha)}, \mathrm{T}(\mathrm{G}_a)\psi_j^{(\beta)}) \\ &= \sum_l \sum_m T_{li}^{(\alpha)*}(\mathrm{G}_a)T_{mj}^{(\beta)}(\mathrm{G}_a)(\phi_l^{(\alpha)}, \psi_m^{(\beta)})\end{aligned}$$

for any element G_a of the group. If we now assume that the basis vectors within each representation have been chosen to be orthonormal then we may use the orthogonality relation (4.23). Hence, averaging over all group elements, we have

$$\begin{aligned}(\phi_i^{(\alpha)}, \psi_j^{(\beta)}) &= \frac{1}{g}\sum_a \sum_l \sum_m T_{li}^{(\alpha)*}(\mathrm{G}_a)T_{mj}^{(\beta)}(\mathrm{G}_a)(\phi_l^{(\alpha)}, \psi_m^{(\beta)}) \\ &= \frac{1}{s_\alpha}\delta_{\alpha\beta}\delta_{ij}\sum_l (\phi_l^{(\alpha)}, \psi_l^{(\alpha)})\end{aligned} \tag{4.38}$$

This shows that any pair of functions which transforms according to unitary irreducible representations are orthogonal unless they belong to (transform according to) the same row of the same (or an equivalent) representation. The power of this result is not so much in the orthogonality of basis functions within the same irreducible representation (the factor δ_{ij} when $\phi \equiv \psi$), since this is largely a matter of choice, but in the orthogonality of basis functions belonging to different rows of equivalent representations or to inequivalent representations (the factor $\delta_{\alpha\beta}$). The latter result holds quite independently of the choice of basis within each representation. The equation (4.38) contains the further result that the scalar product $(\psi_i^{(\alpha)}, \psi_i^{(\alpha)})$ is independent of i showing that, in particular, the functions $\phi_i^{(\alpha)}$ for fixed α have equal normalisation, if they satisfy equation (4.37).

As a particular consequence of the result (4.38) it follows that if $\mathrm{T}^{(\alpha)}$ is the identity representation, then the scalar product vanishes unless $\mathrm{T}^{(\beta)}$ also is the identity representation. Thus if the scalar product is given by an integral over coordinates, as is generally the case, and we take $\phi_i^{(\alpha)} = 1$ as the constant function, then we have

$$\int \phi_j^{(\beta)}\,\mathrm{d}V = 0 \tag{4.39}$$

unless $T^{(\beta)}$ is the identity representation. This shows that the invariant integral of a function which transforms irreducibly will vanish unless the function is invariant. Since an arbitrary function may be expanded into irreducible components it follows that only the invariant component will survive such an integration.

4.17 The direct product of two representations

The *direct product* of an $n \times n$ matrix A and an $m \times m$ matrix B is defined to be the $mn \times mn$ matrix, denoted by $(A \times B)$, with matrix elements

$$(A \times B)_{ij,kl} = A_{ik} B_{jl} \tag{4.40}$$

Each row (and column) of the direct product matrix is labelled by a double suffix ij, the first suffix i referring to a row of A and the second suffix j referring to a row of B. It must be emphasised that $A \times B$ is not the usual product of matrices.

In this way we may construct the direct product representation $T^{(\alpha)} \times T^{(\beta)}$, written as $T^{(\alpha \times \beta)}$, from two representations $T^{(\alpha)}$ and $T^{(\beta)}$,

$$T^{(\alpha \times \beta)}_{ij,kl}(G_a) = T^{(\alpha)}_{ik}(G_a) T^{(\beta)}_{jl}(G_a) \tag{4.41}$$

It may readily be shown that this direct product is in fact a representation, since

$$\begin{aligned}
[T^{(\alpha \times \beta)}(G_a) T^{(\alpha \times \beta)}(G_b)]_{ij,kl} &= \sum_{mn} T^{(\alpha \times \beta)}_{ij,mn}(G_a) T^{(\alpha \times \beta)}_{mn,kl}(G_b) \\
&= \sum_{mn} T^{(\alpha)}_{im}(G_a) T^{(\beta)}_{jn}(G_a) T^{(\alpha)}_{mk}(G_b) T^{(\beta)}_{nl}(G_b) \\
&= T^{(\alpha)}_{ik}(G_a G_b) T^{(\beta)}_{jl}(G_a G_b) \\
&= T^{(\alpha \times \beta)}_{ij,kl}(G_a G_b)
\end{aligned} \tag{4.42}$$

The character of the direct product representation is given by

$$\begin{aligned}
\chi^{(\alpha \times \beta)}(G_a) = \sum_{ij} T^{(\alpha \times \beta)}_{ij,ij}(G_a) &= \sum_{ij} T^{(\alpha)}_{ii}(G_a) T^{(\beta)}_{jj}(G_a) \\
&= \chi^{(\alpha)}(G_a) \chi^{(\beta)}(G_a)
\end{aligned} \tag{4.43}$$

In other words for a given group element the character of the direct product representation $T^{(\alpha \times \beta)}$ is simply the product of the characters of the two representations $T^{(\alpha)}$ and $T^{(\beta)}$. If $T^{(\alpha)}$ and $T^{(\beta)}$ are irreducible then, in general, the representation $T^{(\alpha \times \beta)}$, which has dimension $s_\alpha s_\beta$, will not be irreducible but, knowing its character, we may reduce it by the method of section 4.11. Thus if

$$T^{(\alpha \times \beta)} = \dot{\sum_{\gamma}} m_\gamma T^{(\gamma)} \tag{4.44}$$

then from equations (4.43) and (4.28)

$$m_\gamma = \frac{1}{g} \sum_p c_p \chi_p^{(\gamma)*} \chi_p^{(\alpha)} \chi_p^{(\beta)} \tag{4.45}$$

As an example of this process, consider the product $T^{(3)} \times T^{(3)}$ of the representations of the group D_3 whose characters are given in table 4.2. From equation (4.43) the character of the four-dimensional product representation $T^{(3)} \times T^{(3)}$ is given by

$$\chi^{(3 \times 3)} = (4, 1, 0)$$

the three components referring to the three classes of table 4.2. Now, writing

$$\chi^{(3 \times 3)} = \sum_\alpha m_\alpha \chi^{(\alpha)}$$

and using equation (4.28), we have

$$m_1 = \tfrac{1}{6}(4+2+0) = 1$$
$$m_2 = \tfrac{1}{6}(4+2+0) = 1$$
$$m_3 = \tfrac{1}{6}(8-2+0) = 1$$

so that

$$T^{(3)} \times T^{(3)} = T^{(1)} \dot{+} T^{(2)} \dot{+} T^{(3)}$$

In practice, the direct product representation occurs naturally when considering products of functions. If the set of s_α functions $\phi_k^{(\alpha)}$ transforms according to the representation $T^{(\alpha)}$ and the set $\psi_l^{(\beta)}$ transforms according to $T^{(\beta)}$, then the set of $s_\alpha s_\beta$ products $\{\phi_k^{(\alpha)} \psi_l^{(\beta)}\}$ transforms according to the direct product representation $T^{(\alpha \times \beta)}$. To see this, consider

$$T(G_a)\{\phi_k^{(\alpha)} \psi_l^{(\beta)}\} = \sum_i \sum_j T_{ik}^{(\alpha)}(G_a) T_{jl}^{(\beta)}(G_a) \{\phi_i^{(\alpha)} \psi_j^{(\beta)}\}$$

$$= \sum_{ij} T_{ij,kl}^{(\alpha \times \beta)}(G_a)\{\phi_i^{(\alpha)} \psi_j^{(\beta)}\}$$

using equation (4.41).

Suppose that the direct product representation reduces

$$T^{(\alpha \times \beta)} = \sum_\gamma m_\gamma T^{(\gamma)}$$

then it must be possible, by a change of basis, to choose linear combinations

$$\Psi_k^{(\gamma)t} = \sum_{ij} C(\alpha\beta\gamma t, ijk)\{\phi_i^{(\alpha)} \psi_j^{(\beta)}\} \tag{4.46}$$

which transform irreducibly according to the representation $\mathrm{T}^{(\gamma)}$. Here the subscript k refers to the row of the representations and the label t serves to distinguish functions which have the same γ and k. This will happen wherever the integer m_γ is greater than unity, i.e. when the irreducible representation $\mathrm{T}^{(\gamma)}$ occurs more than once in the reduction (4.44). The coefficients $C(\alpha\beta\gamma t, ijk)$ are usually called the Clebsch–Gordan coefficients for the group. A group which has the property that for any α and β the coefficients m_γ are 0 or 1 is called 'simply reducible'. For such groups, which include $\mathscr{R}_3$, the label t is unnecessary.

In this discussion we have supposed that the functions ϕ and ψ are not identical, in other words we suppose that either $\alpha \neq \beta$ or, if $\alpha = \beta$, then $\phi_i^{(\alpha)} \neq \psi_i^{(\alpha)}$. The case when ϕ and ψ are identical requires some special comments because the set of s_α^2 products $\phi_i^{(\alpha)}\phi_j^{(\alpha)}$ are not linearly independent, clearly $\phi_i^{(\alpha)}\phi_j^{(\alpha)} - \phi_j^{(\alpha)}\phi_i^{(\alpha)} = 0$. Thus it follows that the new basis $\Psi_k^{(\gamma)t}$ will also not be linearly independent and the form which this linear dependence takes is that for certain values of $(\gamma)t$, all the s_γ corresponding basis functions will vanish identically. In other words the basis functions for certain of the representations in the sum (4.44) will vanish identically. The vanishing representations will be those which are antisymmetric in the two factors since these must clearly vanish when the two factors are identical. (We discuss the general problem of symmetrised products of representations in appendix 3 of volume 2.) In the example $\mathrm{T}^{(3)} \times \mathrm{T}^{(3)}$ given above, the basis function of the representation $\mathrm{T}^{(2)}$ in the product will vanish if the two sets of functions are identical.

The Clebsch–Gordan coefficients are usually normalised to satisfy the relation

$$\sum_{ij} |C(\alpha\beta\gamma t, ijk)|^2 = 1$$

This will ensure that the functions $\Psi_k^{(\gamma)t}$ are normalised if the product functions themselves form an orthonormal set. The latter condition is generally satisfied when the two factors ϕ and ψ in the products refer to different coordinates, e.g. particles numbers 1 and 2, for then the natural scalar product is defined by

$$\begin{aligned}(\phi_{i'}^{(\alpha')}(1)\psi_{j'}^{(\beta')}(2), \phi_i^{(\alpha)}(1)\psi_j^{(\beta)}(2)) &= (\phi_{i'}^{(\alpha')}(1), \phi_i^{(\alpha)}(1))(\psi_{j'}^{(\beta')}(2), \psi_j^{(\beta)}(2)) \\ &= \delta_{\alpha'\alpha}\delta_{\beta'\beta}\delta_{i'i}\delta_{j'j} \qquad (4.47)\end{aligned}$$

as a product of the usual scalar products for each particle separately. In this case there can be no linear dependence. From the general orthogonality properties of irreducible representations the functions $\Psi_k^{(\gamma)t}$ are orthogonal with respect to γ and k (see section 4.16) and the additional label t is also assigned to maintain orthogonality. Thus the set $\Psi_k^{(\gamma)t}$ is orthonormal, the transformation (4.46) is unitary and the inverse may be written as

$$\{\phi_i^{(\alpha)}\psi_j^{(\beta)}\} = \sum_{\gamma t k} C^*(\alpha\beta\gamma t, ijk)\Psi_k^{(\gamma)t} \qquad (4.48)$$

with the coefficients satisfying the orthogonality (unitarity) relations

$$\sum_{ij} C^*(\alpha\beta\gamma t, ijk)C(\alpha\beta\gamma' t', ijk') = \delta_{\gamma\gamma'}\delta_{tt'}\delta_{kk'}$$

$$\sum_{\gamma t k} C^*(\alpha\beta\gamma t, ijk)C(\alpha\beta\gamma t, i'j'k) = \delta_{ii'}\delta_{jj'}$$

When the product functions are not such as to ensure the orthonormality (4.47) the same Clebsch–Gordan coefficients may be used in equation (4.46). However, although the functions $\Psi_k^{(\gamma)t}$ will still be orthogonal with respect to γ and k they will not be normalised and will not generally be orthogonal with the same definition of t. The normalisation of some of the $\Psi_k^{(\gamma)t}$ will be zero when, as discussed above, the product functions are not linearly independent. Even in this case the inverse relation (4.48) still holds.

4.18 Reduction of an irreducible representation on restriction to a subgroup

Let $\mathscr{H}$ be a subgroup of $\mathscr{G}$ and let $T^{(\alpha)}(G_a)$ denote an irreducible representation of $\mathscr{G}$. It follows immediately that the set of operators $T^{(\alpha)}(G_a)$ for those elements G_a which lie in the subgroup $\mathscr{H}$ must form a representation of $\mathscr{H}$. However, although $T^{(\alpha)}$ was taken to be irreducible as a representation of $\mathscr{G}$ it will not necessarily be irreducible as a representation of $\mathscr{H}$. Generally, therefore, it will be reducible and we may write

$$T^{(\alpha)} = \dot{\sum_{\tilde{\alpha}}} n_{\alpha\tilde{\alpha}} T^{(\tilde{\alpha})} \tag{4.49}$$

where the sum runs over all irreducible representations $T^{(\tilde{\alpha})}$ of $\mathscr{H}$. To find the coefficients we again use equation (4.28), knowing the characters of both $T^{(\alpha)}$ and $T^{(\tilde{\alpha})}$ from the tables. Notice that equation (4.49) refers entirely to representations of the subgroup $\mathscr{H}$, and $\chi^{(\alpha)}$ may be deduced from the irreducible character table for the group $\mathscr{G}$ by selecting only those elements which belong to the subgroup $\mathscr{H}$.

As an example, consider the restriction from D_3 to its subgroup C_3, the groups which we have met in subsections 2.2(5) and 2.2(6). We first construct the character table for C_3. The irreducible representations, denoted by $\tau^{(\tilde{\alpha})}$, are all one-dimensional since C_3 is an Abelian group. Furthermore, since $R^3 = E$ for all elements, the characters must be the cube roots of unity, $\exp(2k\pi i/3)$ with $k = 0$, 1 or 2 giving table 4.3. The characters of the irreducible representations $T^{(\alpha)}$ of D_3 may now be copied from table 4.2, using only those elements E, R_1 and R_2 of D_3 which belong to the subgroup C_3. It is clear at once that, as representations of C_3,

$$T^{(1)} \equiv T^{(2)} \equiv \tau^{(1)}$$

while the two-dimensional representation reduces

$$T^{(3)} = \tau^{(2)} \dot{+} \tau^{(3)}$$

Table 4.3

C_3	E	R_1	R_2
$\tau^{(1)}$	1	1	1
$\tau^{(2)}$	1	$\exp(2\pi i/3)$	$\exp(4\pi i/3)$
$\tau^{(3)}$	1	$\exp(4\pi i/3)$	$\exp(2\pi i/3)$

Table 4.4

D_3	E	R_1	R_2
$T^{(1)}$	1	1	1
$T^{(2)}$	1	1	1
$T^{(3)}$	2	-1	-1

4.19 Projection operators

In ordinary three-dimensional space the projection of a vector on to the xy-plane is a familiar geometrical idea, the projection of the three-dimensional vector $\boldsymbol{r} = (x, y, z)$ being simply the vector $(x, y, 0)$ which lies in the xy-plane. This concept of projection on to a subspace will carry over into spaces of higher dimension and into function spaces.

Consider a vector space L which is invariant under the transformations $T(G_a)$ induced by the elements G_a of a group $\mathscr{G}$. In general L will not be irreducible and will therefore reduce, in the way described in sections 4.5 and 4.11, into irreducible subspaces. Let us denote by $\boldsymbol{e}_i^{(\alpha)t}$ a set of basis vectors for L where for fixed α and t the subscript i runs through the s_α rows of the irreducible representation $T^{(\alpha)}$ and t is used when there is more than one subspace of L which transforms according to the same (equivalent) representation $T^{(\alpha)}$. Thus, by definition,

$$T(G_a)\boldsymbol{e}_i^{(\alpha)t} = \sum_j T_{ji}^{(\alpha)}(G_a)\boldsymbol{e}_j^{(\alpha)t}$$

We define the subspace $L_{\alpha i}$ of L by the set of basis vectors $\boldsymbol{e}_i^{(\alpha)t}$ in L with fixed α and i. Then we also define $L_\alpha = \sum_{i=1}^{s_\alpha} L_{\alpha i}$.

Having set up the notation we can now show that the operator

$$P_i^{(\alpha)} = \frac{s_\alpha}{g}\sum_a T_{ii}^{(\alpha)*}(G_a)T(G_a) \tag{4.50}$$

projects from L on to $L_{\alpha i}$. To do this, consider

$$\begin{aligned} P_i^{(\alpha)}\boldsymbol{e}_j^{(\beta)t} &= \frac{s_\alpha}{g}\sum_{a,k} T_{ii}^{(\alpha)*}(G_a)T_{kj}^{(\beta)}(G_a)\boldsymbol{e}_k^{(\beta)t} \\ &= \delta_{\alpha\beta}\delta_{ij}\boldsymbol{e}_j^{(\beta)t} \end{aligned}$$

using equation (4.23). Hence $P_i^{(\alpha)}$ is the required projection operator because if we study its effect on an arbitrary vector $\boldsymbol{r} = \sum_{\beta,t,j} c(\beta, t, j)\boldsymbol{e}_j^{(\beta)t}$ of L, then

$$P_i^{(\alpha)}\boldsymbol{r} = \sum_t c(\alpha, t, i)\boldsymbol{e}_i^{(\alpha)t}$$

showing that the operation of $\mathrm{P}_i^{(\alpha)}$ removes from $\boldsymbol{r}$ all components outside $L_{\alpha i}$ and leaves unchanged those components inside $L_{\alpha i}$.

In practice, the operator $\mathrm{P}_i^{(\alpha)}$ of equation (4.50) can be difficult to construct since it involves the diagonal matrix element of $\mathrm{T}^{(\alpha)}$ for all group elements G_a. The operator which projects from L into the larger subspace L_α is much easier to construct since, because $L_\alpha = \sum_i L_{\alpha i}$, it is given by

$$\mathrm{P}^{(\alpha)} = \sum_i \mathrm{P}_i^{(\alpha)} = \frac{s_\alpha}{g} \sum_a \sum_i T_{ii}^{(\alpha)*}(\mathrm{G}_a)\mathrm{T}(\mathrm{G}_a) = \frac{s_\alpha}{g} \sum_a \chi^{(\alpha)*}(\mathrm{G}_a)\mathrm{T}(\mathrm{G}_a) \quad (4.51)$$

and hence requires a knowledge only of the group characters $\chi^{(\alpha)}$. (One may verify, using equation (4.36), that $\sum_\alpha \mathrm{P}^{(\alpha)} = 1$, which is to be expected because $\sum_\alpha L_\alpha = L$.)

The generalised projection operator (or transfer operator)

$$\mathrm{P}_{ij}^{(\alpha)} = \frac{s_\alpha}{g} \sum_a T_{ij}^{(\alpha)*}(\mathrm{G}_a)\mathrm{T}(\mathrm{G}_a) \quad (4.52)$$

is sometimes useful. It is not strictly a projection operator but has instead the property that

$$\mathrm{P}_{ij}^{(\alpha)} \boldsymbol{e}_k^{(\beta)t} = \delta_{\alpha\beta}\delta_{jk}\boldsymbol{e}_i^{(\beta)t} \quad (4.53)$$

showing that it first projects into the subspace $L_{\alpha j}$ and then transforms into $L_{\alpha i}$. Hence, we can show that for any $\boldsymbol{r}$ and fixed α and j, the set of vectors $\boldsymbol{r}_i^{(\alpha)j} \equiv \mathrm{P}_{ij}^{(\alpha)}\boldsymbol{r}$ with $i = 1, 2, \ldots, s_\alpha$ form a basis for the irreducible representation $\mathrm{T}^{(\alpha)}$,

$$\begin{aligned}
\mathrm{T}(\mathrm{G}_b)\boldsymbol{r}_i^{(\alpha)j} &= \frac{s_\alpha}{g} \sum_a T_{ij}^{(\alpha)*}(\mathrm{G}_a)\mathrm{T}(\mathrm{G}_b)\mathrm{T}(\mathrm{G}_a)\boldsymbol{r} \\
&= \frac{s_\alpha}{g} \sum_c T_{ij}^{(\alpha)*}(\mathrm{G}_b^{-1}\mathrm{G}_c)\mathrm{T}(\mathrm{G}_c)\boldsymbol{r}, \text{ where } \mathrm{G}_c = \mathrm{G}_b\mathrm{G}_a \\
&= \frac{s_\alpha}{g} \sum_k T_{ik}^{(\alpha)*}(\mathrm{G}_b^{-1}) \sum_c T_{kj}^{(\alpha)*}(\mathrm{G}_c)\mathrm{T}(\mathrm{G}_c)\boldsymbol{r} \\
&= \sum_k T_{ki}^{(\alpha)}(\mathrm{G}_b)\boldsymbol{r}_k^{(\alpha)j}
\end{aligned}$$

It is soon verified that the operators we have defined satisfy the typical projection operator multiplication rule

$$\mathrm{P}_j^{(\alpha)}\mathrm{P}_k^{(\beta)} = \delta_{\alpha\beta}\delta_{jk}\mathrm{P}_j^{(\alpha)} \quad (4.54)$$

while for the generalised operators

$$\mathrm{P}_{ij}^{(\alpha)}\mathrm{P}_{kl}^{(\beta)} = \delta_{\alpha\beta}\delta_{jk}\mathrm{P}_{il}^{(\alpha)} \quad (4.55)$$

We close this general discussion of projection operators with two warnings.

First, it is not true that the set of vectors $\mathrm{P}_i^{(\alpha)} \boldsymbol{r}$ with given $\boldsymbol{r}$ and fixed α form a basis for $\mathrm{T}^{(\alpha)}$; the generalised operator $\mathrm{P}_{ij}^{(\alpha)}$ is required for this. Secondly, if $\boldsymbol{r}$ is normalised, it does not follow that the projection $\mathrm{P}_i^{(\alpha)} \boldsymbol{r}$ is normalised; the purpose of the numerical factor s_α/g in equation (4.50) is to ensure that those components of $\boldsymbol{r}$ which survive the projection have their magnitudes unchanged.

As an example of projection let us take the function $\psi_1(\boldsymbol{r}) = x^2$ which was used in section 4.4 and analyse it into components which transform irreducibly under the group D_3. Using the character table (table 4.2) for the three irreducible representations $\mathrm{T}^{(1)}$, $\mathrm{T}^{(2)}$ and $\mathrm{T}^{(3)}$, together with equation (4.51) and the expressions given in section 4.4 for $\mathrm{T}(\mathrm{R}_a)\psi_1$,

$$\mathrm{P}^{(1)}x^2 = \tfrac{1}{6}[1+\mathrm{T}(\mathrm{R}_1)+\mathrm{T}(\mathrm{R}_2)+\mathrm{T}(\mathrm{R}_3)+\mathrm{T}(\mathrm{R}_4)+\mathrm{T}(\mathrm{R}_5)]\,x^2$$

$$= \tfrac{1}{6}(x^2+\tfrac{1}{2}x^2+\tfrac{3}{2}y^2+x^2+\tfrac{1}{2}x^2+\tfrac{3}{2}y^2) = \tfrac{1}{2}(x^2+y^2)$$

$$\mathrm{P}^{(2)}x^2 = \tfrac{1}{6}[1+\mathrm{T}(\mathrm{R}_1)+\mathrm{T}(\mathrm{R}_2)-\mathrm{T}(\mathrm{R}_3)-\mathrm{T}(\mathrm{R}_4)-\mathrm{T}(\mathrm{R}_5)]x^2$$

$$= \tfrac{1}{6}(x^2+\tfrac{1}{2}x^2+\tfrac{3}{2}y^2-x^2-\tfrac{1}{2}x^2-\tfrac{3}{2}y^2) = 0$$

$$\mathrm{P}^{(3)}x^2 = \tfrac{2}{6}\,[2x^2-\mathrm{T}(\mathrm{R}_1)x^2-\mathrm{T}(\mathrm{R}_2)x^2]$$

$$= \tfrac{1}{3}(2x^2-\tfrac{1}{2}x^2-\tfrac{3}{2}y^2) = \tfrac{1}{2}(x^2-y^2)$$

Thus we may write $x^2 = \tfrac{1}{2}(x^2+y^2)+\tfrac{1}{2}(x^2-y^2)$, where the first term (x^2+y^2) transforms according to the identity representation $\mathrm{T}^{(1)}$, while the second term transforms according to the representation $\mathrm{T}^{(3)}$.

To proceed further we may use the projection operator (4.50) to project, not simply on to any row of the two-dimensional representation $\mathrm{T}^{(3)}$ but on to particular rows. For this we need the representation matrix elements, given in table 4.1, rather than simply the characters. We find

$$\mathrm{P}_1^{(3)}x^2 = \tfrac{2}{6}[x^2-\tfrac{1}{2}\mathrm{T}(\mathrm{R}_1)x^2-\tfrac{1}{2}\mathrm{T}(\mathrm{R}_2)x^2-\mathrm{T}(\mathrm{R}_3)x^2+\tfrac{1}{2}\mathrm{T}(\mathrm{R}_4)x^2$$
$$+\tfrac{1}{2}\mathrm{T}(\mathrm{R}_5)x^2]$$

$$= \tfrac{1}{3}(x^2-\tfrac{1}{4}x^2-\tfrac{3}{4}y^2-x^2+\tfrac{1}{4}x^2+\tfrac{3}{4}y^2) = 0$$

$$\mathrm{P}_2^{(3)}x^2 = \tfrac{2}{6}[x^2-\tfrac{1}{2}\mathrm{T}(\mathrm{R}_1)x^2-\tfrac{1}{2}\mathrm{T}(\mathrm{R}_2)x^2+\mathrm{T}(\mathrm{R}_3)x^2$$
$$-\tfrac{1}{2}\mathrm{T}(\mathrm{R}_4)x^2-\tfrac{1}{2}\mathrm{T}(\mathrm{R}_5)x^2]$$

$$= \tfrac{1}{3}(x^2-\tfrac{1}{4}x^2-\tfrac{3}{4}y^2+x^2-\tfrac{1}{4}x^2-\tfrac{3}{4}y^2)$$

$$= \tfrac{1}{2}(x^2-y^2)$$

Thus the function x^2 contains no component transforming like the first row of the representation $T^{(3)}$.

Finally in this example we may construct a pair of functions transforming according to the representation $T^{(3)}$. For this we must use the operator (4.52). We already have

$$P_{22}^{(3)}x^2 \equiv P_2^{(3)}x^2 = \tfrac{1}{2}(x^2-y^2)$$

and now we calculate

$$\begin{aligned} P_{12}^{(3)}x^2 &= \tfrac{2}{6}\sum_a T_{12}^{(3)}(R_a)\,T(R_a)x^2 \\ &= \tfrac{1}{3}\big[-(\tfrac{3}{4})^{\frac{1}{2}}T(R_1)x^2+(\tfrac{3}{4})^{\frac{1}{2}}T(R_2)x^2-(\tfrac{3}{4})^{\frac{1}{2}}T(R_4)x^2 \\ &\quad +(\tfrac{3}{4})^{\frac{1}{2}}T(R_5)x^2\big] \\ &= \tfrac{1}{3}(\tfrac{3}{2}xy+\tfrac{3}{2}xy) = xy \end{aligned}$$

Thus the pair of functions $f_1 = xy, f_2 = \frac{1}{2}(x^2-y^2)$ will transform according to the matrix representation $T^{(3)}$ given in section 4.8. To test this let us calculate $T(R_1)f_1$ which, from the matrix, should give $T(R_1)f_1 = -\frac{1}{2}f_1+(\frac{3}{4})^{\frac{1}{2}}f_2$. We have explicitly, from subsection 3.8.6,

$$\begin{aligned} T(R_1)f_1 &= [-\tfrac{1}{2}x+(\tfrac{3}{4})^{\frac{1}{2}}y]\,[-(\tfrac{3}{4})^{\frac{1}{2}}x-\tfrac{1}{2}y] \\ &= -\tfrac{1}{2}xy+\tfrac{1}{2}(\tfrac{3}{4})^{\frac{1}{2}}(x^2-y^2) = -\tfrac{1}{2}f_1+(\tfrac{3}{4})^{\frac{1}{2}}f_2 \end{aligned}$$

which verifies the earlier result.

From a physical point of view one important use of the projection operation is in defining what are called 'symmetry (adapted) coordinates' for a system. Thus the example in subsection 1.2.3 became much simpler in terms of the symmetry coordinates $Q_1 = x_1+x_2$ and $Q_2 = x_1-x_2$. The relevant group is the symmetric group $\mathscr{S}_2$ [see section 2.2(10)] with elements E and the permutation P_{12}. The characters are given in table 4.5 and it is clear that Q_1 transforms according to the $T^{(1)}$, while Q_2 transforms according to $T^{(2)}$. The use of group theory is, of course, unnecessary in such a simple problem but in a more realistic example (see section 6.5) of the vibrations of the ammonia molecule NH_3 in which there are twelve coordinates, the use of symmetry brings a welcome simplification.

Table 4.5

	E	P_{12}
$T^{(1)}$	1	1
$T^{(2)}$	1	-1

4.20 Irreducible sets of operators and the Wigner–Eckart theorem

The most important concept of this chapter has been that of a vector space which is both invariant and irreducible with respect to the transformations induced by a group $\mathscr{G}$. We have seen in particular that the dimensions and transformation properties of such spaces are strictly limited and must correspond to one of the irreducible representations of $\mathscr{G}$. This led to the idea of classifying functions according to their transformation properties and of analysing an arbitrary function into components, each of which transforms according to a particular row i of a particular irreducible representation $T^{(\alpha)}$. We now extend this concept to the classification of operators. It is a process completely analogous to the classification of functions and is of direct physical interest in quantum mechanics where physical observables are described by operators. A study of their transformation properties leads directly (see section 5.4) to an understanding of selection rules in transition processes (see also subsection 1.2.5).

Consider the transformations $T(G_a)$ in some space L, where G_a is an element of the group $\mathscr{G}$. If S is some arbitrary operator in L then the definition of the transformed operator S′ is, from section 3.4, $S' = T(G_a)ST(G_a)^{-1}$. In general, S′ will be quite different from S in the same way that a transformed function $\phi' = T(G_a)\phi$ is generally quite different from the original function ϕ. However we have seen that the functions in an irreducible invariant space transform 'among themselves' in the sense of equation (4.37). We now define an 'irreducible set of operators' $S_i^{(\alpha)}$ in a similar way, by the property

$$S_i^{(\alpha)\prime} \equiv T(G_a)S_i^{(\alpha)}T(G_a)^{-1} = \sum_j T_{ji}^{(\alpha)}(G_a)S_j^{(\alpha)} \tag{4.56}$$

A set of operators $S_i^{(\alpha)}$ which satisfies equation (4.56) is said to transform according to the irreducible representation $T^{(\alpha)}$. The number of operators in such a set is clearly equal to the dimension s_α of $T^{(\alpha)}$. In particular, a scalar operator, for which $S' = S$ will transform according to the identity representation. By an argument exactly like that given in section 4.17, the product of two operators $S_i^{(\alpha)}S_j^{(\beta)}$ transforms according to the row ij of the direct product representation $T^{(\alpha\times\beta)}$.

In preparation for a study of the matrix elements of operators, consider the result of an operation $S_i^{(\alpha)}$ on a function $\phi_j^{(\beta)}$ where, as the notation implies, both the operator and the function transform according to irreducible representations of the same group. Again, following section 4.17, the set of $s_\alpha s_\beta$ functions ψ_{ij} defined by

$$\psi_{ij} = S_i^{(\alpha)}\,\phi_j^{(\beta)} \tag{4.57}$$

will transform according to the direct product representation $T^{(\alpha\times\beta)}$, since

$$\begin{aligned} T(G_a)\psi_{ij} &= T(G_a)S_i^{(\alpha)}T(G_a)^{-1}T(G_a)\phi_j^{(\beta)} \\ &= \sum_{k,m} T_{ki}^{(\alpha)}(G_a)T_{mj}^{(\beta)}(G_a)S_k^{(\alpha)}\,\phi_m^{(\beta)} \\ &= \sum_{k,m} T_{km,ij}^{(\alpha\times\beta)}(G_a)\psi_{km} \end{aligned} \tag{4.58}$$

We may therefore expand each ψ_{ij} into irreducible components, using equation (4.48),

$$\psi_{ij} = \sum_{\gamma',t,k'} C^*(\alpha\beta\gamma' t, ijk')\,\Psi_{k'}^{(\gamma')t} \tag{4.59}$$

Let us now study the matrix element (see subsection 3.3) of an irreducible operator $S_i^{(\alpha)}$ between basis functions $\phi_j^{(\beta)}$ and $\phi_k^{(\gamma)}$. We have

$$\begin{aligned} (\phi_k^{(\gamma)}, S_i^{(\alpha)}\,\phi_j^{(\beta)}) &= (\phi_k^{(\gamma)}, \psi_{ij}) \\ &= \sum_{\gamma',t,k'} C^*(\alpha\beta\gamma' t, ijk')(\phi_k^{(\gamma)}, \Psi_{k'}^{(\gamma')t}) \\ &= \sum_t C^*(\alpha\beta\gamma t, ijk)(\phi_k^{(\gamma)}, \Psi_k^{(\gamma)t}) \end{aligned} \tag{4.60}$$

using equation (4.38).

This shows, firstly, that the matrix element of the operator $S_i^{(\alpha)}$ will vanish unless the irreducible representation $T^{(\gamma)}$ occurs in the reduction of the product $T^{(\alpha)}\times T^{(\beta)}$. Those combinations of α, β and γ for which the matrix element vanishes are deduced from equation (4.45) using the known character tables. Secondly, it shows that the matrix elements for fixed α, β and γ and various i, j and k are all related to a much smaller number of constants. To see this we note that, from equation (4.38), the scalar product $(\phi_k^{(\gamma)}, \Psi_k^{(\gamma)t})$ is independent of k and it is also independent of i and j.

Since the Clebsch–Gordan coefficients C are known from the group theory, equation (4.60) contains only one constant for each term in the sum over t. In particular, for a simply-reducible group, there is only one constant. The expression 'reduced matrix elements' is used for these constants, with the notation

$$\langle \phi^{(\gamma)} \| \mathbf{S}^{(\alpha)} \| \phi^{(\beta)} \rangle_t = (\phi_k^{(\gamma)}, \Psi_k^{(\gamma)t}) \tag{4.61}$$

the suffixes i, j and k being dropped from the operator and functions because the constant is independent of them. With this notation the result (4.60) becomes

$$(\phi_k^{(\gamma)}, \mathbf{S}_i^{(\alpha)} \phi_j^{(\beta)}) = \sum_t C^* (\alpha\beta\gamma t, ijk) \; \langle \phi^{(\gamma)} \| \mathbf{S}^{(\alpha)} \| \phi^{(\beta)} \rangle_t \tag{4.62}$$

which is known as the Wigner–Eckart theorem and shows that the dependence on i, j and k is contained entirely in the Clebsch–Gordan coefficients.

From a physical point of view, the diagonal matrix elements of an operator are the expectation values of the corresponding observable, while the off-diagonal matrix elements govern the transition probabilities from one state to another (see section 5.1). The results of this section are therefore of great significance since in a complex physical system the functions ϕ and the operator S may be very complicated, but from equation (4.62) the symmetry alone enables us to deduce that certain matrix elements vanish and to predict relationships between the others.

For an invariant operator, α is the identity representation and the coefficients C from equation (4.46) are then trivial, namely t is superfluous, $\gamma = \beta$, $j = k$ and $C(\alpha\beta\gamma, ijk) = \delta_{\beta\gamma}\delta_{jk}$. Thus equation (4.62) becomes

$$(\phi_k^{(\gamma)}, \mathbf{S}\phi_j^{(\beta)}) = \langle \phi^{(\beta)} \| \mathbf{S} \| \phi^{(\beta)} \rangle \, \delta_{\beta\gamma} \delta_{jk} \tag{4.63}$$

which says simply that an invariant has zero matrix elements between any two irreducible representations and that within a representation it is diagonal with equal matrix elements on the diagonal. In other words, within a representation it is a multiple of the unit matrix and between representations it is zero. This statement is in fact just Schur's lemma re-emerging as a special case.

To illustrate the idea of irreducible sets of operators consider the operators (see subsection 3.8.4) $\partial/\partial x$, $\partial/\partial y$ and $\partial/\partial z$ in a vector space of continuous functions $\phi(\boldsymbol{r})$. In this context it is simpler to regard $\phi(\boldsymbol{r}) \equiv \phi(x, y, z)$ as a function of the coordinates of $\boldsymbol{r}$. From equation (3.38) we have the transformed function, for rotations R,

$$\phi'(x, y, z) = \mathrm{T}(\mathrm{R})\phi(x\; y\; z) = \phi(\bar{x}\; \bar{y}\; \bar{z}) \tag{4.64}$$

where $\bar{x}, \bar{y}, \bar{z}$ are the coordinates of $\bar{\boldsymbol{r}} = \mathrm{R}^{-1}\boldsymbol{r}$. The corresponding transformed operators are clearly $\partial/\partial\bar{x}$, $\partial/\partial\bar{y}$ and $\partial/\partial\bar{z}$, since

$$\frac{\partial}{\partial \bar{x}} \mathrm{T}(\mathrm{R})\phi(x, y, z) = \frac{\partial}{\partial \bar{x}} \phi(\bar{x}, \bar{y}, \bar{z})$$

while, if we define $\tilde{\phi}(x, y, z) = \frac{\partial}{\partial x}\phi(x, y, z)$ then

$$\mathrm{T}(\mathrm{R})\frac{\partial}{\partial x}\phi(x, y, z) = \mathrm{T}(\mathrm{R})\tilde{\phi}(x, y, z) = \tilde{\phi}(\bar{x}, \bar{y}, \bar{z}) = \frac{\partial}{\partial \bar{x}}\phi(\bar{x}, \bar{y}, \bar{z})$$

so that $(\partial/\partial\bar{x})\mathrm{T}(\mathrm{R}) = \mathrm{T}(\mathrm{R})(\partial/\partial x)$ and hence $\partial/\partial\bar{x} = \mathrm{T}(\mathrm{R})(\partial/\partial x)\mathrm{T}^{-1}(\mathrm{R})$. The expression for $\partial/\partial\bar{x}$ in terms of the original set of operators is obtained from the usual chain rule for partial derivatives

$$\frac{\partial}{\partial\bar{q}} = \sum_{q' = x, y, z}\frac{\partial q'}{\partial\bar{q}}\frac{\partial}{\partial q'}$$

which shows that the three operators form an invariant set with respect to rotations. The coefficients in this transformation are easily related to the transformation matrix for the basis vectors $\boldsymbol{e}_q$ since, if we write

$$\mathbf{R}\boldsymbol{e}_q = \sum_{q'} R_{q'q}\boldsymbol{e}_{q'}$$

then

$$q' = \boldsymbol{e}_{q'}.\boldsymbol{r} = \boldsymbol{e}_{q'}.\mathbf{R}\bar{\boldsymbol{r}} = \sum_q \bar{q}\,\boldsymbol{e}_{q'}.\mathbf{R}\boldsymbol{e}_q = \sum_q \bar{q}\,R_{q'q}$$

so that $\partial q'/\partial\bar{q} = R_{q'q}$ and hence

$$\mathrm{T}(\mathrm{R})\frac{\partial}{\partial q}\mathrm{T}^{-1}(\mathrm{R}) = \sum_{q'} R_{q'q}\frac{\partial}{\partial q'}$$

showing that the three operators transform exactly like the basis vectors $\boldsymbol{e}_x$, $\boldsymbol{e}_y$ and $\boldsymbol{e}_z$.

This set of operators is therefore clearly irreducible for the group $\mathscr{R}_3$ of all rotations but if we consider only the group D_3 then, as we have seen in section 4.11 for the vectors $\boldsymbol{e}_q$, the set of operators is not irreducible but will break up into $\partial/\partial z$ which transforms like $\mathrm{T}^{(2)}$ and an irreducible pair $\partial/\partial x$ and $\partial/\partial y$ transforming according to $\mathrm{T}^{(3)}$ in the notation of table 4.2

As a further illustration of an irreducible set of operators let us consider multiplication by a function, as in subsection 3.8.3. In particular the functions x and y form an irreducible set transforming according to the representation $T^{(3)}$ of the group D_3, since the transformed operators are simply $\bar{x}$ and $\bar{y}$ which are given by equation (3.39). The fact that the complex conjugate enters that equation is immaterial in this case because the matrix is real.

4.21 Representations of direct product groups

The concept of a direct product group $\mathscr{G} \times \mathscr{H}$ was introduced in section 2.5

and it was shown in section 2.8 that the classes of such a group are labelled by every pair of classes, one from $\mathscr{G}$ and one from $\mathscr{H}$. Now given two irreducible matrix representations $T^{(\alpha)}(G_a)$ of $\mathscr{G}$ and $U^{(\beta)}(H_b)$ of $\mathscr{H}$, it is simply shown that the direct product matrices defined by

$$T^{(\alpha\times\beta)}(G_aH_b) = T^{(\alpha)}(G_a)\times U^{(\beta)}(H_b) \tag{4.65}$$

form an irreducible representation of the group $\mathscr{G}\times\mathscr{H}$.

The proof that $T^{(\alpha\times\beta)}$ is a representation is almost identical to that leading to equation (4.42). The proof that $T^{(\alpha\times\beta)}$ is irreducible follows from the character, which as in equation (4.43) is found to be the product of the characters of $T^{(\alpha)}$ and $U^{(\beta)}$,

$$\chi^{(\alpha\times\beta)}(G_aH_b) = \chi^{(\alpha)}(G_a)\chi^{(\beta)}(H_b) \tag{4.66}$$

From this we have, summing over the group elements,

$$\sum_{ab}|\chi^{(\alpha\times\beta)}(G_aH_b)|^2 = \sum_a|\chi^{(\alpha)}(G_a)|^2\sum_b|\chi^{(\beta)}(H_b)|^2 = gh$$

using equation (4.29) for $\mathscr{G}$ and for $\mathscr{H}$. But, since gh is the order of the product group $\mathscr{G}\times\mathscr{H}$ this equation shows that $T^{(\alpha\times\beta)}$ is irreducible.

It may further be shown that the direct product representations $T^{(\alpha\times\beta)}$ exhaust all irreducible representations of $\mathscr{G}\times\mathscr{H}$ as α and β run over all irreducible representations of $\mathscr{G}$ and $\mathscr{H}$. This is done most simply by summing the squares of the dimensions of the $T^{(\alpha\times\beta)}$,

$$\sum_{\alpha,\beta}(s_\alpha s_\beta)^2 = \sum_\alpha s_\alpha^2\sum_\beta s_\beta^2 = gh$$

using equation (4.33). Hence again, using equation (4.33) for the product group $\mathscr{G}\times\mathscr{H}$, this shows that the $T^{(\alpha\times\beta)}$ exhaust all inequivalent irreducible representations of $\mathscr{G}\times\mathscr{H}$. The character table of a product group is obtained simply, labelling both rows and columns by a pair of indices referring to the separate groups $\mathscr{G}$ and $\mathscr{H}$ and by multiplying the corresponding entries in the character tables for each separate group.

As an example, consider the group $C_2\times S_2$ introduced in section 2.5 and usually denoted by C_{2h}. It is an Abelian group so that each element is in a class by itself. The characters for the two groups C_2 and S_2 are given in table 4.6 and the table for C_{2h} is deduced immediately from them, where, for example, $T^{(1\times 2)}$ denotes the representation constructed from the representations $T^{(1)}$ of C_2 and $T^{(2)}$ of S_2.

As a further example, consider the group $D_{3h} = D_3\times S_1$, introduced in section 2.2 (7). From the characters for D_3 given in table 4.2 and the table for S_1 which is isomorphic with S_2, the characters for D_{3h} are constructed in table 4.7.

Table 4.6

C_2	E	R
$T^{(1)}$	1	1
$T^{(2)}$	1	-1

S_2	E	I
$T^{(1)}$	1	1
$T^{(2)}$	1	-1

$C_{2h} = C_2 \times S_2$	E	R	I	$RI = \sigma$
$T^{(1\times 1)}$	1	1	1	1
$T^{(1\times 2)}$	1	1	-1	-1
$T^{(2\times 1)}$	1	-1	1	-1
$T^{(2\times 2)}$	1	-1	-1	1

Table 4.7

D_{3h}	$\mathscr{C}_1(E)$	$\mathscr{C}_2(R_1, R_2)$	$\mathscr{C}_3(R_3, R_4, R_5)$	$\mathscr{C}'_1(\sigma_h)$	$\mathscr{C}'_2(R_1\sigma_h, R_2\sigma_h)$	$\mathscr{C}'_3(\sigma_3, \sigma_4, \sigma_5)$
$T^{(1)}$	1	1	1	1	1	1
$T^{(2)}$	1	1	-1	1	1	-1
$T^{(3)}$	2	-1	0	2	-1	0
$T^{(1)'}$	1	1	1	-1	-1	-1
$T^{(2)'}$	1	1	-1	-1	-1	1
$T^{(3)'}$	2	-1	0	-2	1	0

Bibliography

A more rigorous mathematical treatment of group representations may be found in

Boerner, H. (1963). *Representations of Groups* (North-Holland, Amsterdam)

Problems

4.1 Verify that the 3×3 matrices $T(R_i)$ in subsection 4.3.1 have the same multiplication table (table 2.5 of section 2.2) as the group elements R_i.

4.2 Construct the 3×3 matrix representation of the group D_4 (see problem 2.3) using the basis vectors $\boldsymbol{e}_x$, $\boldsymbol{e}_y$ and $\boldsymbol{e}_z$, where the z-axis is the four-fold symmetry axis.

4.3 Continue subsection 4.3.3 by deducing the 6×6 matrices $T(R_4)$ and $T(R_5)$ and verifying that $T(R_1)T(R_4) = T(R_5)$.

4.4 Starting with the function $\psi_4 = yz$ and the group D_3, generate an invariant subspace of the set of six quadratics discussed in section 4.4. Show that the representation of D_3 provided by this subspace is equivalent to the representation $T^{(3)}$ described in subsection 4.3.1.

4.5 Find the character of the three-dimensional representation of the group D_4 obtained in problem 4.2 and use the character tables in appendix 1 to show that it reduces to a two-dimensional and a one-dimensional representation. Verify this result directly from the matrices found in problem 4.2.

4.6 Show that, for the cyclic group of order n, the characters of the irreducible representations are given by

$$\chi^{(m)}(C_n^p) = \exp(2\pi imp/n)$$

with $m = 0, 1, 2, \ldots, (n-1)$. Hence construct the character table for the group C_4 of four-fold rotations about the z-axis.

4.7 Show that any change of basis will make the matrix $T(R_3)$, in the representation $T^{(3)}$ of table 4.1 (section 4.8) non-diagonal and hence that $T^{(3)}$ is irreducible, in agreement with the conclusion in section 4.12.

4.8 Verify that the reduced representations of problem 4.5 satisfy the orthogonality relations (4.23) and the criterion (4.29) for irreducibility.

4.9 Using the matrices found in problem 4.3, deduce the character of the representation of D_3 provided by the six-dimensional space of subsection 4.3.3 and use it to show that this representation reduces into $2T^{(1)} \dotplus 2T^{(3)}$ as remarked at the end of section 4.7.

4.10 Using the methods of section 4.15,construct the character table for the group D_4. Identify the irreducible representations already found in problem 4.5 with those in the table.

4.11 Show that the function $f(x, y) = x^2 + y^2$ transforms according to the identity representation for the group D_4 of problem 2.3, with the four-fold axis along the z-direction. For each representation, find a linear or quadratic function of x, y and z which transforms accordingly.

4.12 Using the result of problem 4.10,calculate the character of the direct product of the two-dimensional representation of the group D_4 with itself and find the irreducible representations into which it reduces.

4.13 The coordinates x,y of a particle transform according to the two-dimensional irreducible representation T of D_4, so that the four products x_1x_2, x_1y_2, y_1x_2, y_1y_2 for a pair of particles transform according to the product representation $T \times T$. By projection, find the four combinations of these products which transform irreducibly under D_4. Hence write down the Clebsch–Gordan coefficients.

4.14 Find the representations of C_4 to which the irreducible representations of D_4 reduce on restriction to the subgroup C_4. (Use the results of problems 4.6 and 4.10.)

4.15 Use the projection operator (4.51) to show that the function x^3 transforms according to the representation T of the group D_4. (Take the character from appendix 1.) Further, show that it transforms according to the first row of the representation T in the basis used in problem 4.5. (Use the projection operator (4.50).) Finally, use the projection operator (4.52) together with the matrices of problem 4.5 to construct a partner function which transforms according to the second row.

5

Symmetry in Quantum Mechanics

We begin this chapter with a review, in section 5.1, of the basic concepts of quantum mechanics. It is assumed that the reader has met this subject before (see bibliography) and the main purpose of the review is to present those features which we shall need in illustrating the consequences of symmetry. Following a careful definition of symmetry in section 5.2 we proceed to study, one by one, these different consequences. By way of illustration, in section 5.6 we take a single particle moving in a fixed field with various simple symmetries, C_3, D_3, S_2 and $\mathscr{R}_2$. The use of symmetry in approximate methods is described towards the end of the chapter. The later chapters of the book will deal with more complicated groups and more realistic physical systems but all the important consequences of symmetry in quantum systems are introduced in this chapter.

5.1 Brief review of the framework of quantum mechanics

In quantum mechanics the behaviour of a system with n degrees of freedom described by some coordinates $(r_1 r_2, \ldots, r_n) \equiv \boldsymbol{r}$ is completely specified by its wave function (or state vector) $\psi(\boldsymbol{r}, t)$. This wave function takes on a value, which is generally a complex number, for each value of $\boldsymbol{r}$ and t. The real number $|\psi(\boldsymbol{r}, t)|^2 \, \mathrm{d}V$ is interpreted as the probability that, at time t, the coordinates have values lying in volume $\mathrm{d}V$ at $\boldsymbol{r}$. The normalisation of the wave functions must therefore be chosen such that the total probability of being in some position is equal to unity, i.e.

$$\int |\psi(\boldsymbol{r}, t)|^2 \, \mathrm{d}V = 1 \tag{5.1}$$

where the integral runs over all values of the coordinates.

The wave functions are found by solving the Schrodinger equation

$$\mathrm{H}(\boldsymbol{r}, t)\psi(\boldsymbol{r}, t) = i\hbar \frac{\partial}{\partial t}\psi(\boldsymbol{r}, t) \tag{5.2}$$

where $\mathrm{H}(\boldsymbol{r}, t)$ is the Hamiltonian (operator) corresponding to the classical Hamiltonian $\mathrm{H} = \mathrm{T} + \mathrm{V}$, where T and V are the kinetic and potential energies, respectively. There are general rules for obtaining a quantum mechanical operator from its classical analogue—for example the Cartesian coordinate x of a particle is replaced by the operator x, while the conjugate momentum p_x is replaced by the operator $-i\hbar(\partial/\partial x)$. The boundary conditions imposed on the differential equation (5.2), for example that ψ should vanish at the spatial boundaries, are sufficient to ensure that $\psi(\boldsymbol{r}, t)$ is uniquely determined for all time by equation (5.2) if the function $\psi(\boldsymbol{r}, t_0)$ is known at some time t_0.

Many important problems in quantum mechanics are described by a time-independent Hamiltonian. In such cases it follows immediately from equation (5.2) that there are solutions in which the time dependence appears as a factor

$$\psi_E(\boldsymbol{r}, t) = \psi_E(\boldsymbol{r})\exp(-iEt/\hbar) \tag{5.3}$$

where $\psi_E(\boldsymbol{r})$ is a solution (eigenfunction) of the time-independent eigenvalue equation

$$\mathrm{H}\psi_E(\boldsymbol{r}) = E\psi_E(\boldsymbol{r}) \tag{5.4}$$

and E is the corresponding eigenvalue. The infinite set of eigenfunctions $\psi_E(\boldsymbol{r})$ is a complete set in the sense that any continuous function with the same boundary conditions may be expanded as an infinite seriesin the $\psi_E(\boldsymbol{r})$. Hence any solution of the Schrodinger equation may be expanded

$$\psi(\boldsymbol{r}, t) = \sum_E a_E \psi_E(\boldsymbol{r})\exp(-iEt/\hbar) \tag{5.5}$$

The simple solutions (5.3) are called 'stationary' states, since the probability density $|\psi_E(\boldsymbol{r}, t)|^2$ is time independent. The eigenvalue E is interpreted physically as the energy of the system.

The set of all continuous functions satisfying the boundary conditions imposed on equation (5.2) form a function space, usually called the Hilbert space, and it is convenient to define a scalar product of two functions ψ and ϕ by the integral

$$(\phi, \psi) = \int \phi^*(\boldsymbol{r})\psi(\boldsymbol{r})\mathrm{d}V \tag{5.6}$$

taken over all possible values of the coordinates $\boldsymbol{r}$. The Hamiltonian must be Hermitian with respect of this definition of scalar product in order that the energies E shall be real. As a consequence, the set of eigenfunctions $\psi_E(\boldsymbol{r})$ are orthogonal with respect to this scalar product and therefore provide a convenient basis. In accordance with equation (5.1) it is usual to define the $\psi_E(\boldsymbol{r})$ to be normalised to unity. Notice that for a stationary state the condition (5.1) becomes $(\psi_E, \psi_E) = 1$.

To each classical observable O, like the energy, momentum, angular momentum, etc., there is a corresponding Hermitian operator O (usually time independent) which operates in the Hilbert space of the wave functions. Each operator will have its own set of eigenfunctions and eigenvalues $\mathrm{O}\psi^O_\lambda(\boldsymbol{r}) = \lambda\psi^O_\lambda(\boldsymbol{r})$ and each set $\psi^O_\lambda(\boldsymbol{r})$ will form an orthonormal basis. The only definite values resulting from a measurement of the observable O are the eigenvalues λ and a particular value λ will be observed only if the wave function of the system at that particular time is precisely the corresponding eigenfunction $\psi^O_\lambda(\boldsymbol{r})$ of O. In general, the wave function $\psi(\boldsymbol{r}, t)$ will not coincide with any particular eigenfunction of O but, because of the completeness of the $\psi^O_\lambda(\boldsymbol{r})$, we can always expand

$$\psi(\boldsymbol{r}, t) = \sum_\lambda c_\lambda(t)\psi^O_\lambda(\boldsymbol{r}) \tag{5.7}$$

The result of a measurement of the observable O in such a state is not completely determined but $|c_\lambda(t)|^2$ is the probability of obtaining the value λ. The mean value or, as it is usually called, 'the expectation value' of O in the state ψ is given by the sum $\sum_\lambda |c_\lambda(t)|^2\lambda$. This expectation value may be expressed very conveniently as a scalar product since

$$(\psi, \mathrm{O}\psi) = \sum_{\lambda'}\sum_\lambda c^*_{\lambda'}(t)\, c_\lambda(t)\,\lambda(\psi^O_{\lambda'}, \psi^O_\lambda) = \sum_\lambda |c_\lambda(t)|^2\lambda \tag{5.8}$$

using the orthonormality of the eigenfunctions $\psi^O_\lambda(\boldsymbol{r})$. It is customary in quantum mechanics to use the notation $\langle\psi|\mathrm{O}|\psi\rangle$ for the scalar product $(\psi, \mathrm{O}\psi)$. Notice that the stationary states give precise values E for a measurement of the energy, the corresponding operator being the Hamiltonian. To be consistent in our notation we should have used ψ^H_E for the stationary states but for brevity we used simply ψ_E, in view of the special significance of the energy operator H.

Assuming that the spectrum of an operator (the set of eigenvalues) is discrete we have the phenomenon of quantisation in which the measured values of the operator are restricted to that discrete set of eigenvalues. For some operators it may happen that the spectrum is continuous, in which case the expansion (5.7) becomes

$$\psi(\boldsymbol{r}, t) = \int c(\lambda, t)\psi_{\lambda}^{0}(\boldsymbol{r})\mathrm{d}\lambda$$

where the integration runs over the possible range of eigenvalues λ.

From the experimental point of view one may investigate a quantum system by measuring its energies and the expectation values of various operators. These can then be compared with the values calculated from some assumed Hamiltonian. Another very important kind of experiment is the measurement of the transition rate, or transition probability, for a process in which a system changes from an initial state ψ_i to a final state ψ_f. It may be argued that the transition probability W_{if} is proportional to the square of an off-diagonal matrix element

$$W_{if} \propto |(\psi_f, \mathrm{O}\psi_i)|^2 \tag{5.9}$$

where O is the operator relevant to the particular process. It is, of course, by the measurement of energies, expectation values and transition probabilities that one learns about the symmetry of a quantum system and in the subsequent sections of this chapter we shall show how these quantities are affected by symmetry. The interaction of a system, for example an atom or atomic nucleus, with an electromagnetic field is a particularly convenient and fruitful source of information. It provides both expectation values, through the perturbation of energies, and transition probabilities through the absorption and emission of radiation. Without going into the theory in any detail, we now describe the main features of the electromagnetic interaction and list the various operators which occur.

As one might expect from simple classical ideas, the interaction of a system with an electromagnetic field depends on the positions and momenta of the constituent charges. Next to the total charge of the system, the most important quantities are what are known as the electric dipole and magnetic dipole moments, which for a system of particles i with charges e_i are given by $\sum_i e_i\boldsymbol{r}_i$ and $\sum_i (e_i/2Mc)\boldsymbol{r}_i \wedge \boldsymbol{p}_i$, respectively. (For particles with spin there is an additional contribution to the magnetic moment, see section 8.4.) In an atom, where all the Z electrons have the same charge, these expressions simplify to $-Ze\boldsymbol{R}$ and $-(e/2Mc)\boldsymbol{L}$, where $\boldsymbol{R}$ is the centre-of-mass vector and $\boldsymbol{L}$ the total angular momentum. These operators are involved both in the energy perturbations and in the transition probabilities for the quantum system. In particular, the energy shift due to a uniform magnetic field $\boldsymbol{B}$ is given by $e\boldsymbol{L}.\boldsymbol{B}/2Mc$. For transitions, the electric dipole process is dominant in atoms, and for unpolarised radiation all three components of $\boldsymbol{R}$ will contribute. However, for the absorption of polarised light, only certain components of $\boldsymbol{R}$ are relevant. For example, in what is known as plane-polarised light, the electric field vector points in a fixed direction $\boldsymbol{p}$ which is perpendicular to the light beam. The relevant operator is then the component of $\boldsymbol{R}$ in the direction $\boldsymbol{p}$ and we refer to this case as polarisation in the $\boldsymbol{p}$-direction.

5.2 Definition of symmetry in a quantum system

Consider a system governed by a time-independent Hamiltonian $\mathrm{H}(\boldsymbol{r})$ with $\psi(\boldsymbol{r})$ denoting an arbitrary wave function—not necessarily an eigenfunction. As in the previous section, the coordinate vector $\boldsymbol{r}$ describes the n degrees of freedom of the system. Any group $\mathscr{G}$ of coordinate transformations $\mathrm{G}_a\boldsymbol{r} = \boldsymbol{r}'$ will define a corresponding set of induced transformations $\mathrm{T}(\mathrm{G}_a)$ in the space of wave functions through the definition (3.37):

$$\mathrm{T}(\mathrm{G}_a)\psi(\boldsymbol{r}) = \psi'(\boldsymbol{r}) = \psi(\mathrm{G}_a^{-1}\boldsymbol{r})$$

It also defines, through equation (3.24), the transformed Hamiltonian operator $\mathrm{T}(\mathrm{G}_a)\mathrm{H}\mathrm{T}^{-1}(\mathrm{G}_a) = \mathrm{H}'$. If the Hamiltonian is invariant under these transformations, i.e. if

$$\mathrm{T}(\mathrm{G}_a)\mathrm{H}\mathrm{T}^{-1}(\mathrm{G}_a) = \mathrm{H} \tag{5.10}$$

for all G_a in the group $\mathscr{G}$, then $\mathscr{G}$ is said to be a symmetry group of the Hamiltonian. We shall soon see how the existence of a symmetry group leads to a variety of important consequences. By multiplying on the right by T, the condition (5.10) takes the equivalent form $\mathrm{T}(\mathrm{G}_a)\mathrm{H} - \mathrm{H}\mathrm{T}(\mathrm{G}_a) = 0$, or $[\mathrm{T}(\mathrm{G}_a), \mathrm{H}] = 0$, i.e. the Hamiltonian commutes with all the induced transformations of the group.

The group elements G_a of a symmetry group are called symmetry elements. In practice, if one can find one or two symmetry elements of a Hamiltonian, a symmetry group may be generated by multiplying them together and taking powers until no new elements are produced. (It follows immediately from equation (5.10) that the product of two symmetry elements is another symmetry element.) It is, however, rather difficult to be sure that one has found *all* the symmetry elements of a given Hamiltonian and not simply the elements belonging to a subgroup of the full symmetry group.

For many transformations, such as rotations, reflections and translations, the kinetic energy operator is an invariant so that the invariance of the Hamiltonian rests on the condition

$$V(\boldsymbol{r}) = V(\mathrm{G}_a\boldsymbol{r}) \tag{5.11}$$

on the potential energy V.

In the presentation above, we have been fairly explicit in the definition of the induced transformation but it is possible to imagine a more abstract set of transformations $\mathrm{T}(\mathrm{G}_a)$ in the function space which are not related to coordinate transformations but which nevertheless satisfy the multiplication rules for a group and therefore define a symmetry group.

5.3 Degeneracy and the labelling of energies and eigenfunctions

The first two consequences of the existence of a symmetry group $\mathcal{G}$ of the Hamiltonian are intimately related and we discuss them together. First we state the results and then show how they are derived.

(1) The eigenfunctions ψ and eigenvalues E of H may be labelled by the irreducible representations $T^{(\alpha)}$ of the symmetry group $\mathcal{G}$, so that we write $\psi^{(\alpha)}$ and $E^{(\alpha)}$.

(2) The energy $E^{(\alpha)}$ will be (at least) s_α-fold degenerate, where s_α is the dimension of $T^{(\alpha)}$.

To derive these results we first note that the set of all degenerate eigenfunctions at some chosen eigenvalue E forms a vector space V†. (If Φ and ψ both have energy E then so does any linear combination of them.) Next, we argue that V must be an invariant space with respect to the transformations $T(G_a)$ induced by $\mathcal{G}$. To show the invariance we simply define the transformed function $\psi' = T(G_a)\psi$ in the usual way and then deduce that, if ψ is an eigenfunction of H with energy E, then so is ψ', because

$$H\psi' = HT(G_a)\psi = T(G_a)H\psi = ET(G_a)\psi = E\psi' \tag{5.12}$$

where we have made use of equation (5.10). The vector space V therefore provides a representation of the symmetry group through the transformations $T(G_a)$. This representation is either irreducible or may be reduced into its irreducible constituents so that in either case a basis for V may be chosen in which the basis vectors are written as $\psi_i^{(\alpha)}$, being labelled by an irreducible representation label α of the group $\mathcal{G}$ and by a row label i. The degeneracy is given by the dimension of V which is at least s_α and will be greater if V is reducible.

The argument above justifies the results given at the beginning of this section but leaves one important point to be clarified. If the representation provided by V is not irreducible then that energy will carry not one but a number of labels α corresponding to the irreducible representations which occur in its reduction. In practice, if one is using the full symmetry group, it is very rare to find such a situation—as rare as it would be to find equal eigenvalues for an arbitrary matrix. One therefore uses the expression 'normal degeneracy' for one described by an irreducible representation and 'accidental degeneracy' when two or more irreducible representations occur at the same energy. An accidental degeneracy can be produced by the variation of some parameter in H (such as a magnetic or electric field strength) so that, for some particular values of this parameter, two normally non-degenerate levels cross. If, for a particular Hamiltonian, one finds systematic accidental degeneracies when using a symmetry group $\mathcal{G}$ there is usually an explanation in the existence of a larger symmetry group. As we have seen in section 4.18, an irreducible

† The symbol V used on this page for the vector space should not be confused with the same symbol V which was used for the potential energy on the previous page.

representation of a group $\mathscr{K}$ will generally reduce into a sum of several irreducible representations of any subgroup. Thus the accidentally degenerate levels are brought together within a single irreducible representation of the larger group. Some famous examples of this are given in chapter 19 of volume 2. Examples of the much more important case of normal degeneracy are given in section 5.6.

5.4 Selection rules and matrix elements of operators

In a system with a symmetry group $\mathscr{G}$ we have seen that the eigenfunctions ψ of the Hamiltonian may be labelled by the irreducible representations of $\mathscr{G}$. In section 4.20 we saw that if an operator $O^{(\alpha)}$ was defined to transform according to an irreducible representation $T^{(\alpha)}$ then its matrix elements, taken between functions which also transform irreducibly under the same group, have some simple features. It is therefore clear that we shall be able to draw some simple conclusions about the physically interesting matrix elements if we analyse the operators into their irreducible parts. The most striking result is that if a particular transition process is governed by an operator $O^{(\alpha)}$ then a transition from a state $\psi^{(\beta)}$ may take place only to those final states $\psi^{(\gamma)}$ for which the representation $T^{(\gamma)}$ occurs in the reduction (4.44):

$$T^{(\alpha)} \times T^{(\beta)} = \sum_{\gamma} m_{\gamma} T^{(\gamma)}$$

Thus transitions from the state $\psi^{(\beta)}$ via the operator $O^{(\alpha)}$ select those final states $\psi^{(\gamma)}$ which transform according to any one of the representations on the right-hand side of equation (4.44). For other final states, the matrix element vanishes and no transition can take place. One refers to this phenomenon as a 'selection rule' and transitions which cannot take place are said to be 'forbidden'. Notice that a selection rule does not tell you that a transition will occur but only that, so far as the symmetry is concerned, it may occur. In particular cases there can always be other reasons which cause particular matrix elements to vanish. To find the full set of selection rules one must take the irreducible representations $T^{(\beta)}$ one by one and deduce the coefficients m_{γ} from equation (4.45) using the known character table for the symmetry group. The transition may then take place for every γ for which $m_{\gamma} \neq 0$.

So far we have assumed that the transition is governed by an operator $O^{(\alpha)}$ transforming according to an irreducible representation α. In practice it may happen that the operator O transforms according to a reducible representation. In this case we may either analyse it into its irreducible parts

$$O = \sum_{\alpha} O^{(\alpha)}$$

and deduce the transitions allowed separately for each irreducible component, or, if we know the character $\chi(G_a)$ of the 'reducible' representation directly, then the coefficients m_γ are given by

$$m_\gamma = \frac{1}{g} \sum_a \chi^{(\gamma)}(G_a)^* \, \chi(G_a) \chi^{(\beta)}(G_a)$$

in place of equation (4.45).

The operators $\mathbf{R} = (X, Y, Z)$ for electric dipole transitions transform like a vector under rotations and are odd under inversion. They thus form a three-dimensional representation of any group of rotations and inversions. For a proper rotation about the z-axis, through an angle θ, the character is simply $(2\cos\theta + 1)$ from equation (4.6) and, since the trace is independent of basis, this is also the character for rotation about any axis. In the same way for magnetic dipole transitions, the relevant operator $\mathbf{L} = \sum_i \mathbf{r}_i \wedge \mathbf{p}_i$ is also a vector and therefore has the same character for rotations. However, whereas $\mathbf{R}$ changes sign under inversion, $\mathbf{L}$ does not, since both $\mathbf{r}$ and $\mathbf{p}$ change sign, so that when improper rotations are considered, the characters for $\mathbf{R}$ and $\mathbf{L}$ differ in sign. One often refers to these two three-dimensional representations of any group of rotations and inversions as the vector and pseudo- (or axial) vector representations, respectively.

The Wigner–Eckart theorem, equation (4.62), goes beyond the selection rule to relate matrix elements which have the same α, β and γ but differ in their row labels i, j and k—see equation (4.62). In other words it tells how a matrix element changes when one of the wave functions is replaced by another member of the degenerate multiplet. Since the simple transition processes involve a sum over any degenerate final states, this feature of symmetry is of less importance. We leave any further discussion of applications of the Wigner–Eckart theorem until chapter 8.

5.5 Conservation laws

An observable O is said to be conserved in a given system if its mean value in any state $\psi(\mathbf{r}, t)$ of the system does not vary with time. An equivalent statement is that if, at some time, the wave function of the system coincides with an eigenfunction of O, then it will remain an eigenfunction of O with the same eigenvalue for all time. We now show that an operator O has this conservation property if it commutes with the Hamiltonian and is also time independent. We have

$$\begin{aligned}
\frac{\mathrm{d}}{\mathrm{d}t}(\psi, \mathrm{O}\psi) &= \left(\frac{\partial\psi}{\partial t}, \mathrm{O}\psi\right) + \left(\psi, \mathrm{O}\frac{\partial\psi}{\partial t}\right) \\
&= \{-(\mathrm{H}\psi, \mathrm{O}\psi) + (\psi, \mathrm{OH}\psi)\}/i\hbar \\
&= (\psi, [\mathrm{O}, \mathrm{H}]\psi)/i\hbar = 0
\end{aligned} \tag{5.13}$$

using the Schrodinger equation (5.2) and the assumed commutation property $[\mathrm{H}, \mathrm{O}] = 0$.

Now, if the system possesses symmetry, the operators $\mathrm{T}(\mathrm{G}_a)$ for any G_a in $\mathscr{G}$ will commute with the Hamiltonian and are therefore conserved. At first sight this suggests a large number of conserved quantities but they are not independent. The conservation of $\mathrm{T}(\mathrm{G}_c)$ is clearly not independent of the conservation of $\mathrm{T}(\mathrm{G}_a)$ and $\mathrm{T}(\mathrm{G}_b)$ if $\mathrm{G}_a\mathrm{G}_b = \mathrm{G}_c$. There is therefore a minimum number of independent conserved quantities for each group corresponding to the minimum number of group elements needed to generate all group elements by multiplication amongst themselves. Thus for a cyclic group, like C_3, there is only one conserved quantity since, by definition of a cyclic group, all elements are obtained as powers of a single element. For the group D_3, see table 2.5, one needs two elements R_1 and R_3 to generate all others and so there are two independent conserved quantities. For the continuous group $\mathscr{R}_2$ it will be seen in subsection 7.3.5 that all group elements may be generated from a single operator.

It should also be remembered from the postulates of quantum mechanics that only Hermitian operators may represent physical observables. Thus for a physically observable conserved quantity one must construct a Hermitian operator from the $\mathrm{T}(\mathrm{G}_a)$. We shall see that only for the continuous groups do these conserved quantities correspond to familiar classical concepts like momentum and angular momentum.

5.6 Examples

To illustrate these consequences of symmetry we now consider four examples, corresponding to symmetry groups C_3, D_3, S_2 and $\mathscr{R}_2$. Although it is not necessary to specify the wave function in detail (this is the strength of the symmetry argument) one may imagine these examples to relate to the motion of a single particle in a potential $V(\boldsymbol{r})$. The kinetic energy operator ∇^2, being a scalar product, is invariant with respect to any rotation and is therefore an invariant in all four examples. On the other hand, the potential may be chosen to be invariant only with respect to one of the more restricted rotation groups listed above. This simple problem is not entirely artificial since it can represent the motion of an electron around the nucleus of an atom. Although the Coulomb attraction to the nucleus is spherically symmetric one may imagine an externally applied field which reduces the symmetry. Alternatively, if the atom is in a crystal then the presence of other atoms of the crystal will reduce the symmetry from $\mathscr{R}_3$ to some finite group of rotations.

5.6.1 Symmetry group C_3

For this group the potential $V(\boldsymbol{r})$ is invariant under three-fold rotations R_1 and $\mathrm{R}_2 = \mathrm{R}_1^2$ about the z-axis, using the notation of section 2.2(5). The group C_3 is a cyclic group and the three one-dimensional irreducible representations $\tau^{(1)}$, $\tau^{(2)}$ and $\tau^{(3)}$, given in table 4.3 of section 4.18, may be written

$$\tau^{(\alpha)}(R_1) = \exp[2\pi(\alpha-1)i/3] \tag{5.14}$$

with $\alpha = 1, 2$ and 3. From the general results of section 5.3 we can therefore deduce that the eigenstates will be non-degenerate, can be labelled by the irreducible representation label α and will have the property $T(R_1)\psi^{(\alpha)} = \exp[2\pi(\alpha-1)i/3]\psi^{(\alpha)}$. In our particular example of a single particle this implies, from the definition (3.37), that

$$\psi^{(\alpha)}\left(\phi - \frac{2\pi}{3}\right) = \exp[2\pi(\alpha-1)i/3]\psi^{(\alpha)}(\phi) \tag{5.15}$$

where ϕ is the usual polar coordinate around the z-axis. (Without the direct use of group theory we could have derived this result by noticing that the Hamiltonian H commutes with the three-fold rotation operator $\tau(R_1)$ and hence the eigenstates of H can be chosen to make $\tau(R_1)$ diagonal and can be labelled by the eigenvalues of $\tau(R_1)$ which are given by equation (5.14). The operator $\tau(R_1)$ represents the conserved quantity for C_3 but it has no simple physical interpretation.)

The symmetry properties of functions which satisfy equation (5.15) are seen more clearly by writing

$$\psi^{(\alpha)}(\phi) = u_\alpha(\phi)\exp[-i(\alpha-1)\phi] \tag{5.16}$$

where $u_\alpha(\phi)$ is a periodic function of ϕ with period $2\pi/3$. This may be justified by substituting (5.16) in equation (5.15) whence

$$u_\alpha\left(\phi - \frac{2\pi}{3}\right)\exp\{-i(\alpha-1)[\phi-(2\pi/3)]\}$$
$$= \exp[2\pi(\alpha-1)i/3]u_\alpha(\phi)\exp[-i(\alpha-1)\phi]$$

i.e.

$$u_\alpha\left(\phi - \frac{2\pi}{3}\right) = u_\alpha(\phi) \tag{5.17}$$

Equation (5.16) gives $\psi^{(\alpha)}(\phi)$ for all ϕ if it is known for the range $0 \leqslant \phi < 2\pi/3$. Symmetry arguments thus lead to the conclusion that, to solve Schrodinger's equation for a system with C_3 symmetry, it is only necessary to solve an equivalent equation for $u_\alpha(\phi)$ for the range $0 \leqslant \phi < 2\pi/3$ with periodic boundary conditions.

We will now study the selection rules for electric dipole transitions between the eigenstates $\psi^{(\alpha)}(\phi)$. For light polarised along the z-axis the appropriate operator (see section 5.1) is $\boldsymbol{ez}$. Clearly this is invariant under the three-fold rotations and hence belongs to the identity representation $\tau^{(1)}$. Since the product representation $\tau^{(1)} \times \tau^{(\alpha)} = \tau^{(\alpha)}$ this operator can only cause transitions between states with the same irreducible representation label α. For light polarised in a direction in the xy-plane the relevant operator is a mixture of x and y. Since $x \pm iy = r \sin\theta \exp(\pm i\phi)$, it follows by comparison with equation (5.16) that the operator transforms according to a mixture of $\tau^{(2)}$ and $\tau^{(3)}$. From the character table (table 4.3) we see that the products of representations give $\tau^{(1)} \times \tau^{(\alpha)} = \tau^{(\alpha)}$, $\tau^{(2)} \times \tau^{(2)} = \tau^{(3)}$, $\tau^{(2)} \times \tau^{(3)} = \tau^{(1)}$, and $\tau^{(3)} \times \tau^{(3)} = \tau^{(2)}$ Hence the allowed transitions for light polarised in a direction in the xy-plane are $(1) \leftrightarrow (2)$, $(1) \leftrightarrow (3)$ and $(2) \leftrightarrow (3)$, so that all transitions which change the label α are possible.

5.6.2 Symmetry group D_3

This group was described in section 2.2(6) and the potential is now invariant under three-fold rotations about the z-axis and also under two-fold rotations about three axes in the xy-plane. The characters of D_3 were given in table 4.2. The eigenstates will therefore be either non-degenerate carrying the label $\alpha = 1$ or 2, or doubly degenerate with $\alpha = 3$. States $\psi^{(1)}$ will be invariant under all group operations. States $\psi^{(2)}$ will be invariant under the three-fold rotations R_1 and R_2 but change sign under the two-fold rotations R_3, R_4 and R_5. Pairs of states $\psi_1^{(3)}$, $\psi_2^{(3)}$ will in general be mixed under the group operations.

Selection rules for electric dipole transitions can easily be found. The relevant operator $O = e\mathbf{R}$ is a vector with character $(2\cos\theta + 1)$, see section 5.4, so that from the D_3 characters reproduced in table 5.1 we see that it transforms like $T^{(2)} \dotplus T^{(3)}$.

Table 5.1

D_3	$\mathscr{C}_1(E)$	$\mathscr{C}_2(R_1, R_2)$	$\mathscr{C}_3(R_3, R_4, R_5)$	
$T^{(1)}$	1	1	1	
$T^{(2)}$	1	1	-1	
$T^{(3)}$	2	-1	0	
O	3	0	-1	$= T^{(2)} \dotplus T^{(3)}$
$O \times T^{(1)}$	3	0	-1	$= T^{(2)} \dotplus T^{(3)}$
$O \times T^{(2)}$	3	0	1	$= T^{(1)} \dotplus T^{(3)}$
$O \times T^{(3)}$	6	0	0	$= 2T^{(3)} \dotplus T^{(2)} \dotplus T^{(1)}$

Thus we deduce that the following transitions are allowed $T^{(1)} \leftrightarrow T^{(2)}$, $T^{(1)} \leftrightarrow T^{(3)}$, $T^{(2)} \leftrightarrow T^{(3)}$, $T^{(3)} \leftrightarrow T^{(3)}$, while the transitions $T^{(1)} \leftrightarrow T^{(1)}$, $T^{(2)} \leftrightarrow T^{(2)}$ are forbidden. Again by analysing O into its component $T^{(2)}$ (along z) and $T^{(3)}$ in the xy-plane we can deduce that, for polarised light, the following transitions are allowed:

(1) Polarisation along z $\quad T^{(1)} \leftrightarrow T^{(2)}, T^{(3)} \leftrightarrow T^{(3)}$
(2) Polarisation in xy-plane $\quad T^{(1)} \leftrightarrow T^{(3)}, T^{(2)} \leftrightarrow T^{(3)}, T^{(3)} \leftrightarrow T^{(3)}$

5.6.3 Symmetry group S_2

This very simple group, see section 2.2(4), contains only the identity and the inversion I with $I^2 = E$. It has two one-dimensional irreducible representations, the identity representation $T^{(1)}$ and a representation $T^{(2)}$ which associates -1 with the inversion. The eigenstates of a Hamiltonian which is invariant under inversion will therefore belong to $T^{(1)}$ or $T^{(2)}$. We refer to them as even or odd parity, respectively, and they are often distinguished by the labels $\pm$. Because the representations are one-dimensional, there is no degeneracy. Selection rules are obtained very simply from the product rules $T^{(1)} \times T^{(2)} = T^{(2)}$ and $T^{(1)} \times T^{(1)} = T^{(2)} \times T^{(2)} = T^{(1)}$. Thus, an odd parity operator, like the electric dipole, will need a change of parity between initial and final states, while the magnetic dipole operator (see section 5.4) will allow transitions only between states of the same parity.

5.6.4 Symmetry group $\mathscr{R}_2$

This is the continuous group of all rotations about the z-axis so that the potential $V(\mathbf{r})$ must be a function of r and θ only, if it is to be invariant under $\mathscr{R}_2$. The group elements are $R(a)$ and the operators $T(R(a))$ are defined, following equation (3.37), as $T(R(a))\psi(r, \theta, \phi) = \psi(r, \theta, \phi - a)$ when acting on a wave function for a single particle. The coordinates r, and θ, play no part in this discussion and we shall omit them. Now we may use a Taylor series expansion to write

$$\psi(\phi - a) = \psi(\phi) - a\frac{\partial}{\partial \phi}\psi(\phi) + \frac{1}{2!}a^2\frac{\partial^2}{\partial \phi^2}\psi(\phi) + \dots$$

$$= \exp\left(-a\frac{\partial}{\partial \phi}\right)\psi(\phi)$$

so that the operator $T(R(a))$ may be given the explicit form

$$T(R(a)) = \exp\left(-a\frac{\partial}{\partial \phi}\right) \tag{5.18}$$

The exponential of an operator may be interpreted as an abbreviation for the corresponding series. It is apparent therefore that all rotations about the z-axis may be generated from the single operator $\partial/\partial\phi$.

Since the group elements commute, the group is Abelian and hence (see section 4.8) the irreducible representations must be one-dimensional. We shall discuss the irreducible representations of $\mathscr{R}_2$ more carefully in section 7.3 but it is clear from equation (5.18) that functions of the kind $\psi^{(m)}(\phi) = A\exp(im\phi)$, where m is any integer, will serve as basis vectors for one-dimensional irreducible representations given by $T^{(m)}(R(a)) = \exp(-ima)$. By choosing the label m to be real we make $T^{(m)}$ unitary and by choosing m to be an integer we ensure the continuity of $\psi^{(m)}(\phi)$ as ϕ goes through 2π.

Selection rules for electric dipole transitions can be found by noting that z is invariant under all rotations about the z-axis and hence belongs to the representation with $m = 0$. For light polarised in the xy-plane the ϕ-dependence of the relevant operator is, as in previous examples, a combination of $\exp(i\phi)$ and $\exp(-i\phi)$ and hence is a mixture of the representations $m = 1$ and $m = -1$. The rule for forming product representations in $\mathscr{R}_2$ is particularly simple since

$$T^{(m)}(R(a)) \times T^{(m')}(R(a)) = \exp(-ima - im'a) = T^{(m+m')}(R(a))$$

Light which is polarised along the z-axis thus gives rise to transitions in which the m-value does not change, $\Delta m = 0$, whereas light polarised in the xy-plane gives rise to transitions in which the m-value changes by unity, $\Delta m = \pm 1$.

From the discussion of section 5.5 one would expect to find a conservation law in the case of $\mathscr{R}_2$ symmetry if we can construct a Hermitian operator from the rotation operator (5.18) and it is soon verified that the exponent $\partial / \partial\phi$ will serve such a purpose if multiplied by i. In fact if we write $l_z = -i(\partial / \partial\phi)$ and transform to Cartesian coordinates we have

$$l_z = -i\frac{\partial}{\partial\phi} = -i\left(x\frac{\partial}{\partial y} - y\frac{\partial}{\partial x}\right) = (xp_y - yp_x)/\hbar \tag{5.19}$$

where $p_x = -i\hbar\,\partial/\partial x$ is the usual quantum mechanical expression for the momentum. This shows that the conserved quantity l_z represents the angular momentum about the z-axis. In an eigenstate $\psi^{(m)}(\phi)$ the operator l_z has value m.

5.7 Use of group theory in a variational approximation

In most physically realistic problems in quantum mechanics the Schrodinger equation cannot be solved exactly because of the mathematical complexity, especially when a large number of particles are involved. In such cases one often uses the variational method to find an approximate solution for the ground state and possibly the first few excited states. The essence of the method is to construct a relatively simple normalised trial function ψ which contains a number of variable parameters and then to calculate the expectation value $(\psi, H\psi)$ of the Hamiltonian. It can be shown that this value is an upper bound to the ground state energy for any values of the parameters. Consequently by minimising the expectation value with respect to the parameters in the wave function we find the lowest upper bound possible with the chosen form of trial function. As more parameters are introduced, the expectation value $(\psi, H\psi)$ approaches closer to the ground state energy and the trial function ψ approaches the ground state wave function. As a second step, one chooses trial functions orthogonal to the ground state to obtain, by the same procedure, an approximation to the first excited state and so on.

If the system has symmetry then, without any calculation, one can say that the eigenfunctions transform irreducibly under the symmetry group. This is already a helpful guide in choosing the variational trial function since it tells us not to take a trial function which is a mixture of pieces which transform according to different irreducible representations. Further than this, it is generally true to say that a system governed by attractive forces between its constituents has a ground state with maximum symmetry. In terms of representations, this says that the ground state should transform like the identity representation—that it should be invariant under the group. For example, it may be shown rigorously that the ground state for a single particle moving in a spherically symmetric potential $V(r)$ is itself spherically symmetric, with $l = 0$. One may therefore restrict to spherically symmetric trial functions in approximating the ground state. To find the first excited state one would build a trial function transforming according to some other representation and again it is usually the representation with next greatest symmetry, in some sense. For example in the harmonic oscillator potential the first excited state has $l = 1$. There will also be excited states with maximum symmetry, see for example the spectrum of the hydrogen atom in section 8.5.

One form of the variational method which is often used is to take as trial function a linear combination

$$\psi = \sum_{i=1}^{n} c_i \phi_i \tag{5.20}$$

of n fixed functions ϕ_i with the coefficients c_i as the variational parameters. This is sometimes called the Rayleigh–Ritz method and is equivalent to diagonalising the Hamiltonian in the finite n-dimensional space of the chosen functions ϕ_i. If the ϕ_i are chosen to be orthonormal, then the Schrodinger equation $\mathrm{H}\psi = E\psi$ reduces to a matrix equation as follows. From $\mathrm{H}\psi = E\psi$ and using (5.20) we have

$$\sum_i c_i \mathrm{H}\phi_i = E\sum_i c_i \phi_i$$

i.e.

$$\sum_i c_i(\phi_j, \mathrm{H}\phi_i) = Ec_j$$

and, writing H_{ji} for the matrix element $(\phi_j, \mathrm{H}\phi_i)$, we have

$$\sum_i (H_{ji} - E\delta_{ji})c_i = 0\,. \tag{5.21}$$

showing that the energy E and coefficients c_i emerge as the eigenvalues and eigenvector components of the matrix H_{ji}. This does not mean that the exact eigenvalues of the Schrodinger equation may be deduced from a finite $n \times n$

matrix. It is an approximation (upper bound) to E and will be exact only if the exact wave function ψ may be expanded in the finite series (5.20). Generally this is not possible unless n is allowed to extend to infinity, making the functions ϕ_i a complete set, in which case the matrix to be diagonalised also becomes infinite.

If the functions ϕ_i are not orthonormal and we denote their scalar products by $S_{ji} = (\phi_j, \phi_i)$ then equation (5.21) is modified to

$$\sum_i (H_{ji} - ES_{ji})c_i = 0 \tag{5.22}$$

and the energies then satisfy the determinantal equation

$$|\mathrm{H} - E\mathrm{S}| = 0 \tag{5.23}$$

One can make use of symmetry in the Rayleigh–Ritz method in two ways. In the first place the sum (5.20) should only include basis functions ϕ_i which transform according to the same irreducible representation $\mathrm{T}^{(\alpha)}$. If the representation has dimension greater than one, as is usually the case when approximating excited states, then we may further restrict the terms in the sum (5.20) to functions which transform according to the same row of the representation $\mathrm{T}^{(\alpha)}$. The reason for this is clear from equation (4.63) which showed that an invariant operator, like the Hamiltonian, has zero matrix elements between functions transforming like different rows of a representation. Furthermore equation (4.63) showed that the matrix elements were independent of the row label. It follows therefore that the matrix of H set up in a basis of functions which all transform like row i of an irreducible representation $\mathrm{T}^{(\alpha)}$, will be precisely the same as the matrix of H set up in the corresponding basis where all functions transform like some other row of that representation. Consequently it is only necessary to set up and diagonalise the matrix for one choice of i. The fact that every other choice j gives the same matrix and therefore the same eigenvalues simply brings out the known degeneracy of order s_α, the dimension of $\mathrm{T}^{(\alpha)}$.

We shall see an example of the use of symmetry in the Rayleigh–Ritz (or matrix) method in the next section.

5.8 Symmetry-breaking perturbations

In many physical problems the Hamiltonian is composed of one dominant contribution and one (or more) smaller contributions. Let us write

$$\mathrm{H} = \mathrm{H}_0 + \mathrm{H}_1 \tag{5.24}$$

where H_1 is small. It is often true that H_0 is much simpler than H so that an approximate solution may be found by treating H_1 as a perturbation on the exact solution for H_0 only. In fact, a perturbation series in powers of H_1 may be developed but here we restrict our attention to the first-order corrections. We shall be particularly interested in the effect of symmetry.

Let us write the Schrodinger equation for the unperturbed system as $H_0\psi_k = E_k\psi_k$, for a level E_k which is non-degenerate. Then, if we treat ψ_k as a trial function for H, the expectation value of the energy will be $(\psi_k, H\psi_k) = E_k + (\psi_k, H_1\psi_k)$. This shows the first-order correction to the unperturbed energy E_k to be $\Delta E = (\psi_k, H_1\psi_k)$.

If the energy level E_n is degenerate, with a set of independent wave functions ϕ_{ni}, so that $H_0\phi_{ni} = E_n\phi_{ni}$ then, to find a first approximation to the correction due to H_1 we must take a trial function of the form (5.20), as a linear combination of the degenerate states. Thus from the results of the previous section, the energy shifts are given by the eigenvalues of the matrix of H_1 in the set ϕ_{ni}.

Let us now suppose that, in the separation (5.24), H_0 is a Hamiltonian of high symmetry with symmetry group $\mathscr{G}_0$ and H_1 has a lower symmetry group $\mathscr{G}_1$ which is only a subgroup of $\mathscr{G}_0$. The full Hamiltonian will, of course, also have the symmetry group $\mathscr{G}_1$. We will consider the case where H_1 is a small perturbation which does not affect H_0 sufficiently to mix functions corresponding to different energy levels of H_0. (Mathematically we require $|\int\psi_n^* H_1\psi_m| \ll |E_n - E_m|$.) If we consider some particular energy eigenvalue E_n of H_0 then its eigenfunctions will be the basis functions $\psi_i^{(\alpha)}$ for some irreducible representation α of $\mathscr{G}_0$ by the arguments of section 5.3 and the level will have s_α-fold degeneracy. We can see intuitively that the effect of H_1 is likely to cause a splitting of this degeneracy, since $\mathscr{G}_1$ may have no irreducible representation of dimension s_α and therefore will not have eigenstates with degeneracy s_α. The space $L^{(\alpha)}$ of eigenfunctions of the level E_n will be reduced with respect to the subgroup $\mathscr{G}_1$ into subspaces which are irreducible with respect to $\mathscr{G}_1$, i.e. in the notation of section 4.18,

$$L^{(\alpha)} = \sum_{\tilde{\alpha}, q} L^{(\tilde{\alpha}), q} \tag{5.25}$$

The label q serves to distinguish spaces which transform according to equivalent irreducible representations $T^{(\tilde{\alpha})}$ of $\mathscr{G}_1$. With the help of the character tables for $\mathscr{G}_0$ and $\mathscr{G}_1$ it is then a simple matter to deduce which labels $\tilde{\alpha}$ of $\mathscr{G}_1$ occur in equation (5.25) for a given representation $T^{(\alpha)}$ of $\mathscr{G}_0$.

5.8.1 Examples

Examples of this process were given in section 4.18 for the restriction $D_3 \to C_3$. It was shown that the two-dimensional representation $T^{(3)}$ of D_3 reduced to the sum of two one-dimensional representations of C_3. In physical terms, however, this particular reduction does not usually lead to a splitting of the energy doublet since the two representations of C_3, being complex conjugates, have the same energy for a real potential $V(\boldsymbol{r})$ (or, more generally, for any time-reversal invariant Hamiltonian, see subsection 15.7.4 of volume 2.)

In the restriction $\mathscr{R}_2 \to C_3$ there is again no splitting of degeneracy because the group $\mathscr{R}_2$ has no degeneracies. In this case the states which, under $\mathscr{R}_2$, were labelled by $m = 0, \pm 1, \pm 2 \ldots$ would, under C_3, be labelled by $m = 0, \pm 1$ by the addition or subtraction of any multiple of 3.

For a more interesting example we need to go to a larger group. The octahedral group O described in subsection 9.3.1 has a three-dimensional irreducible representation T_1, and, on restriction to the subgroup D_3, one sees from the character tables in appendix 1 that it reduces $T_1 = A_2 \dot{+} E$ into two irreducible representations of D_3 with dimensions one and two, respectively. Thus the triply degenerate T_1 level of a Hamiltonian with symmetry group O will split, when a perturbation of symmetry D_3 is added, into an A_2-level and an E-doublet. Further examples of this process will be given in subsection 9.9.2 in connection with the crystal-field splitting of atomic levels.

5.8.2 Magnitude of the splitting

The use of the character table above has enabled us to deduce the way in which the degeneracies will split but it does not tell us the magnitude of the splitting or even the ordering of the perturbed levels. Assuming that the wave functions for the unperturbed levels are known, these details of the splitting may be found by evaluating the matrix elements of the perturbing Hamiltonian H_1 and if necessary diagonalising a small matrix as described in section 5.7. In choosing a basis for this matrix the use of symmetry again simplifies the calculation, as we shall now show.

The argument at the end of the previous section showed that, following the reduction (5.25), the matrix of H_1 has no coupling between functions transforming according to inequivalent irreducible representation $\mathrm{T}^{(\tilde{\alpha})}$ and that, within the set $\psi_i^{(\tilde{\alpha})}$ with $i = 1, 2, \ldots, s_{\tilde{\alpha}}$, the matrix of H_1 is a multiple of the unit matrix. If a particular representation $\mathrm{T}^{(\tilde{\alpha})}$ occurs only once in the reduction (5.25) then it is only necessary to calculate the expectation value

$$(\psi_i^{(\tilde{\alpha})}, \mathrm{H}_1 \psi_i^{(\tilde{\alpha})}) \tag{5.26}$$

for any row label i and this gives the energy shift due to H_1. Given the wave functions for the unperturbed level in some arbitrary basis, the perturbed wave function $\psi_i^{(\tilde{\alpha})}$ may be found by using the projection operator (4.50) or, in this case, the simpler operator (4.51). The energy shift (5.26) may then be evaluated directly. If a representation $\mathrm{T}^{(\tilde{\alpha})}$ occurs more than once in the reduction (5.25), say a number $m_{\tilde{\alpha}}$ times, then there will be a number $m_{\tilde{\alpha}}$ of linearly independent functions of the type $\psi_i^{(\tilde{\alpha})}$ which may be labelled in any convenient way $\psi_i^{(\tilde{\alpha})q}$ with $q = 1, 2, \ldots, m_{\tilde{\alpha}}$. In general, the perturbation H_1 will have non-zero matrix elements between members of this set and the energy shifts are found by setting up the $m_{\tilde{\alpha}} \times m_{\tilde{\alpha}}$ matrix in some convenient basis and then diagonalising. The basis functions are again found by using the projection operator (4.50) acting on wave functions of the unperturbed degenerate multiplet. An application of the general method described here will be found in section 8.5.

5.9 The indistinguishability of particles

In classical mechanics it is possible to have a system of identical particles and yet to distinguish between them. One can imagine selecting a particular particle at some instant of time and then following its path. In quantum mechanics the motion of particles is described only by wave functions so that it is not possible to speak of a precise path—the uncertainty principle. One therefore supposes that in quantum mechanics, identical particles are indistinguishable in the sense that if one of them is observed, it is not possible to tell which one. The mathematical statement of indistinguishability is that all observables are represented by operators which are symmetric with respect to permutation of the particle number labels.

The Hamiltonian for a system of n interacting identical particles is necessarily invariant under permutations so that the group $\mathscr{S}_n$ of all permutations, see section 2.2(10), is a symmetry group of the system. Except for small n, there are many different irreducible representations of $\mathscr{S}_n$, the description of which is left until chapter 17 of volume 2. However, if a certain eigenstate were to belong to an irreducible representation $T^{(\alpha)}$ of $\mathscr{S}_n$ with dimension $s_\alpha > 1$ then, because of the symmetry of all physical operators, we should have a degeneracy which was completely unobservable. No physical operator could produce any splitting or induce any transitions from one of the degenerate states to any other. In nature, this rather bizarre situation seems not to arise. Experiment shows that the only representations of $\mathscr{S}_n$ which occur are the two one-dimensional representations corresponding to totally symmetric or totally antisymmetric states. (The first of these is the identity representation and the corresponding state ψ_S satisfies $P\psi_S = \psi_S$ for all permutations P. An antisymmetric state ψ_A has the property $P_{ij}\psi_A = -\psi_A$ for all pairs ij, where P_{ij} interchanges i and j leaving all other particles unchanged. We discuss these representations in detail in section 17.4 of volume 2.). For example, the whole of the theory of atomic structure including the understanding of the periodic table would collapse without the assumption of antisymmetric wave functions for the electrons. Generally, one finds that particles with integer spin have totally symmetric wave functions while those with half-integer spin have totally antisymmetric wave functions. In the latter case it is clearly impossible for any two identical particles to occupy the same single-particle state, a feature known as the Pauli principle. One refers to particles with symmetric wave functions as 'bosons' and those with antisymmetric wave functions as 'fermions', saying that they satisfy 'Bose–Einstein' and 'Fermi–Dirac' statistics, respectively.

In subsection 16.3.4 of volume 2 we discuss the spin-statistics theorem which, on the basis of some very general field-theoretic assumptions, reaches the conclusion that particles with half-integer spin cannot have symmetric wave functions and those with integer spin cannot have antisymmetric wave functions. This theorem is therefore consistent with the observations mentioned above but does not exclude the mixed-symmetry representations for either integer or half-integer spin particles. Indeed there has been some speculation, see section 12.3, that the as yet unobserved quarks may belong to mixed symmetry representations, a situation referred to as parastatistics.

Finally, we comment that, if the particle-number labels are unobservable, it should be possible to formulate quantum mechanics without introducing them. Indeed this may be done, in what is called the 'occupation number' or 'second quantisation' method described in subsection 16.3.1 of volume 2. However for many problems it is simpler to introduce the number labels, as we shall do, as a way of ensuring the symmetry or antisymmetry of the wave function, knowing that no physical significance is to be attached to them.

5.10 Complex conjugation and time-reversal

If $\psi(\boldsymbol{r}, t)$ is a solution of the time-dependent Schrodinger equation (5.2) with a real, time-independent Hamiltonian, then it follows directly from that equation that $\phi(\boldsymbol{r}, t) = \psi^*(\boldsymbol{r}, -t)$ is also a solution. In other words, if $\psi(\boldsymbol{r}, t)$ describes a possible state of motion of the system then so does $\phi(\boldsymbol{r}, t)$. The physical relation between $\psi(\boldsymbol{r}, t)$ and $\phi(\boldsymbol{r}, t)$ is understood by comparing expectation values of operators like $\boldsymbol{r}$ and $\boldsymbol{p}$ in the two states. We have, on using the definition (5.6) of scalar product,

$$\langle \phi(\boldsymbol{r}, t) | \mathrm{x} | \phi(\boldsymbol{r}, t) \rangle = \langle \psi(\boldsymbol{r}, -t) | \mathrm{x} | \psi(\boldsymbol{r}, -t) \rangle$$
$$\langle \phi(\boldsymbol{r}, t) | \mathrm{p}_x | \phi(\boldsymbol{r}, t) \rangle = -\langle \psi(\boldsymbol{r}, -t) | \mathrm{p}_x | \psi(\boldsymbol{r}, -t) \rangle$$

But, classically, if we replace $x(t)$ by $x(-t)$ and $p_x(t)$ by $-p_x(-t)$ we obtain a trajectory in which the motions of the particles are all reversed. A particle projected upwards at $t = 0$ and reaching its highest point A at $t = 1$ would be replaced by a particle falling from A at $t = -1$ and reaching the ground at $t = 0$. We therefore refer to the transformation from $\psi(\boldsymbol{r}, t)$ to $\phi(\boldsymbol{r}, t)$ as time-reversal and introduce the time-reversal operator Υ defined by $\phi(\boldsymbol{r}, t) = \Upsilon\psi(\boldsymbol{r}, t) = \psi^*(\boldsymbol{r}, -t)$. A more general definition which also reverses the direction of spin is given in subsection 15.7.4 of volume 2.

For a stationary state (5.3), the factor $\exp(-iEt/\hbar)$ is unchanged by the time-reversal operation so that ψ and ψ^* have the same energy, a result which also follows from the eigenvalue equation (5.4). Hence, if ψ is complex, ψ and ψ^* are independent and a two-fold degeneracy occurs. This is another example of a degeneracy arising from a symmetry in the Hamiltonian, with respect to time-reversal in the present case†.

Since the successive operation of two time-reversals restores the original physical situation we have a group isomorphic to S_2 and one might expect consequences analogous to those described in section 5.6 for the group S_2, but this is not the case. The reason is that the operator Υ which we have to use to represent time-reversal is neither linear nor unitary. We cannot therefore use the representation theory built up in chapter 4 and we defer further discussion

† In this paragraph ψ refers to the $\psi_E(\boldsymbol{r})$ of equation (5.3) rather than the $\psi(\boldsymbol{r}, t)$.

of time-reversal until subsection 15.7.4 of volume 2. There it will be shown that a 'parity' under time-reversal may be given to an operator but not to a wave function. Whether the addition of time-reversal to an existing symmetry group leads to an increased degeneracy or not is a detailed question whose answer depends on the group. We return to this question briefly in section 9.8 and in subsection 15.7.4. Finally, we remark that it is not difficult to find a Hamiltonian which does not have time-reversal symmetry; for example the operator **L** introduced in section 5.1 in connection with the interaction with a magnetic field, is odd under time-reversal, while the kinetic energy is even.

Bibliography

There are so many books on non-relativistic quantum mechanics that selection is difficult, but for a very simple introduction we suggest

Mathews, P. T. (1963). *Introduction to Quantum Mechanics* (McGraw-Hill, London)

and at a more advanced level

Landau, L. D. and Lifschitz, E. M. (1958). *Quantum Mechanics* (Pergamon, London)

Messiah, A. (1961). *Quantum Mechanics* (North-Holland, Amsterdam)

Problems

We provide only a few problems in this chapter since all the concepts introduced here will reappear in later chapters in applications to a variety of physical systems.

5.1 Show that the groups C_3 and D_3, respectively, are symmetry groups for the two potentials (a) $V = \cos\theta \sin^3\theta \cos 3\phi$; (b) $V = \sin^3\theta \cos 3\phi$. Do either of these potentials have further reflection or inversion symmetries?

5.2 According to the arguments given in section 5.3 the eigenstates of a Hamiltonian with D_4 symmetry may be labelled by the irreducible representations of that group, which were found in question 4.10. What degeneracies, if any, would you expect? Deduce selection rules for dipole transitions in such a system.

5.3 Show that, for a particle moving in a potential with C_3 symmetry, the operator $\cos(2\pi l_z/3)$ is conserved.

5.4 Consider a particle constrained to move on a circle (so that there is only one coordinate ϕ) under the influence of a potential $V(\phi)$ with C_3 symmetry. In seeking an eigenfunction in the form of a Fourier series $\psi(\phi) = \sum_n a_n \exp(in\phi)$ show that the sum may be restricted to those integers $n = m + 3k$, where $m = 0$, 1 or 2 is fixed and k runs over all integers.

5.5 The functions xe^{-r^2}, ye^{-r^2} and ze^{-r^2} are eigenfunctions for a particle moving in a spherically symmetric harmonic oscillator potential. These three states are obviously degenerate. If a perturbing potential with D_3 symmetry is switched on, deduce how this three-fold degeneracy will be split. (Find the character of the three-dimensional representation of D_3 and reduce it.)

The consequences of spherical symmetry will be discussed in detail in chapters 7 and 8 and the restriction to the finite point-group symmetries is described in chapter 9.

6

Molecular Vibrations

In this chapter we come to the first detailed application of symmetry to a physical system. We study the vibration of the constituent atoms of a molecule about their equilibrium positions. The atoms are treated as point particles and we assume that their motion is governed by a potential energy which is quadratic in their displacements from equilibrium. The equilibrium configuration of the atoms usually has symmetry with respect to a group of rotations and reflections which interchange identical atoms and both kinetic and potential energies are invariants under the symmetry group. Although the dimensions of a molecule are sufficiently small that one must use quantum mechanics to describe the system it is nevertheless convenient first to derive the classical solution in terms of normal modes and then to quantise the Hamiltonian, which separates very simply when expressed in terms of the normal coordinates. The representation theory developed in chapter 4 is used to classify the normal modes according to irreducible representations of the symmetry group. The quantum mechanical solution provides an excellent example of the general theory of symmetry in a quantum system given in chapter 5, including the selection rules governing infrared and Raman absorption. As particular examples we choose the molecules H_2O and NH_3 whose equilibrium configurations are shown later in figures 6.1 and 6.2.

6.1 The harmonic approximation

We shall treat the N atoms of a molecule as point particles moving in a potential which has a minimum when they are in their equilibrium positions. When the atoms are displaced from their equilibrium positions the potential increases and for small displacements it can be expanded in a power series in the displacements of the individual atoms. To specify the displacement of a particular atom requires three components for the three mutually perpendicular directions x, y and z, and a general displacement of the molecule will therefore require $3N$ components. We denote these components by $q_1, q_2, q_3, q_4, \ldots, q_{3N}$, where q_1, q_2 and q_3 denote, respectively, the x, y and z displacements of atom number one, q_4, q_5 and q_6 denote the x, y and z displacements of atom number two, and so on. Expanding the potential in a Taylor series in the q_i, the linear term must disappear because the q_i are measured from the equilibrium position and the first non-vanishing term is

$$V = V_0 + \sum_{i,j} \tfrac{1}{2} B_{ij} q_i q_j + \ldots \tag{6.1}$$

where $B_{ij} = (\partial^2 V/\partial q_i \partial q_j)$, and V_0, the potential for no displacement, can be chosen to be zero. The harmonic approximation for the vibrations of the molecule is obtained by ignoring terms of order higher than quadratic in the expansion of V and is thus appropriate when the displacements are 'small'.

The Hamiltonian for the molecule in the harmonic approximation can therefore be written as

$$H = \sum_{i=1}^{3N} \tfrac{1}{2} M_i \dot{q}_i^2 + \sum_{i,j=1}^{3N} \tfrac{1}{2} B_{ij} q_i q_j \tag{6.2}$$

where M_i is the mass of the particle associated with the component q_i. The kinetic energy term here has the simple form of a sum of squares of the velocities $\dot{q}_i$, while the potential energy involves cross terms of the form $q_i q_j$. The solution of the problem either classically or in quantum mechanics is made much simpler if we define new coordinates Q_k by a transformation

$$q_i = \sum_k A_{ik} Q_k$$

which is chosen to make both T and V free from cross terms so that they take the simple forms:

$$T = \tfrac{1}{2} \sum_k \dot{Q}_k^2 \, , \; V = \tfrac{1}{2} \sum_k \omega_k^2 \, Q_k^2 \tag{6.3}$$

This transformation is usually achieved in two stages by first introducing mass-weighted coordinates $\alpha_i = (M_i)^{\frac{1}{2}} q_i$ so that the kinetic energy term has the simple form $T = \frac{1}{2} \sum_i \dot{\alpha}_i^2$. The potential energy is now

$$V = \tfrac{1}{2} \sum_{ij} D_{ij} \alpha_i \alpha_j$$

with
$$D_{ij} = \frac{B_{ij}}{(M_i M_j)^{\frac{1}{2}}} = \frac{\partial^2 V}{\partial \alpha_i \, \partial \alpha_j} \tag{6.4}$$

The potential energy may be reduced to the required diagonal form by finding the eigenvectors of the $3N \times 3N$ matrix D_{ij}. Let us denote the eigenvalues by $\omega_k^2 (k = 1, 2, \ldots, 3N)$ and the corresponding eigenvector by $a_{jk} (j = 1, 2, \ldots, 3N)$ so that

$$\sum_j D_{ij} a_{jk} = \omega_k^2 \, a_{ik} \tag{6.5}$$

Since the matrix D_{ij} is symmetric it follows (see section 3.6) that the eigenvectors are orthogonal and we are free to normalise them, so that

$$\sum_i a_{ik} a_{il} = \delta_{kl} \tag{6.6}$$

We now define the new coordinates Q_k by the transformation

$$q_i M_i^{\frac{1}{2}} = \alpha_i = \sum_k a_{ik} Q_k \tag{6.7}$$

from which it follows, using equations (6.5) and (6.6), that

$$V = \tfrac{1}{2} \sum_{i,j} D_{ij} \alpha_i \alpha_j = \tfrac{1}{2} \sum_{i,j,k,l} D_{ij} \, a_{ik} a_{jl} Q_k Q_l$$

$$= \tfrac{1}{2} \sum_{i,k,l} \omega_l^2 \, a_{ik} a_{il} Q_k Q_l = \tfrac{1}{2} \sum_k \omega_k^2 \, Q_k^2$$

and
$$T = \tfrac{1}{2} \sum_i \dot{\alpha}_i^2 = \tfrac{1}{2} \sum_{i,k,l} a_{ik} a_{il} \dot{Q}_k \dot{Q}_l = \tfrac{1}{2} \sum_k \dot{Q}_k^2$$

Thus the simple forms (6.3) have been achieved by a transformation with

$$A_{ik} = a_{ik}/(M_i)^{\frac{1}{2}} \tag{6.8}$$

6.2 Classical solution

It is now a simple problem to solve the classical equations of motion if we use the Lagrange equations, see subsection 16.1.1, of volume 2,

$$\frac{\mathrm{d}}{\mathrm{d}t}\left(\frac{\partial L}{\partial \dot{Q}_k}\right) = \frac{\partial L}{\partial Q_k} \tag{6.9}$$

where $L = T - V$.

Using the form (6.3) of T and V and equation (6.9) we have $\ddot{Q}_k = -\omega_k^2 Q_k$.

This shows that the new coordinates Q_k are uncoupled, each one behaving as a simple harmonic oscillator with solutions

$$Q_k = c_k \cos(\omega_k t + \epsilon_k) \tag{6.10}$$

The Q_k are called the 'normal' coordinates of the system. The original coordinates are given by

$$q_i = \sum_k A_{ik} c_k \cos(\omega_k t + \epsilon_k) \tag{6.11}$$

where the $6N$ constants c_k and ϵ_k would be determined from the initial conditions.

A movement in which only one normal coordinate Q_p departs from zero is called a normal mode. It can be achieved by setting all $c_k = 0$ except for one $k = p$ (say). In such a mode we have from equation (6.11)

$$q_i(\text{mode } p) = A_{ip} c_p \cos(\omega_p t + \epsilon_p) \tag{6.12}$$

showing that the displacements q_i all change with the same frequency ω_p and their ratios remain fixed by the coefficients A_{ip} which, apart from the mass factors M_i, are the components of the eigenvector of the matrix D_{ij} corresponding to the eigenvalue ω_p^2. In a general vibration the displacements (6.11) are superpositions of all modes, each with its own frequency.

6.3 Quantum mechanical solution

The quantum mechanical solution is also straightforward since the Hamiltonian for the molecule now has the form

$$\mathrm{H} = -\frac{\hbar^2}{2}\sum_{k=1}^{3N}\frac{\partial^2}{\partial Q_k^2} + \tfrac{1}{2}\sum_{k=1}^{3N}\omega_k^2 Q_k^2 = \sum_{k=1}^{3N}\mathrm{H}_k \tag{6.13}$$

where $$\mathrm{H}_k = -\frac{\hbar^2}{2}\frac{\partial^2}{\partial Q_k^2} + \tfrac{1}{2}\omega_k^2 Q_k^2$$

is a one-dimensional simple harmonic oscillator. We have quantised the Hamiltonian in the usual way by replacing $\dot{Q}_k^2$ by $-\hbar^2\partial^2/\partial Q_k^2$ (Strictly, this step should be justifed by first quantising (6.2) in the original Cartesian coordinates q_i and then transforming to the new coordinates Q_k.). If $\psi_{n_k}(Q_k)$ denotes an eigenfunction of H_k with energy ε_{n_k} then the eigenfunctions Ψ of H are products of the ψ_{n_k}

$$\Psi(n_1 n_2, \ldots, n_{3N}) = \prod_k \psi_{n_k}(Q_k) \tag{6.14}$$

This follows immediately since

$$\begin{aligned} H\Psi &= \sum_k H_k \prod_{k'} \psi_{n_{k'}}(Q_{k'}) \\ &= \sum_k \varepsilon_{n_k} \prod_{k'} \psi_{n_{k'}}(Q_{k'}) \\ &= \sum_k \varepsilon_{n_k} \Psi \end{aligned}$$

so that the total energy of the molecule in the state Ψ is

$$E(n_1 n_2, \ldots, n_{3N}) = \sum_k \varepsilon_{n_k}$$

The solutions of Schrodinger's equation for a one-dimensional simple harmonic oscillator can be found in any standard quantum mechanics book,

$$\begin{aligned} \psi_{n_k} &= A H_{n_k}(\mu_k Q_k) \exp(-\tfrac{1}{2}\mu_k^2 Q_k^2) \\ \varepsilon_{n_k} &= (n_k + \tfrac{1}{2})\hbar\omega_k \end{aligned} \tag{6.15}$$

where n_k is zero or a positive integer, H_{n_k} is a Hermite polynomial, $\mu_k = (\omega_k/\hbar)^{\frac{1}{2}}$ and A a normalising constant. The total energy is

$$E = \sum_k (n_k + \tfrac{1}{2})\hbar\omega_k \tag{6.16}$$

We can also regard this result as 'quantising' the classical calculation by giving the normal mode of frequency ω_k a number n_k of quanta of excitation so that its energy is $(n_k + \frac{1}{2})\hbar\omega_k$.

6.4 Effects of symmetry in molecular vibrations

In discussing symmetry in quantum mechanics in section 5.3, it was shown that the eigenfunctions of a symmetrical Hamiltonian could be labelled by the irreducible representation labels of the symmetry group of the Hamiltonian and that degeneracies with orders equal to the dimension of those representations would occur. For a molecule with symmetry both the kinetic and potential energies are unchanged by a symmetry operation and hence these general results apply. Furthermore we show that the normal coordinates Q_k may be labelled by the irreducible representations of the symmetry group and that for a representation $T^{(\alpha)}$ of dimension s_α there will be a set of s_α linearly independent normal coordinates all with the same frequency. The proof will be essentially identical to that given in section 5.3 but it should be noted that the present result for vibrations has significance both in classical and quantum mechanics. The reason is that even in classical mechanics the vibration problem leads to an eigenvalue equation (6.5).

In this and the following two sections we describe the properties of the normal modes. The structure of the wave functions, the energy spectrum and the selection rules for electromagnetic transitions will be described in sections 6.6 and 6.7.

To deduce the properties of normal modes we regard a general displacement $\boldsymbol{q}$ as a vector in a $3N$-dimensional space with the mass weighted coordinates $\alpha_i = q_i M_i^{\frac{1}{2}}$, introduced in section 6.1. The basis vector $\boldsymbol{e}_j$ represents a displacement in which all coordinates except α_j are zero and $\alpha_j = 1$, i.e. $q_j = M_j^{-\frac{1}{2}}$. Thus

$$\boldsymbol{q} = \sum_{i=1}^{3N} \alpha_i \boldsymbol{e}_i \tag{6.17}$$

It will be convenient to define the scalar product of two displacements $\boldsymbol{q}$ and $\boldsymbol{q}'$ by

$$(\boldsymbol{q}, \boldsymbol{q}') = \sum_{i=1}^{3N} \alpha_i \alpha_i' \tag{6.18}$$

so that in particular $(\boldsymbol{e}_i, \boldsymbol{e}_j) = \delta_{ij}$. A potential energy operator D may be defined in this space by the equation

$$\mathrm{D}\boldsymbol{e}_j = \sum_i D_{ij} \boldsymbol{e}_i \tag{6.19}$$

where D_{ij} are the coefficients introduced in equation (6.4). The value of the potential energy in a displacement $\boldsymbol{q}$ is then given by

$$V(\boldsymbol{q}) = \tfrac{1}{2}(\boldsymbol{q}, \mathrm{D}\boldsymbol{q}) = \tfrac{1}{2} \sum_{i,j} \alpha_i \alpha_j (\boldsymbol{e}_i, \mathrm{D}\boldsymbol{e}_j)$$

$$= \tfrac{1}{2} \sum_{i,j} \alpha_i \alpha_j D_{ij} \tag{6.20}$$

agreeing with equation (6.4)

The coordinates q_i in a normal mode p were shown, equation (6.12), to be proportional to the coefficients A_{ip} and hence, apart from the time factor, the displacement in a normal mode p is represented by the vector with $\alpha_i = A_{ip} M_i^{\frac{1}{2}}$

$$\boldsymbol{u}_p = \sum_i A_{ip} M_i^{\frac{1}{2}} \boldsymbol{e}_i = \sum_i a_{ip} \boldsymbol{e}_i \tag{6.21}$$

using equation (6.8). Hence, in this framework, the normal displacements are orthogonal

$$(\boldsymbol{u}_p, \boldsymbol{u}_{p'}) = \sum_{i,j} a_{ip} a_{jp'} (\boldsymbol{e}_i, \boldsymbol{e}_j) = \sum_i a_{ip} a_{ip'} = \delta_{pp'}$$

using equation (6.6). The normal coordinates Q_k are the coordinates of a general displacement in this basis since, from equations (6.17) and (6.7),

$\boldsymbol{q} = \sum_k Q_k \boldsymbol{u}_k$. The most important property of the normal displacements $\boldsymbol{u}_p$ is that they are eigenvectors of the operator D for

$$\mathrm{D}\boldsymbol{u}_p = \sum_{i,j} a_{ip} D_{ji} \boldsymbol{e}_j = \sum_j \omega_p^2 a_{jp} \boldsymbol{e}_j = \omega_p^2 \boldsymbol{u}_p \qquad (6.22)$$

using equation (6.5).

A symmetry operation G_a on the molecule will carry a displacement $\boldsymbol{q}$ into a new displacement $\boldsymbol{q}'$. In particular it will carry each basis vector $\boldsymbol{e}_i$ into a new displacement $\boldsymbol{e}_i'$ and we define the transformations $\mathrm{T}(\mathrm{G}_a)$ induced in the $3N$-dimensional space by G_a by $\boldsymbol{q}' = \mathrm{T}(\mathrm{G}_a)\boldsymbol{q}$, with

$$\boldsymbol{e}_i' = \mathrm{T}(\mathrm{G}_a)\boldsymbol{e}_i = \sum_j T_{ji}(\mathrm{G}_a)\boldsymbol{e}_j \qquad (6.23)$$

where G_a is an element of the symmetry group $\mathcal{G}$. The scalar product $(\boldsymbol{q}, \boldsymbol{q}) = \sum_i M_i q_i^2$ of $\boldsymbol{q}$ with itself is the mass-weighted sum of squares of the magnitudes of the displacements of each atom and hence is unchanged by a symmetry operation. Thus $(\boldsymbol{q}', \boldsymbol{q}') = (\boldsymbol{q}, \boldsymbol{q})$ so that the transformation $\mathrm{T}(\mathrm{G}_a)$ is unitary and being real it is therefore orthogonal. Since the potential energy is unchanged by a symmetry operation it follows that the operator D is an invariant, with respect to the transformations $\mathrm{T}(\mathrm{G}_a)$, i.e. $\mathrm{D} = \mathrm{TDT}^{-1}$.

We may now follow precisely the arguments given in section 5.3 in connection with degeneracies in quantum mechanics, drawing on the analogy between equations (5.12) and (6.22). The arguments are set out briefly below.

(1) If $\boldsymbol{u}_p$ is a normal displacement with frequency ω_p then it follows that $\boldsymbol{u}_p' = \mathrm{T}(\mathrm{G}_a)\boldsymbol{u}_p$ is also a normal displacement with the same frequency ω_p for

$$\mathrm{D}\boldsymbol{u}_p' = \mathrm{DT}(\mathrm{G}_a)\boldsymbol{u}_p = \mathrm{T}(\mathrm{G}_a)\mathrm{D}\boldsymbol{u}_p = \omega_p^2 \mathrm{T}(\mathrm{G}_a)\boldsymbol{u}_p = \omega_p^2 \boldsymbol{u}_p'$$

(2) Hence, the set of normal displacements with the same frequency form a basis for a representation of the symmetry group $\mathcal{G}$.

(3) Thus, barring accidental degeneracy, each frequency will be labelled by some irreducible representation $\mathrm{T}^{(\alpha)}$ of $\mathcal{G}$ and will contain s_α linearly independent normal modes.

There is one very significant difference between this problem of classifying the normal modes and the problem of classifying the eigenfunctions of a quantum mechanical Hamiltonian. The total number of normal modes is finite and equal to $3N$, whereas the number of eigenfunctions is generally infinite. In fact by calculating the character of the $3N$-dimensional representation of all displacements using any convenient basis we may use the methods of section 4.11 to study its reduction to irreducible representations and hence classify all the normal modes of the molecule.

6.5 Classification of the normal modes

The $3N$-dimensional space of all possible displacements provides a representation of the symmetry group $\mathscr{G}$ through equation (6.23). Let us call it the representation $T^{(3N)}$. We have also shown that the normal displacements with the same normal frequency form the basis of an irreducible representation $T^{(\alpha)}$ of $\mathscr{G}$. But since the $3N$-dimensional space of all displacements breaks up into a sum of spaces corresponding to the different normal frequencies it follows that the classification of the normal frequencies may be found by studying the reduction

$$T^{(3N)} = \dot{\sum_{\alpha}} m_\alpha T^{(\alpha)} \tag{6.24}$$

of the representation $T^{(3N)}$ into its irreducible constituents $T^{(\alpha)}$, using the techniques of section 4.11. If, for some α, the coefficient $m_\alpha > 1$ then there will be a number m_α of different frequencies which all carry the same label α and each have a set of s_α linearly independent normal coordinates.

The reduction (6.24) is straightforward if we can find the character $\chi_p^{(3N)}$ of the representation $T^{(3N)}$ for each of the classes $\mathscr{C}_p$ of the group $\mathscr{G}$. We can then use equation (4.28) to determine the number of times each irreducible representation will appear in the reduction.

$$m_\alpha = \frac{1}{g} \sum_p c_p \chi_p^{(3N)} \chi_p^{(\alpha)*} \tag{6.25}$$

In order to find the character $\chi_p^{(3N)}$ we write equation (6.23) in a slightly more detailed form by relabelling the basis vectors $\boldsymbol{e}_i$ as $\boldsymbol{e}_{ti}$, so that we can distinguish the individual atoms $t = 1, 2, \ldots, N$ and their Cartesian coordinates $i = x, y$ or z,

$$T(G_a)\boldsymbol{e}_{ti} = \sum_{t', i'} T^{(3N)}_{t'i', ti}(G_a)\boldsymbol{e}_{t'i'} \tag{6.26}$$

The character is then

$$\chi_p^{(3N)} = \chi^{(3N)}(G_a) = \sum_{t, i} T^{(3N)}_{ti, ti}(G_a) \tag{6.27}$$

for any element G_a in the class $\mathscr{C}_p$. However, the only contributions to $\chi_p^{(3N)}$ come from those atoms t which are unmoved by G_a since, if G_a moves atom t to site t', the diagonal element $T^{(3N)}_{ti, ti}$ in equation (6.26) is zero. The contribution to the character $\chi_p^{(3N)}$ from a particular atom t which is unmoved by G_a is $\sum_i T^{(3N)}_{ti, ti}(G_a)$, but this must be just the character for the rotation G_a of an ordinary three-dimensional vector about a fixed point; the vector representation discussed in section 5.4. If G_a is a proper rotation $R(\theta)$ through an angle θ then this character is simply given by $(2 \cos \theta + 1)$, as in section 5.4, and if $N_{R(\theta)}$ denotes the number of atoms left unmoved by $R(\theta)$ then the total character is given by

$$\chi^{(3N)}(\mathrm{R}(\theta)) = N_{R(\theta)}(2\cos\theta + 1)$$

For an improper rotation $\mathrm{S}(\theta)$ which we can describe as a proper rotation $\mathrm{R}(\theta)$ followed by reflection σ_h in a plane perpendicular to the axis of $\mathrm{R}(\theta)$, the character will be given by

$$\chi^{(3N)}(\mathrm{S}(\theta)) = N_{S(\theta)}(2\cos\theta - 1)$$

where $N_{S(\theta)}$ is the number of atoms unmoved by $\mathrm{S}(\theta)$. We can see this result by considering an improper rotation about the z-axis. The improper rotation matrix in a Cartesian basis is easily seen to be

$$\begin{pmatrix} \cos\theta & -\sin\theta & 0 \\ \sin\theta & \cos\theta & 0 \\ 0 & 0 & -1 \end{pmatrix}$$

where the -1 appears because of the reflection σ_h. The character for this is given by the trace $(2\cos\theta - 1)$.

Having found $\chi^{(3N)}$ we could now use equation (6.25) to find the coefficients m_α in equation (6.24). It is, however, convenient first of all to remove from $\chi^{(3N)}$ the contribution which comes from the six normal modes of zero frequency which are not vibrational modes but correspond to overall translation and rotation of the molecule with no relative motion of the atoms. In these modes the molecule moves as a rigid body and the three translational modes can be chosen as displacements along the x, y and z Cartesian axes. These clearly transform under rotation according to the vector representation and have a character

$$\chi_{\text{trans}}(\mathrm{R}(\theta)) = (2\cos\theta + 1)$$
$$\chi_{\text{trans}}(\mathrm{S}(\theta)) = (2\cos\theta - 1)$$

The three rotational modes again transform like a vector under rotation but they are even under inversion, i.e. they form a pseudo-vector, see section 5.4. (To justify this statement we note (see subsection 7.4.1) that the displacement $d(\boldsymbol{r})$ of any point $\boldsymbol{r}$ due to a small-angle rigid rotation defined by the vector $\boldsymbol{a}$ is given by $\boldsymbol{a} \wedge \boldsymbol{r}$. If $\boldsymbol{d}'$ denotes the displacement obtained from $\boldsymbol{d}$ by a finite rotation R then geometrically we see that $\boldsymbol{d}'(\boldsymbol{r}) = \mathrm{R}(\boldsymbol{a} \wedge \mathrm{R}^{-1}\boldsymbol{r}) = \mathrm{R}\boldsymbol{a} \wedge \boldsymbol{r}$ and hence that the displacement transforms like $\boldsymbol{a}$, a vector. In a similar way if $\tilde{\boldsymbol{d}}$ denotes the displacement obtained from $\boldsymbol{d}$ by inversion then $\tilde{\boldsymbol{d}} = \mathrm{I}(\boldsymbol{a} \wedge \mathrm{I}\boldsymbol{r}) = \boldsymbol{a} \wedge \boldsymbol{r}$, which justifies that $\boldsymbol{d}$ is a pseudo-vector.) Hence for a proper rotation the character of the three rotational modes is again the character of the vector representation, while for an improper element there is a change of sign,

$$\chi_{rot}(R(\theta)) = (2\cos\theta + 1), \qquad \chi_{rot}(S(\theta)) = -(2\cos\theta - 1)$$

After subtracting these from $\chi^{(3N)}$ we are left with the character for the remaining non-zero frequency modes $\chi_{vib} = \chi^{(3N)} - \chi_{trans} - \chi_{rot}$,

$$\chi_{vib}(R(\theta)) = (N_{R(\theta)} - 2)\,(2\cos\theta + 1)$$
$$\chi_{vib}(S(\theta)) = N_{S(\theta)}(2\cos\theta - 1) \qquad (6.28)$$

We shall use these results to classify the normal vibrations of the water molecule and the ammonia molecule.

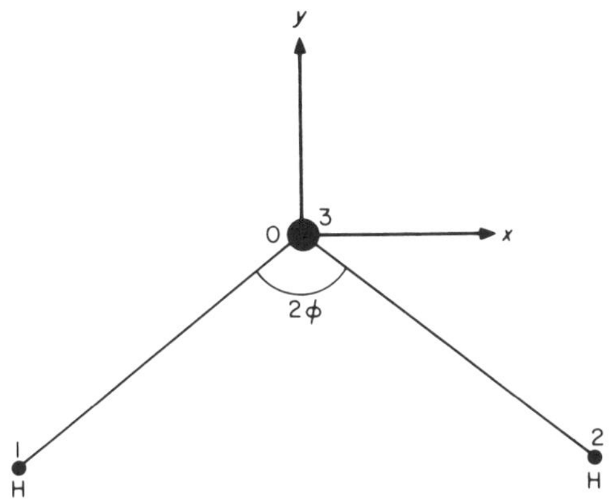

Figure 6.1

6.5.1 The water molecule

The water molecule H_2O has the shape shown in figure 6.1 and symmetry group C_{2v} (see chapter 9 for details). The value of the 'bond angle' 2ϕ is unimportant for the discussion of symmetry but is known to be about 105°. This group has four symmetry elements: E, the identity; C_2, a two-fold rotation about an axis through the oxygen atom and in the plane of the molecule; σ_v, a reflection in the plane of the molecule; and σ'_v, a reflection in the plane perpendicular to the plane of the molecule and bisecting the bond angle. Each element is in a class of its own. The vibrational characters are found from equation (6.28) as follows:

for	E(proper, $\theta = 0$, $N_E = 3$),	$\chi_{vib}(E) = 3$
	C_2(proper, $\theta = 180°$, $N_{C_2} = 1$),	$\chi_{vib}(C_2) = 1$
	σ_v(improper, $\theta = 0$, $N_{\sigma_v} = 3$),	$\chi_{vib}(\sigma_v) = 3$
	σ'_v(improper, $\theta = 0$, $N_{\sigma'_v} = 1$),	$\chi_{vib}(\sigma'_v) = 1$

The characters are given in table 6.1 and decomposition is trivial by inspection in this case. (The notation for the irreducible representations is conventional and is described in appendix 1.) We conclude that there are two non-degenerate modes transforming according to the representation A_1 and one non-degenerate mode transforming according to representation B_1.

Table 6.1

C_{2v}	E	C_2	σ_v	σ'_v	
A_1	1	1	1	1	
A_2	1	1	-1	-1	
B_1	1	-1	1	-1	
B_2	1	-1	-1	1	
χ_{vib}	3	1	3	1	$= 2A_1 \dot{+} B_1$

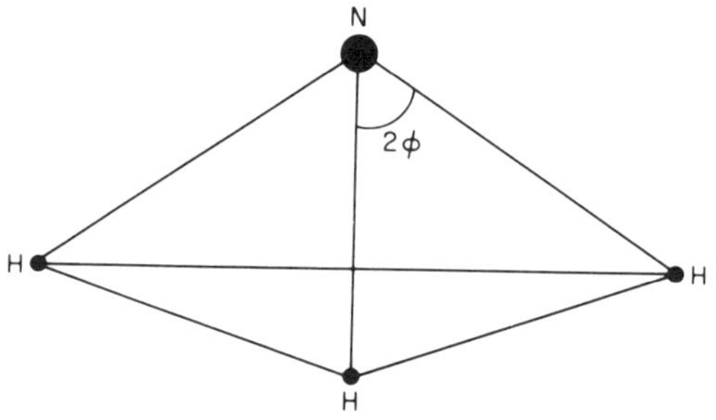

Figure 6.2

6.5.2 The ammonia molecule

The ammonia molecule NH_3 is three-dimensional with the three hydrogen atoms forming the equilateral base of a tetrahedron, as shown in figure 6.2, and its symmetry group is C_{3v} (see subsection 9.3.2). Again, the value of the bond angle, known to be $2\phi \approx 108°$, plays no part in the discussion of symmetry. It is convenient to refer to the plane of the three H atoms as 'horizontal'. The six group elements fall into three classes, the identity E, the two three-fold rotations about the 'vertical' axis through the N atom and the reflections in the three vertical planes containing the N atom and one of the H atoms. The vibrational characters from equation (6.28) are,

for E(proper, $\theta = 0$, $N_E = 4$), $\chi_{vib}(E) = 6$

C_3(proper, $\theta = 2\pi/3$, $N_{C_3} = 1$), $\chi_{vib}(C_3) = 0$

σ_v(improper, $\theta = 0$, $N_{\sigma_v} = 2$), $\chi_{vib}(\sigma_v) = 2$

From the character table 6.2 we then deduce by inspection or from the formula (6.25) that there are two A_1 modes and two two-fold degenerate E modes—a total of four different frequencies for the six modes. (The notation $2C_3$ records that there are two elements in the class containing C_3. The symbol E is again conventional for a two-dimensional representation, see appendix 1. This italic E is not to be confused with the roman typeface E used for the identity element.)

Table 6.2

C_{3v}	E	$2C_3$	$3\sigma_v$	
A_1	1	1	1	
A_2	1	1	−1	
E	2	−1	0	
χ_{vib}	6	0	2	$= 2A_1 \dot{+} 2E$

6.6 Vibrational energy levels and wave functions

The classification of normal modes has immediate relevance to the classical vibration problem since it enables us to deduce the shape of the modes from symmetry arguments alone, i.e. without knowing details of the potential. We return to this point in more detail in section 6.8. In the present section and in section 6.7 we discuss the significance of the symmetry classification of normal modes in the quantum mechanical treatment of molecular vibrations.

The general vibrational state is labelled by the set of quantum numbers n_k giving the number of quanta of excitation in each mode and, as shown in section 6.3, can be written

$$\begin{aligned}\Psi(n_1 n_2, \ldots, n_{3N-6}) &= B \prod_k H_{n_k}(\mu_k Q_k) \exp\left(-\tfrac{1}{2}\mu_k^2 Q_k^2\right)\\ &= B \prod_k H_{n_k}(\mu_k Q_k) \exp\left(-\tfrac{1}{2}\sum_{k'} \mu_{k'}^2 Q_{k'}^2\right)\end{aligned} \tag{6.29}$$

where B is a normalisation constant.

Since we are interested only in the internal structure of the molecule we have excluded the six zero-frequency modes (translation and rotation) from the wave function (6.29). The exponent in equation (6.29) is an invariant under the symmetry group $\mathscr{G}$ and this may be shown by introducing the more detailed notation

$$\sum_k \mu_k^2 Q_k^2 = \hbar^{-1} \sum_k \omega_k Q_k^2 = \hbar^{-1} \sum_\alpha \sum_{p=1}^{m_\alpha} \omega_{\alpha p} \sum_{i=1}^{s_\alpha} Q_{\alpha p i}^2 \tag{6.30}$$

where α labels an irreducible representation of $\mathscr{G}$, i labels the rows of that representation and p serves to distinguish different frequencies with the same label α. This notation exhibits the fact that to each frequency $\omega_{\alpha p}$ there is a set of s_α coordinates labelled by i. We now show that the sum $\sum_{i=1}^{s_\alpha} Q_{\alpha pi}^2$ is invariant for every α and p and hence that the exponent (6.30) is invariant. Consider the transformation of coordinate $Q_{\alpha pi}$ by a symmetry operation T

$$Q'_{\alpha pi} = \mathrm{T} Q_{\alpha pi} = \sum_j T_{ij}^{(\alpha)} Q_{\alpha pj} \tag{6.31}$$

using equation (3.20), so that

$$\sum_i (Q'_{\alpha pi})^2 = \sum_i \sum_{jk} T_{ij}^{(\alpha)} T_{ik}^{(\alpha)} Q_{\alpha pj} Q_{\alpha pk} = \sum_j Q_{\alpha pj}^2 \tag{6.32}$$

since the transformation T is orthogonal (unitary and real). Thus the exponential in the wave function (6.29) is invariant and the symmetry of Ψ is given by the symmetry of the remaining product of Hermite polynomials.

The ground state of the molecule will have all $n_k = 0$ and, since $H_0(\mu_k Q_k)$ is a constant, $\Psi(0, 0, \ldots, 0)$ is an invariant. Thus the ground state transforms according to the identity representation A_1 for all molecules.

The lowest excited states are those in which only one mode is excited and with only one quantum. In the notation of equation (6.30) $n_{\alpha_1 p_1} = 1, n_{\alpha p} = 0$ for $\alpha p \neq \alpha_1 p_1$. They are called 'fundamental' states. The symmetry of such a state is thus the symmetry of $H_1(\mu_{\alpha_1 p_1} Q_{\alpha_1 p_1 i})$ and since $H_1(x) \propto x$ this is just the symmetry of the set of normal coordinates $Q_{\alpha_1 p_1 i}$ with fixed $\alpha_1 p_1$ and $i = 1, 2, \ldots, s_{\alpha_1}$ which transform according to $\mathrm{T}^{(\alpha_1)}$. There will therefore be a total of $3N - 6$ fundamental states labelled exactly like the normal modes discussed in section 6.5. The fundamental energies are given by $\hbar\omega_{\alpha p}$ and their degeneracies are given by s_α.

Excited states in which two or more modes of different frequency are excited are called 'combination levels', and states in which a single mode is excited with more than one quantum are called 'overtone levels'. For example, a simple combination level in which $n_{\alpha_1 p_1} = 1$, $n_{\alpha_2 p_2} = 1$ with all other $n_{\alpha p} = 0$ will have a wave function with the symmetry of $H_1(\mu_{\alpha_1 p_1} Q_{\alpha_1 p_1 i}) H_1(\mu_{\alpha_2 p_2} Q_{\alpha_2 p_2 j})$, i.e. the symmetry of the product $Q_{\alpha_1 p_1 i} Q_{\alpha_2 p_2 j}$ which belongs to the direct product representation $\mathrm{T}^{(\alpha_1)} \times \mathrm{T}^{(\alpha_2)}$. These simple combination levels may therefore be labelled by the product representation which generally is not irreducible. This is an example of an 'accidental' degeneracy. In fact, if anharmonic terms in the potential are taken into account without disturbing the symmetry the degeneracy would be lifted and the resulting sets of states would correspond to the irreducible components of the product representation in the usual way. For example in NH_3 if the two E doublets are both singly excited, with the product representation $E \times E$ reducing to $E \dotplus A_1 \dotplus A_2$, the anharmonic terms would split the four-fold degenerate level into a doublet E and two singlets A_1 and A_2.

The structure of an overtone level is slightly more complicated since the states no longer transform simply according to product representations. For instance if $Q_{\alpha p1}$ and $Q_{\alpha p2}$ are the two degenerate coordinates of a doublet at energy $\hbar\omega_{\alpha p}$ then we can form only the three functions $H_2(\mu_{\alpha p} Q_{\alpha p1})$, $H_1(\mu_{\alpha p} Q_{\alpha p1}) H_1(\mu_{\alpha p} Q_{\alpha p2})$, $H_2(\mu_{\alpha p} Q_{\alpha p2})$ for the excited states at energy $2\hbar\omega_{\alpha p}$, rather than four. The point is that the products $H_1(\mu_{\alpha p} Q_{\alpha p1}) H_1(\mu_{\alpha p} Q_{\alpha p2})$ and $H_1(\mu_{\alpha p} Q_{\alpha p2}) H_1(\mu_{\alpha p} Q_{\alpha p1})$ are identical. Thus in general the overtones given by a double excitation of a mode with symmetry $T^{(\alpha)}$ are not simply described by the product representation of $T^{(\alpha)}$ with itself $T^{(\alpha)} \times T^{(\alpha)}$, as one might have expected from the discussion of combination levels. Only those functions which are symmetric in the two factors will exist, producing what is called the 'symmetrised product representation' whose character for a group element G_a is given by

$$\chi_{\text{sym}}^{(\alpha\times\alpha)}(G_a) = \tfrac{1}{2}[\chi^{(\alpha)}(G_a)]^2 + \tfrac{1}{2}\chi^{(\alpha)}(G_a^2) \tag{6.33}$$

This formula is a special case of a more general result described in appendix 3.1 of volume 2 but it may also be deduced simply as follows. The possible independent product functions may be labelled by ii or ij with $i < j$ where i and j denote row labels of $T^{(\alpha)}$. For a function of the type ii the contribution to the character is simply $[T_{ii}^{(\alpha)}(G_a)]^2$ while for one of the type ij it is $\{T_{ii}^{(\alpha)}(G_a) T_{jj}^{(\alpha)}(G_a) + T_{ji}^{(\alpha)}(G_a) T_{ij}^{(\alpha)}(G_a)\}$, the second of these two terms arising from transformation to the product function ji which is identical to ij. The character for the symmetrised product is therefore given by the sum

$$\sum_i [T_{ii}^{(\alpha)}(G_a)]^2 + \sum_{i<j} (T_{ii}^{(\alpha)}(G_a) T_{jj}^{(\alpha)}(G_a) + T_{ji}^{(\alpha)}(G_a) T_{ij}^{(\alpha)}(G_a))$$

$$= \tfrac{1}{2}\sum_{i,j} (T_{ii}^{(\alpha)}(G_a) T_{jj}^{(\alpha)}(G_a) + T_{ji}^{(\alpha)}(G_a) T_{ij}^{(\alpha)}(G_a))$$

$$= \tfrac{1}{2}[\chi^{(\alpha)}(G_a)]^2 + \tfrac{1}{2}\sum_j T_{jj}^{(\alpha)}(G_a^2) = \tfrac{1}{2}[\chi^{(\alpha)}(G_a)]^2 + \tfrac{1}{2}\chi^{(\alpha)}(G_a^2)$$

Thus in the example of NH_3 we find the character for the symmetrised product $E \times E$ to be

	E	$2C_3$	$3\sigma_v$	
$\chi_{\text{sym}}^{(E\times E)} =$	3	0	1	$= E \dotplus A_1$

showing that the first overtone of the fundamental excitation E is composed of an A_1 state and an E doublet.

The rotation of a molecule will also give rise to discrete energy levels. To each vibrational level there is a band of rotational levels. However, the energy scale of the rotational motion is about a factor ten less than that of the vibrations. It is therefore usually possible to distinguish the two effects, although strictly one should also consider the rotation–vibration interaction—see Wilson *et al.* (1955) in the bibliography. We shall not consider the rotational motion here.

6.7 Infrared and Raman absorption spectra of molecules

Energy levels of a quantum system will in general decay to the ground state by the emission of radiation. In the opposite process, incident radiation may be absorbed by the system which then moves from its ground state into an excited state. In such a transition between a ground state with energy E_0 and an excited state with energy E_1, the absorbed radiation must carry the energy difference $E_1 - E_0$ and through the Planck relation $E_1 - E_0 = \hbar\omega$; the frequency ω of the absorbed radiation is related to the energy level differences of the system. (In molecular spectroscopy it is the practice to write $\omega = 2\pi c\bar{v}$ where c is the velocity of light, and to quote the value of $\bar{v}$, the inverse of the wavelength (or wave number), in cm^{-1}. Thus an excitation energy given in units of cm^{-1} must be multiplied by hc to obtain strict energy units.)

6.7.1 Infrared spectra

A typical excitation energy for a fundamental vibrational level of a molecule is of the order of 2000 cm^{-1}, corresponding to radiation in the infrared region—see Herzberg (1945; 1966) in the bibliography. The dominant mechanism for absorption or emission of radiation is electric dipole whose operator transforms like the vector representation (see section 5.4) of the group $\mathscr{R}_3$. But since the ground state of the molecule is an invariant it follows immediately from the discussion of selection rules in section 5.4 that the only fundamentals excited by infrared absorption will be those labelled by the irreducible representations which occur in the reduction of the vector representation into irreducible representations of the symmetry group of the molecule. The character of the vector representation as we have seen in section 6.5 is given by

$$\chi^{(V)}(\mathrm{R}(\theta)) = 2\cos\theta + 1$$

$$\chi^{(V)}(\mathrm{S}(\theta)) = 2\cos\theta - 1$$

so that for the symmetry group C_{2v} of the water molecule

$$\chi^{(V)} = A_1 \dot{+} B_1 \dot{+} B_2$$

while for the group C_{3v} in ammonia,

$$\chi^{(V)} = A_1 \dot{+} E$$

Thus in both the water and ammonia molecules all fundamentals will be excited.

6.7.2 Raman spectra

In the infrared absorption considered above, a single quantum of light is absorbed during the vibrational transition. In a Raman experiment the sample is illuminated by monochromatic light and the scattered light observed in a spectrograph. The observed spectrum consists of a strong line of the same frequency as the incident light with weaker lines symmetrically on either side. The lines of lower frequency are called Stokes lines and are weaker in intensity than the anti-Stokes lines on the high frequency side. The lines are interpreted to be due to molecular vibrational transitions in which the difference between the vibrational energy levels is equal to the change in energy of the photon. The Stokes lines correspond to excitation of molecular vibration and the anti-Stokes lines to de-excitation. At low temperatures the Stokes lines will clearly dominate since the probability of finding a molecule in an excited state is small.

Quantum mechanically the Raman transition involves a second order interaction with the electromagnetic field and can be thought of as a transition first to some virtual intermediate state and then from this to the final state—a careful discussion of the Raman scattering process may be found in Birman (1974). In most cases of interest the intermediate state is at a high excitation energy corresponding to a different electron configuration. In this way it may be shown that the Raman absorption process is governed by the quadratic operators xy, yz, zx, x^2, y^2 and z^2 evaluated between the two vibrational states involved. This contrasts with the vector operator x, y and z for the infrared absorption. To deduce selection rules for the Raman process one therefore needs to deduce the character of this set of quadratic operators. In fact it is the symmetrised product of the vector representation with itself and is given by the formula (6.33):

$$\chi_{\text{sym}}^{(V\times V)}(\mathrm{G}_a) = \tfrac{1}{2}\chi^{(V)}(\mathrm{G}_a^2) + \tfrac{1}{2}[\chi^{(V)}(\mathrm{G}_a)]^2$$

This character may then be used to reduce the Raman operator into its irreducible parts and hence deduce the Raman selection rules. One finds for the group C_{2v}

$$\chi_{\text{sym}}^{(V\times V)} \to 3A_1 \dot{+} A_2 \dot{+} B_1 \dot{+} B_2$$

and for the group C_{3v}

$$\chi_{\text{sym}}^{(V\times V)} \to 2A_1 \dot{+} 2E$$

Thus, referring back to the examples of the water molecule and the ammonia molecule discussed in section 6.5 we see that the Raman process will excite all fundamentals in both cases. We find no forbidden transition in this example because of the simplicity of the symmetry group. As the symmetry increases one finds more forbidden transitions—see problem 6.3.

We notice that the Raman operator has even parity while the infrared dipole operator has odd parity, so that for molecules with a centre of inversion, lines which appear in the infrared spectrum are excluded from the Raman spectrum and vice versa.

6.8 Displacement patterns and frequencies of the normal modes

We have been able to deduce a number of properties of the energy levels and transitions in a molecule simply from the symmetry of the molecule. To deduce the classical frequencies ω_k and hence the excitation energies $\hbar\omega_k$ of the fundamental levels it is, of course, necessary to know the potential precisely. In this section we describe the calculation of the frequencies assuming that the potential energy function is known. In practice, however, one is often working in the opposite direction, knowing the frequencies from experiment and thereby deducing something about the potential between the atoms of the molecule. We shall also deduce the displacement patterns in the normal modes, a step which helps one to picture the motion of the atoms. These calculations are straightforward when the particular mode being studied is labelled by a representation $T^{(\alpha)}$ which occurs only once. We will not give details for the more general case when the representation $T^{(\alpha)}$ occurs more than once.

If in the reduction (6.24) a particular representation $T^{(\alpha)}$ appears only once (i.e., $m_\alpha = 1$) then it is possible to construct the displacement patterns in that particular mode by using the projection operators $P_{ij}^{(\alpha)}$ defined by equation (4.52). If $\boldsymbol{q}$ is an arbitrary displacement then, apart from overall normalisation and for any fixed j,

$$\boldsymbol{u}_i^{(\alpha)} = P_{ij}^{(\alpha)}\boldsymbol{q} \tag{6.34}$$

will be the normal displacement, corresponding to row i of the representation $T^{(\alpha)}$. This step again calls for no details of the potential beyond its symmetry. To find the frequency naturally involves the magnitude of the potential and from equations (6.22) and (6.20) the frequency is given by

$$\omega_\alpha^2 = \frac{(\boldsymbol{u}_i^{(\alpha)}, D\boldsymbol{u}_i^{(\alpha)})}{(\boldsymbol{u}_i^{(\alpha)}, \boldsymbol{u}_i^{(\alpha)})} = \frac{2V(\boldsymbol{u}_i^{(\alpha)})}{(\boldsymbol{u}_i^{(\alpha)}, \boldsymbol{u}_i^{(\alpha)})} \tag{6.35}$$

It is therefore sufficient to calculate the value of the potential energy in the normal displacement $\boldsymbol{u}_i^{(\alpha)}$ to find its frequency. Notice that because of the invariance of D, the frequency ω_α will be independent of the row label i, exhibiting the expected s_α-fold degeneracy. In fact for evaluating the frequency it is sufficient to use the much simpler projection operator $P^{(\alpha)}$ defined in equation (4.51), which involves only the group characters, in place of the more complicated $P_{ij}^{(\alpha)}$ which involves the representation matrix elements.

For the water molecule illustrated in figure 6.1 we found in section 6.5 that there were two vibrational A_1 modes and one B_1 mode. The single B_1 mode is

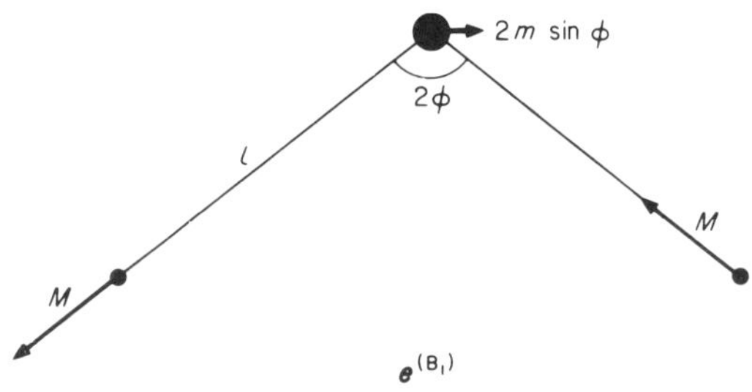

Figure 6.3

easily found by projection. In the notation of section 6.5 the three displacements

$$e_{3x},\ e_{1x}+e_{2x} \quad \text{and} \quad e_{1y}-e_{2y}$$

are all quickly found to be of B_1 type by projection on e_{1x}, e_{1y} and e_{3x}. The vibrational mode must be a linear combination of them which is free from translation and rotation and this is the displacement which is shown in figure 6.3, namely

$$\sin\phi\{2m^{\frac{1}{2}}e_{3x}-M^{\frac{1}{2}}(e_{1x}+e_{2x})\}-M^{\frac{1}{2}}\cos\phi(e_{1y}-e_{2y}) \tag{6.36}$$

For the two A_1 modes a more extensive calculation yields a linear combination of the patterns shown in figure 6.4, the amount of each pattern depending on the details of the potential.

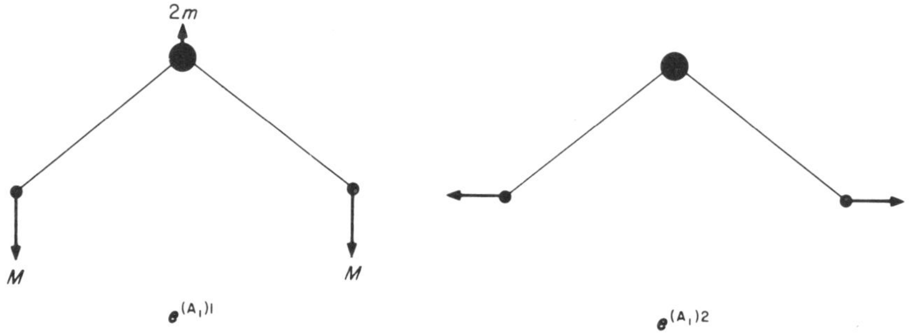

Figure 6.4

Another example of the calculation of displacement patterns can be found in problem 6.1 whose solution is also given.

Bibliography

For further reading on molecular vibrations we suggest:

Wilson, E. B., Decius, J. C. and Cross, P. C. (1955). *Molecular Vibrations* (McGraw-Hill, New York)

and for more details of spectra either of the following:

Hertzberg, G. (1945). *Molecular Spectra and Molecular Structure II*, Infra-red and Raman Spectra of Polyatomic Molecules (Van Nostrand, Princeton)
Hertzberg, G. (1966). *Molecular Spectra and Molecular Structure III*, Electronic Spectra of Polyatomic Molecules (Van Nostrand, Princeton)

Birman, J. L. (1974). Theory of space groups and infra-red and Raman lattice processes in insulating crystals, *Handb. Phys.*, **XXV** (2b); Light and Matter (1b).

Problems

6.1 A hypothetical molecule consists of four identical atoms at the corners of a plane rectangle (symmetry group D_{2h}, see chapter 9). Find the characters for the vibrational modes and hence classify them by symmetry. Are there any degeneracies in this problem? Draw a diagram indicating the displacements of the atoms in the normal mode of symmetry B_3^-, in the notation of the character table in appendix 1.

6.2 The coplanar molecule boron trifluoride, BF_3, has three fluorine atoms at the vertices of an equilateral triangle and the boron atom lies at the centre, thus having symmetry group D_{3h} (see chapter 9). Classify the normal modes of vibration and point out any degeneracies.

6.3 Discuss the infrared and Raman spectra in the first problem above.

Further problems on molecular vibrations are given at the end of chapter 9, after a more complete discussion of the possible point groups.

7

Continuous Groups and their Representations, including Details of the Rotation Groups $\mathscr{R}_2$ and $\mathscr{R}_3$

In chapter 2 the distinction was made between finite groups and infinite groups and the phrase 'continuous group' was used for an infinite group whose elements are defined by the values of a set of continuous parameters. In the examples of section 2.2, the group C_3 is finite while the group $\mathscr{R}_2$ is a continuous group with one parameter. Throughout chapters 4 and 5 we have made no distinction between finite and continuous groups except that at a number of points in chapter 4 the process of summing over all group elements was used and this is clearly not meaningful for a continuous group. In this chapter we take a closer look at the structure of continuous groups and show that these sums may be replaced by integrals over the group parameters. In particular we study the groups $\mathscr{R}_2$ and $\mathscr{R}_3$. The unitary groups will be described in later chapters.

7.1 General remarks

The group elements of a continuous group will be denoted by $G(a_1 a_2, \ldots, a_r)$ where the r continuous, real parameters a_q are all essential in the sense that the group elements cannot be distinguished by any smaller number. The number r is called the dimension of the group. Each parameter will have a well-defined range of values. For the elements to satisfy the group postulate a multiplication law must be defined and the product of two elements

$$G(a_1 a_2, \ldots, a_r)G(b_1 b_2, \ldots, b_r) = G(c_1 c_2, \ldots, c_r) \tag{7.1}$$

must be another group element. Thus the new parameters c_q must be expressible as functions of the a's and b's,

$$c_q = \phi_q(a_1 a_2, \ldots, a_r; b_1 b_2, \ldots, b_r) \tag{7.2}$$

It is usual to define the parameters in such a way that the identity element has all parameters zero. The r functions ϕ_q must satisfy many conditions in order for the group postulates to be satisfied. The functions ϕ_q will be assumed to be differentiable functions of the parameters since this is sufficiently general for the physical applications in this book. Groups of this kind are sometimes called 'Lie groups' after the Norwegian mathematician Sophus Lie (1842–1899). (The Lie groups are classified in section 20.4 of volume 2.)

In deriving the orthogonality relations for irreducible representations and characters in chapter 4 the process of summing over all group elements was frequently used. For a finite group this sum presents no difficulty but for a continuous group not only is there an infinite number of group elements but they are continuously distributed. Thus any such sum will not simply contain an infinite number of terms but will involve an integral over the parameters (like the angle of rotation) of the group. For an arbitrary group such integrals may not converge, in which case the orthogonality relations will not exist, but it is fortunate that for most of the continuous groups of physical interest it is possible to define an appropriate integral which is finite. In such cases, usually referred to as 'compact' groups, all the group sums $\sum_{a=1}^{g} f(a)$ used in chapter 4 may be replaced by group integrals

$$\int\int\ldots\int f(a_1 a_2, \ldots, a_r)\rho(a_1 a_2, \ldots, a_r)\mathrm{d}a_1\,\mathrm{d}a_2, \ldots, \mathrm{d}a_r$$

over the parameters $a_1, a_2, \ldots, a_r$ of the continuous group with a carefully chosen 'weight function' $\rho(a_1 a_2, \ldots, a_r)$. The number g of group elements is replaced by the group volume obtained by integrating over all values of the parameters. The choice of $\rho(a_1 a_2, \ldots, a_r)$ and the calculation of the volume is described in appendix 4.3 of volume 2. The orthogonality relations and all the results which followed from them in chapters 4 and 5 will then be true also for compact continuous groups. In particular, the concepts of representation, irreducibility and character are unchanged. The matrix elements and character of a representation are now continuous functions $T_{ij}^{(\alpha)}(a_1 a_2, \ldots, a_r)$, $\chi^{(\alpha)}(a_1 a_2, \ldots, a_r)$ of the group parameters. For a continuous group one no longer has a finite character table. Since the number of classes of a continuous group is infinite the table would not be expected to be finite and, like the elements themselves, the classes are continuously distributed. The number of inequivalent irreducible representations is also infinite although the dimension of the irreducible representations is (generally) finite.

The fact that there is an infinite number of irreducible representations means that in the expansion of an arbitrary function into functions belonging to irreducible representations there is an infinite number of terms. Thus for example the complex Fourier series

$$f(\phi) = \sum_{m=-\infty}^{+\infty} c_m \exp(im\phi)$$

for an arbitrary function f of angle ϕ is such an expansion, since as we shall see in section 7.3 each function $\exp(im\phi)$ transforms according to a one-dimensional irreducible representation of the group $\mathscr{R}_2$.

7.2 Infinitesimal operators

A continuous group of dimension r with parameters $a_1, a_2, \ldots, a_r$ has an infinite number of group elements but most of the properties of the group may be deduced from a finite number r of operators, called the infinitesimal operators. It will be convenient to use the symbol $\boldsymbol{a}$ for the set of parameters a_1, $a_2, \ldots, a_r$. Consider a representation $\mathrm{T}(\boldsymbol{a})$ of the group $\mathscr{G}$ in a space L. By convention the parameters are chosen such that the identity element has all $a_q = 0$, so that

$$\mathrm{T}(0, 0, \ldots, 0) = 1 \tag{7.3}$$

If all parameters a_q are small then to first order in these parameters,

$$\mathrm{T}(\boldsymbol{a}) \approx 1 + \sum_{q=1} a_q \mathrm{X}_q \tag{7.4}$$

where the X_q are some fixed linear operators, independent of the parameters a_q. These operators X_q are called the 'infinitesimal operators' of the representation T and from equation (7.4) they are given explicitly as partial derivatives

$$\mathrm{X}_q = \lim_{a_q \to 0} \{\mathrm{T}(0, 0, \ldots, a_q \ldots 0) - 1\}/a_q = \left[\frac{\partial}{\partial a_q}\mathrm{T}(\boldsymbol{a})\right]_{\boldsymbol{a}=0} \tag{7.5}$$

Before developing the properties of these infinitesimal operators in general let us first examine the special case when $\mathscr{G}$ is a one-parameter group with the multiplication law $\mathrm{G}(c) = \mathrm{G}(a)\mathrm{G}(b)$ with $c = a + b$. In other words the parameter is additive as in the group $\mathscr{R}_2$. Then for any integer n, we may write $\mathrm{T}(a) = \{\mathrm{T}(a/n)\}^n$. For large n, the parameter a/n is small and in the limit $n \to \infty$ we need keep only the first term in equation (7.4) so that, making use of the limit definition of an exponential function,

$$T(a) = \lim_{n \to \infty} \{1 + (a/n)X\}^n = \exp(aX) \tag{7.6}$$

in which the exponential is to be interpreted as the usual infinite series $\exp(aX) = \sum_{n=0}^{\infty} a^n X^n/n!$

This example illustrates how the finite operator $T(a)$ can be built up from the infinitesimal operator X. In fact a similar proof may be constructed in the general case to show that $T(\boldsymbol{a})$ is determined uniquely from the parameters a_q and the infinitesimal operators X_q. The vital properties of the infinitesimal operators are contained in the following three theorems which we do not prove, but first we prove the simple result that if the representation T is unitary then the operators X_q are skew-Hermitian, i.e. $X_q^\dagger = -X_q$. This follows immediately from equation (7.4) since, a_q being real, the unitary condition implies that, for small a_q,

$$1 = T(\boldsymbol{a})T^\dagger(\boldsymbol{a}) \approx \left(1 + \sum_q a_q X_q\right)\left(1 + \sum_q a_q X_q^\dagger\right)$$

$$\approx \left\{1 + \sum_q a_q (X_q + X_q^\dagger)\right\}$$

Equating the first-order term to zero gives $X_q^\dagger = -X_q$. If a factor $i = (-1)^{\frac{1}{2}}$ is removed from X_q then it will become Hermitian and this is sometimes done.

Theorem 1. If two representations of a group $\mathscr{G}$ have the same infinitesimal operators then they are the same representation.

Theorem 2. For any representation T of $\mathscr{G}$ the set of infinitesimal operators X_q satisfy commutation relations

$$[X_q, X_p] = \sum_t c_{qp}^t X_t \tag{7.7}$$

where the numbers c_{qp}^t, called 'structure constants', are the same for all representations T of $\mathscr{G}$.

Theorem 3. Any set of operators X_q, defined in a space L, will be the infinitesimal operators of a representation T of $\mathscr{G}$ in L if they satisfy the commutation relations (7.7).

The essence of theorem 1 is that the representation is uniquely determined by its infinitesimal operators, a generalisation of the result (7.6) for a one-parameter group.

The second theorem gives the 'law of multiplication' of the infinitesimal operators. In general, the infinitesimal operator for the product of two group elements is just the sum of the infinitesimal operators for each factor,

$$\mathrm{T}(\boldsymbol{a})\mathrm{T}(\boldsymbol{b}) \approx \left(1+\sum_q a_q \mathrm{X}_q\right)\left(1+\sum_p b_p \mathrm{X}_p\right) \approx 1+\sum_q (a_q+b_q)\mathrm{X}_q$$

for small $\boldsymbol{a}$ and $\boldsymbol{b}$. However, for the product $\mathrm{T}(\boldsymbol{a})\,\mathrm{T}(\boldsymbol{b})\mathrm{T}^{-1}(\boldsymbol{a})\mathrm{T}^{-1}(\boldsymbol{b})$, all the first order terms vanish although this product does not represent the identity. In fact

$$\mathrm{T}(\boldsymbol{a})\mathrm{T}(\boldsymbol{b})\mathrm{T}^{-1}(\boldsymbol{a})\mathrm{T}^{-1}(\boldsymbol{b}) \approx 1 + \sum_{q,\,p} a_q b_p [\mathrm{X}_q, \mathrm{X}_p] + \text{terms of order} > 2 \qquad (7.8)$$

Now, from the group properties,

$$\mathrm{T}(\boldsymbol{a})\mathrm{T}(\boldsymbol{b})\mathrm{T}^{-1}(\boldsymbol{a})\mathrm{T}^{-1}(\boldsymbol{b}) = \mathrm{T}(\boldsymbol{c}) \approx 1 + \sum_t c_t \mathrm{X}_t$$

for some $\boldsymbol{c}$, so that comparison with the previous expression leads us to conclude that $\boldsymbol{c}$ must be of order ab and also that the commutator $[\mathrm{X}_q, \mathrm{X}_p]$ must be a linear combination of the X_t. Hence we arrive at equation (7.7), that the commutator of any two infinitesimal operators must be a linear combination of the set of infinitesimal operators. It may further be shown that the structure constants c^t_{qp} are given uniquely from the multiplication law for the group elements so that the structure constants are clearly independent of the particular representation.

As a result of these theorems the study of continuous groups is in some ways simpler than the study of finite groups since one has only to study the algebra of the infinitesimal operators. The group multiplication table is in a sense replaced by the set of structure constants.

One further result of these theorems is that if a subset of the infinitesimal operators of some group $\mathscr{G}$ is closed under commutation, in the sense that the commutator of any two members of the subset is a linear combination of members of the subset, then that subset forms the infinitesimal operators of a subgroup of $\mathscr{G}$.

A set of functions $\phi^{(\alpha)}_i$ is said to transform according to the irreducible representation $\mathrm{T}^{(\alpha)}$ if the transformed functions may be expressed by equation (4.37),

$$\mathrm{T}\,\phi^{(\alpha)}_i = \sum_j T^{(\alpha)}_{ji}\,\phi^{(\alpha)}_j$$

where $\mathrm{T}^{(\alpha)}$ is the known matrix of the representation. For a continuous group it is sufficient to show only that the infinitesimal changes in the $\phi^{(\alpha)}_i$ are given by the known matrices of the infinitesimal operators. Thus with $\mathrm{T} = 1 + \sum_q a_q \mathrm{X}_q$ the condition (4.37) becomes, for each q,

$$\mathrm{X}_q \phi^{(\alpha)}_i = \sum_j (\mathrm{X}_q)^{(\alpha)}_{ji}\,\phi^{(\alpha)}_j \qquad (7.9)$$

where $(\mathrm{X}_q)^{(\alpha)}_{ji}$ are the known matrix elements of the operator X_q for the representation $T^{(\alpha)}$. In particular if ϕ is an invariant, it transforms like the identity representation for which $T = 1$. Thus for all q the infinitesimal operators are zero so that $X_q\phi = 0$.

In the same way, to show that a set of operators $S_i^{(\alpha)}$ transform according to $T^{(\alpha)}$ one needs to show that, for each q, the infinitesimal change in the operators is given by the same known matrices. Using equation (4.56) for small a_q,

$$S' = TST^{-1} \approx \left(1 + \sum_q a_q X_q\right) S \left(1 - \sum_q a_q X_q\right) \approx S + \sum_q a_q [X_q, S]$$

showing that, for operators, the infinitesimal change is given by a commutator. Hence the condition analogous to (7.9) is, for each q,

$$[X_q, S_i^{(\alpha)}] = \sum_j (X_q)_{ji}^{(\alpha)} S_j^{(\alpha)} \tag{7.10}$$

In practice it is usually simpler to use equations (7.9) and (7.10) in identifying the transformation properties of functions and operators rather than equations (4.37) and (4.56) which refer to finite changes. For example an invariant operator S will satisfy $[X_q, S] = 0$ for all q; in other words S commutes with all the infinitesimal operators.

In a product representation the infinitesimal operators are the sums of the infinitesimal operators for each factor. The matrix element (4.41) becomes, for infinitesimals,

$$T_{ij,kl}^{(\alpha\times\beta)}(\boldsymbol{a}) = \left\{\delta_{ik} + \sum_q a_q (X_q^{(\alpha)})_{ik}\right\}\left\{\delta_{jl} + \sum_p a_p (X_p^{(\beta)})_{jl}\right\}$$

$$= \delta_{ik}\delta_{jl} + \sum_q a_q \{(X_q^{(\alpha)})_{ik}\delta_{jl} + (X_q^{(\beta)})_{jl}\delta_{ik}\}$$

In terms of the product functions $\phi_k^{(\alpha)}$ and $\psi_l^{(\beta)}$ which were used as a basis for the product representation in section 4.17 one may therefore write, for each infinitesimal operator,

$$X_q = X_q(1) + X_q(2) \tag{7.11}$$

where $X_q(1)$ denotes the product of the infinitesimal operator X_q for the $\phi_k^{(\alpha)}$ with the unit operator for the $\psi_l^{(\beta)}$, while $X_q(2)$ has the opposite role.

7.3 The group $\mathscr{R}_2$

$\mathscr{R}_2$ is an Abelian group with a single parameter, the angle of rotation a in the range $0 \le a < 2\pi$. The parameter is additive since if $R(c) = R(a)R(b)$ then $c = a + b$ (or $c = a + b - 2\pi$ if $c \ge 2\pi$). There will be no ambiguity by addition or subtraction of multiples of 2π so long as we restrict our attention to periodic functions.

7.3.1 Irreducible representations

Since $\mathscr{R}_2$ is an Abelian group its irreducible representations are one-dimensional, see section 4.8. Thus to find the possible irreducible representations of $\mathscr{R}_2$ we must find functions $T(a)$ which satisfy $T(a+b) = T(a)\,T(b)$ and $T(0) = 1$. Differentiating with respect to b, for fixed a, we have $T'(a+b) = T(a)T'(b)$, where T' denotes the derivative, and then setting $b = 0$, $T'(a) = T(a)T'(0)$. But this is a simple differential equation for the function $T(a)$ with solution $T(a) = \exp[aT'(0)]$ using $T(0) = 1$. Thus we deduce that the irreducible representations of $\mathscr{R}_2$ are exponentials of the rotation angle a. The coefficient $T'(0)$ may take any value but the representation will be continuous, $T(a) = T(a+2\pi)$, only if $T'(0)$ is a purely imaginary integer. It is usual to define $m = iT'(0)$ with m a positive or negative integer, so that the continuous irreducible representations of $\mathscr{R}_2$ are given by

$$T^{(m)}(a) = \exp(-ima) \tag{7.12}$$

and labelled by $m = 0, \pm 1, \pm 2, \ldots$. Such representations are clearly unitary. A vector transforming according to the irreducible representation $T^{(m)}$ is thus denoted by $\boldsymbol{e}_m$ and has the property $\mathrm{T}(a)\boldsymbol{e}_m = \exp(-ima)\boldsymbol{e}_m$.

Notice that by taking $m = \pm\frac{1}{2}, \pm\frac{3}{2}, \ldots$, it is possible to form unitary representations which are continuous only over the extended range $0 \le a < 4\pi$. They would be double-valued over the usual range $0 \le a < 2\pi$. We shall see in section 8.4 that such double-valued representations are necessary in describing spin in quantum mechanics (see also section 7.6).

7.3.2 Character

Since the representations are one-dimensional the character is identical with the representation and, since each element is in a class by itself, the character is

$$\chi^{(m)} \equiv T^{(m)}(a) = \exp(-ima) \tag{7.13}$$

which is a continuous function of the parameter a.

Orthogonality

The orthogonality of characters, given by equation (4.25) for a finite group, is now expressed by the integral

$$\int_{a=0}^{2\pi} \chi^{(m)}\chi^{(m')*}\mathrm{d}a = \int_0^{2\pi} \exp[ia(m'-m)]\mathrm{d}a = 2\pi\delta_{m'm} \tag{7.14}$$

This is an obvious choice of integral to replace the group sum in a finite group but we return to this point in appendix 4.3 of volume 2. In the notation of section 7.1 the weight function $\rho(a)$ has been chosen to be unity. The 'volume' 2π which appears in equation (7.14) takes the place of the number of group elements, g in equation (4.25).

7.3.3 Multiplication of representations

The character of the product of two representations $T^{(m_1)}$ and $T^{(m_2)}$ will be simply $\exp[-i(m_1+m_2)a]$ which is the character of the representation $T^{(m_1+m_2)}$. Thus we have the rather obvious result

$$T^{(m_1)} \times T^{(m_2)} = T^{(m_1+m_2)} \tag{7.15}$$

7.3.4 Examples of basis vectors

(1) Consider the unit vectors $\boldsymbol{e}_x$ and $\boldsymbol{e}_y$ in the x and y directions and the group $\mathscr{R}_2$ of rotations about the z-axis. Then, as in subsection 3.8.1,

$$\begin{aligned} R_z(a)\boldsymbol{e}_x &= \cos a\boldsymbol{e}_x + \sin a\boldsymbol{e}_y \\ R_z(a)\boldsymbol{e}_y &= -\sin a\boldsymbol{e}_x + \cos a\boldsymbol{e}_y \end{aligned}$$

so that

$$R_z(a)(\boldsymbol{e}_x \pm i\boldsymbol{e}_y) = \exp(\mp ia)(\boldsymbol{e}_x \pm i\boldsymbol{e}_y) \tag{7.16}$$

showing that the vectors $(\boldsymbol{e}_x \pm i\boldsymbol{e}_y)$ transform according to the irreducible representations with $m = \pm 1$, respectively. Notice that it was necessary to use complex coefficients to produce irreducible representations. It is clear geometrically that no vector in the xy-plane with real coefficients will be simply multiplied by a constant as the result of an arbitrary rotation about the z-axis.

(2) Next, consider functions $\psi(r, \phi)$ of position in the xy-plane using polar coordinates for convenience. The general definition (3.37) for the induced transformation of functions gives, as in subsection 3.8.5,

$$T(R_z(a))\psi(r, \phi) = \psi(r, \phi - a) \tag{7.17}$$

from which it is clear that a function of the kind $\psi(r, \phi) = \psi(r)\exp(im\phi)$ will transform according to the representation $T^{(m)}$,

$$T(R_z(a))\exp(im\phi) = \exp[im(\phi - a)] = \exp(-ima)\exp(im\phi)$$

The expansion of an arbitrary function in a complex Fourier series

$$\psi(r, \phi) = \sum_{m=-\infty}^{\infty} \psi_m(r)\exp(im\phi)$$

where

$$\psi_m(r) = \frac{1}{2\pi}\int_0^{2\pi} \psi(r, \phi)\exp(-im\phi)\mathrm{d}\phi$$

is therefore an analysis of the function into components each of which transforms according to a specific irreducible representation $T^{(m)}$ of $\mathscr{R}_2$.

7.3.5 Infinitesimal operators

Let us now deduce the single infinitesimal operator for $\mathscr{R}_2$ in these examples. First, in the vector space of $\boldsymbol{e}_x$ and $\boldsymbol{e}_y$ the matrix of $\mathrm{R}_z(a)$ is, see subsection 3.8.1,

$$\mathrm{R}_z(a) = \begin{pmatrix} \cos a & -\sin a \\ \sin a & \cos a \end{pmatrix}$$

which, for small a, becomes

$$1 + \begin{pmatrix} 0 & -a \\ a & 0 \end{pmatrix} = 1 + a\begin{pmatrix} 0 & -1 \\ 1 & 0 \end{pmatrix}$$

so that the infinitesimal matrix operator is

$$\mathrm{X} = \begin{pmatrix} 0 & -1 \\ 1 & 0 \end{pmatrix}$$

We may verify equation (7.6) in this example using the fact that $\mathrm{X}^2 = -1$

$$\begin{aligned} \exp a\mathrm{X} &= 1 + a\mathrm{X} + \tfrac{1}{2}a^2\mathrm{X}^2 + \tfrac{1}{6}a^3\mathrm{X}^3 + \tfrac{1}{24}a^4\mathrm{X}^4 \ldots \\ &= (1 - \tfrac{1}{2}a^2 + \tfrac{1}{24}a^4 + \ldots) + \mathrm{X}(a - \tfrac{1}{6}a^3 + \ldots) \\ &= \cos a + \mathrm{X}\sin a = \begin{pmatrix} \cos a & -\sin a \\ \sin a & \cos a \end{pmatrix} = \mathrm{R}_z(a) \end{aligned}$$

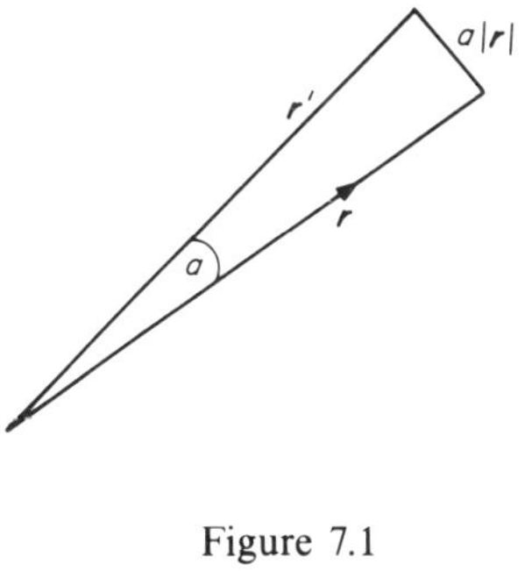

Figure 7.1

Geometrically it is evident that the first-order change in a vector $\boldsymbol{r}$ in the x, y-plane caused by a rotation $\mathrm{R}_z(a)$ about the z-axis is the addition of a vector of length $a|\boldsymbol{r}|$ in a direction perpendicular to $\boldsymbol{r}$, as illustrated in figure 7.1. Thus for small a we have

$$\mathrm{R}_z(a)\boldsymbol{r} \approx \boldsymbol{r} + a(\boldsymbol{e}_z \wedge \boldsymbol{r}) \tag{7.18}$$

where $\boldsymbol{e}_z$ is the unit vector in the z-direction. Thus we may write for the operator

$$\mathrm{R}_z(a) \approx 1 + a(\boldsymbol{e}_z \wedge$$

and for the infinitesimal operator in example (1) of subsection 7.3.4

$$\mathrm{X} = \boldsymbol{e}_z \wedge \tag{7.19}$$

In the function space, example (2), the infinitesimal operator is deduced by using a Taylor series to expand the right-hand side of equation (7.17):

$$\mathrm{T}(\mathrm{R}_z(a))\psi(r, \phi) = \psi(r, \phi) - a\frac{\partial}{\partial \phi}\psi(r, \phi) + \tfrac{1}{2}a^2\frac{\partial^2}{\partial \phi^2}\psi(r, \phi) + \ldots$$

so that we may write for the operator, for small a, $\mathrm{T}(\mathrm{R}_z(a)) \approx 1 - a(\partial/\partial\phi)$. Hence, in this example, the infinitesimal operator is the differential operator

$$\mathrm{X} = -\frac{\partial}{\partial \phi} \tag{7.20}$$

In fact, the Taylor series is an exponential series in the differential operator $\partial/\partial\phi$ so that $\mathrm{T}(\mathrm{R}_z(a)) = \exp[-a(\partial/\partial\phi)]$, in agreement with equation (7.6) again. The link with angular momentum in quantum mechanics has already been brought out in equation (5.19), since

$$\mathrm{X} = -\frac{\partial}{\partial \phi} = -i(\mathbf{r} \wedge \mathbf{p})_z/\hbar = -il_z \tag{7.21}$$

where l_z is the operator for the angular momentum (in units of $\hbar$) about the z-axis.

In an irreducible representation $\mathrm{T}^{(m)}$ of $\mathscr{R}_2$ the matrix element of the infinitesimal operator X is simply $-im$, as may be seen from the second term in the expansion of equation (7.12) for small a. In fact if a function ψ has the property $\mathrm{X}\psi = -im\psi$ then ψ must transform according to the irreducible representation $\mathrm{T}^{(m)}$ of $\mathscr{R}_2$.

The addition of improper elements, reflections and inversion, to the group $\mathscr{R}_2$ is left until section 9.6.

7.4 The group $\mathscr{R}_3$

A rotation in three dimensions is conveniently denoted by $\mathrm{R}_{\boldsymbol{k}}(a)$ with a the angle of rotation $0 \leq a \leq \pi$, and $\boldsymbol{k}$ a unit vector along the axis of rotation, see subsection 2.2(9) and section 3.8. It contains three parameters, the rotation angle a and the two polar angles of $\boldsymbol{k}$. However, for some purposes it is simpler to use instead the three parameters $a_x = ak_x$, $a_y = ak_y$, $a_z = ak_z$, where the k_q are the three components of $\boldsymbol{k}$ referred to a fixed set of axes. One would then use the notation $\mathrm{R}(\boldsymbol{a})$ instead of $\mathrm{R}_{\boldsymbol{k}}(a)$.

$a \in [0, 2\pi)$?

Some elementary geometry shows that the effect of a rotation on an arbitrary vector $\boldsymbol{r}$ is given by

$$\mathrm{R}_{\boldsymbol{k}}(a)\boldsymbol{r} = \cos a\, \boldsymbol{r} + (1-\cos a)(\boldsymbol{r}.\boldsymbol{k})\boldsymbol{k} + \sin a(\boldsymbol{k} \wedge \boldsymbol{r})$$

from which the matrix of $\mathrm{R}_{\boldsymbol{k}}(a)$ in the Cartesian basis is immediately deduced. For example,

$$[R_{\boldsymbol{k}}(a)]_{xx} = \cos a + (1-\cos a)k_x^2,$$
$$[R_{\boldsymbol{k}}(a)]_{yx} = (1-\cos a)k_x k_y + k_z \sin a, \text{ etc.}$$

A rotation preserves both lengths and angles between vectors drawn from the origin. It therefore preserves the scalar product between any two vectors $\boldsymbol{r}_1$ and $\boldsymbol{r}_2$. Hence if we write $\boldsymbol{r}'_1 = \mathrm{R}\boldsymbol{r}_1$, $\boldsymbol{r}'_2 = \mathrm{R}\boldsymbol{r}_2$ then (see section 3.5) $(\boldsymbol{r}'_2, \boldsymbol{r}'_1) = (\mathrm{R}\boldsymbol{r}_2, \mathrm{R}\boldsymbol{r}_1) = (\boldsymbol{r}_2, \mathrm{R}^\dagger\mathrm{R}\boldsymbol{r}_1)$, so that R is unitary, $\mathrm{R}^\dagger\mathrm{R} = 1$, and its determinant has modulus unity.

In the usual Cartesian basis the matrix elements of R are obviously real, so that the matrix is then orthogonal and its determinant has value ± 1. In fact, by choosing an orthonormal basis, the rotations may be defined as the set of orthogonal 3×3 matrices with determinant of $+1$. The identity clearly has a determinant of $+1$ and since $\mathscr{R}_3$ is a continuous group it follows that all rotations must have a determinant of $+1$. There is nevertheless a geometrical role for the matrices with a determinant of -1. Notice that the inversion operation, which changes the sign of all vectors, has a matrix

$$\mathrm{I} = \begin{pmatrix} -1 & 0 & 0 \\ 0 & -1 & 0 \\ 0 & 0 & -1 \end{pmatrix}$$

and so has a determinant of -1. The orthogonal matrices with determinant equal to -1 will therefore correspond to products of rotations and inversion. The symbol O_3^+ is sometimes used for $\mathscr{R}_3$, denoting the group of orthogonal transformations in three dimensions with determinant equal to $+1$. The symbol O_3 then denotes the group of all orthogonal transformations, including the inversions. In fact O_3 is the product group $O_3 = \mathscr{R}_3 \times S_2$ of $\mathscr{R}_3$ with the inversion group S_2. Notice that the transformations with determinant equal to -1 do not form a group by themselves.

7.4.1 Infinitesimal operators

The expression (7.18) for a small-angle rotation about the z-axis immediately generalises for a small-angle rotation $\mathrm{R}_{\boldsymbol{k}}(a)$ about a general axis $\boldsymbol{k}$,

$$\mathrm{R}_{\boldsymbol{k}}(a)\boldsymbol{r} = \boldsymbol{r} + a(\boldsymbol{k} \wedge \boldsymbol{r})$$
$$= \boldsymbol{r} + a \sum_{q=x,y,z} k_q(\boldsymbol{e}_q \wedge \boldsymbol{r}) = \boldsymbol{r} + \sum_q a_q(\boldsymbol{e}_q \wedge \boldsymbol{r}) \qquad (7.22)$$

where $a_q = ak_q$. This exhibits the first-order change in $\boldsymbol{r}$ in the general form of equation (7.4) showing that geometrically the three infinitesimal operators appropriate to the parameters a_q are

$$\mathrm{X}_q = (\boldsymbol{e}_q \wedge \tag{7.23}$$

corresponding to infinitesimal rotations about the x-, y- and z-axes, respectively. The matrices of the infinitesimal operators in the basis $\boldsymbol{e}_x, \boldsymbol{e}_y, \boldsymbol{e}_z$ are thus

$$\mathrm{X}_x = \begin{pmatrix} 0 & 0 & 0 \\ 0 & 0 & -1 \\ 0 & 1 & 0 \end{pmatrix}, \mathrm{X}_y = \begin{pmatrix} 0 & 0 & 1 \\ 0 & 0 & 0 \\ -1 & 0 & 0 \end{pmatrix} \text{ and } \mathrm{X}_z = \begin{pmatrix} 0 & -1 & 0 \\ 1 & 0 & 0 \\ 0 & 0 & 0 \end{pmatrix} \tag{7.24}$$

Some elementary vector algebra now shows that for any $\boldsymbol{r}$

$$\begin{aligned}[\mathrm{X}_x, \mathrm{X}_y]\boldsymbol{r} &= \boldsymbol{e}_x \wedge (\boldsymbol{e}_y \wedge \boldsymbol{r}) - \boldsymbol{e}_y \wedge (\boldsymbol{e}_x \wedge \boldsymbol{r}) \\ &= x\boldsymbol{e}_y - y\boldsymbol{e}_x = (\boldsymbol{e}_z \wedge \boldsymbol{r}) = \mathrm{X}_z\boldsymbol{r}\end{aligned}$$

so that the commutation relations for the infinitesimal operators of the group $\mathscr{R}_3$ are

$$[\mathrm{X}_x, \mathrm{X}_y] = \mathrm{X}_z, \quad [\mathrm{X}_y, \mathrm{X}_z] = \mathrm{X}_x \quad \text{and} \quad [\mathrm{X}_z, \mathrm{X}_x] = \mathrm{X}_y \tag{7.25}$$

From equation (7.23), the three operators X_q form a vector, transforming like the $\boldsymbol{e}_q$ under rotations. Because of the link (7.21) between infinitesimal rotation operators and angular momentum operators in quantum mechanics, these commutation relations are essentially the same as those for the three components of the angular momentum which the reader may have met in his study of quantum mechanics. In fact, much of the algebra of this section has a parallel in the theory of angular momentum, particularly the deduction of the structure of the irreducible representations.

The infinitesimal operators X_q are skew-Hermitian (see section 7.2) and for many purposes it is convenient to extract a factor $i = (-1)^{\frac{1}{2}}$ by defining $\mathrm{J}_q = i\mathrm{X}_q$. We shall use the word infinitesimal operator to describe both X_q and J_q and we hope that no confusion will arise. Then the J_q are Hermitian and satisfy the commutation relations

$$[\mathrm{J}_x, \mathrm{J}_y] = i\mathrm{J}_z, \quad [\mathrm{J}_y, \mathrm{J}_z] = i\mathrm{J}_x, \quad [\mathrm{J}_z, \mathrm{J}_x] = i\mathrm{J}_y \tag{7.26}$$

precisely those of the angular momentum operators in units of $\hbar$. Furthermore, whereas the eigenvalues of X_z were found in section 7.3 to be $-im$ the eigenvalues of J_z will be m. It will also be convenient to work in terms of the linear combinations

$$\mathrm{J}_{\pm} = \mathrm{J}_x \pm i\mathrm{J}_y \tag{7.27}$$

rather than J_x and J_y. The reason for this choice lies in the commutation relations of $J_\pm$ with J_z,

$$[J_z, J_\pm] = iJ_y \pm J_x = \pm(J_x \pm iJ_y) = \pm J_\pm \tag{7.28}$$

They show that if $\psi(m)$ is an eigenfunction of J_z belonging to the representation $T^{(m)}$ of $\mathscr{R}_2$, so that $J_z\psi(m) = m\psi(m)$, then the functions defined by $J_\pm\psi(m)$ belong to $T^{(m\pm1)}$, i.e. $J_\pm$ transform according to $T^{(\pm1)}$, since

$$\begin{aligned} J_z\{J_\pm\psi(m)\} &= (J_\pm J_z \pm J_\pm)\psi(m) = J_\pm(J_z \pm 1)\psi(m) \\ &= J_\pm(m \pm 1)\psi(m) \\ &= (m \pm 1)\{J_\pm\psi(m)\} \end{aligned} \tag{7.29}$$

In consequence, the operators $J_\pm$ are called the 'raising' and 'lowering' operators since they raise or lower the eigenvalue of J_z by one unit. Notice that this property follows from the commutation relations alone, so that it will be true of the infinitesimal operators in any representation of $\mathscr{R}_3$.

7.4.2 Irreducible representations

For the group $\mathscr{R}_2$ it was a trivial matter to deduce not only the irreducible representation labels m but also the representation matrix element $\exp(-ima)$. This simplicity was due to the fact that $\mathscr{R}_2$ is an Abelian group and has only one infinitesimal operator. For more general groups there is a set of infinitesimal operators X_q and the properties of the irreducible representations are deduced from the commutation relations (7.7) of the X_q. For the group $\mathscr{R}_3$ these are given by equation (7.25) or equivalently by equation (7.28) together with

$$[J_+, J_-] = [J_x + iJ_y, J_x - iJ_y] = 2J_z \tag{7.30}$$

Our object in this section is to find the dimensions of the irreducible representations of $\mathscr{R}_3$, to find a means of labelling them, and to deduce the matrix elements of the infinitesimal operators in each irreducible representation. Let D denote an irreducible representation of $\mathscr{R}_3$ in a vector space L. (It is conventional to use the symbol D for irreducible representations of $\mathscr{R}_3$ rather than T which we have used for any group.) We choose a basis in which the infinitesimal operator J_z is diagonal so that the basis vectors $\boldsymbol{e}_m$ carry a label m referring to the eigenvalue m of J_z and hence to the irreducible representation $T^{(m)}$ of $\mathscr{R}_2$ to which they belong. (This notation $\boldsymbol{e}_m$, with m a number, for the basis vectors in representation space should not be confused with the three unit vectors $\boldsymbol{e}_x$, $\boldsymbol{e}_y$, $\boldsymbol{e}_z$ in ordinary physical space, which were used in subsection 7.4.1.)

At this stage it is quite possible that there may be more than one linearly independent basis vector with the same value of m. However we shall see that this does not occur so that the single label m is sufficient to distinguish one basis vector from another within the irreducible representation. Let j denote the greatest value of m for any of the basis vectors of D and let $\boldsymbol{e}_j$ denote a basis vector with $m = j$, called a vector of 'greatest weight'. Necessarily the vector $\boldsymbol{e}_j$ must have the property

$$\mathrm{J}_+ \boldsymbol{e}_j = 0 \tag{7.31}$$

for otherwise the new vector $\mathrm{J}_+ \boldsymbol{e}_j$ would have greater weight than $\boldsymbol{e}_j$. Acting on $\boldsymbol{e}_j$ repeatedly with the lowering operator J_- we construct a set of normalised vectors

$$\begin{aligned} \boldsymbol{e}_{j-1} &= A_{j-1}\,\mathrm{J}_- \boldsymbol{e}_j \\ \boldsymbol{e}_{j-2} &= A_{j-2}\,\mathrm{J}_- \boldsymbol{e}_{j-1} \\ \boldsymbol{e}_{j-3} &= A_{j-3}\,\mathrm{J}_- \boldsymbol{e}_{j-2}, \text{ etc.} \end{aligned} \tag{7.32}$$

The constants A_m, which are as yet undetermined, are included to ensure normalisation of the vectors $\boldsymbol{e}_m$ given that $\boldsymbol{e}_j$ is normalised. This set of vectors all have different m-values and are therefore orthogonal and independent. Since the vector space of a representation must be invariant under the group operations it follows that the set of vectors must all belong to L. Thus, if the space L is to have finite dimension, the set of vectors must terminate and this can happen only if at some stage the further application of the lowering operator produces zero. In other words, for some integer t,

$$\mathrm{J}_- \boldsymbol{e}_{j-t} = 0 \tag{7.33}$$

It is clear that the set

$$\boldsymbol{e}_j, \boldsymbol{e}_{j-1}, \ldots, \boldsymbol{e}_{j-t} \tag{7.34}$$

is invariant under J_z and J_- and we now show that it is also invariant under J_+. This enables us to conclude that the set is invariant under any group element of $\mathscr{R}_3$ and hence that it provides the basis for a representation of $\mathscr{R}_3$. To show the invariance under J_+ we first construct the quadratic $\mathbf{J}^2 = \mathbf{J}\,.\,\mathbf{J} = \mathrm{J}_x^2 + \mathrm{J}_y^2 + \mathrm{J}_z^2$ which, from the commutation relations (7.25), may be seen to satisfy $[\mathbf{J}^2, \mathrm{J}_q] = 0$ for $q = x$, y and z and is therefore, from equation (7.10), an invariant. We note that $\mathbf{J}^2$ can also be written in terms of the raising and lowering operators

$$\mathbf{J}^2 = \tfrac{1}{2}(\mathrm{J}_+\mathrm{J}_- + \mathrm{J}_-\mathrm{J}_+) + \mathrm{J}_z^2$$

and, using equation (7.30), this may be written as

$$\mathbf{J}^2 = \mathrm{J}_-\mathrm{J}_+ + \mathrm{J}_z^2 + \mathrm{J}_z \tag{7.35a}$$

or as

$$\mathbf{J}^2 = \mathrm{J}_+\mathrm{J}_- + \mathrm{J}_z^2 - \mathrm{J}_z \tag{7.35b}$$

From equation (7.35a) it follows that $\boldsymbol{e}_j$ is an eigenvector of $\mathbf{J}^2$ since

$$\mathbf{J}^2 \boldsymbol{e}_j = (\mathrm{J}_-\mathrm{J}_+ + \mathrm{J}_z^2 + \mathrm{J}_z)\boldsymbol{e}_j = j(j+1)\boldsymbol{e}_j$$

using the property (7.31), and since $[\mathbf{J}^2, \mathrm{J}_-] = 0$, we also have

$$\mathbf{J}^2 \boldsymbol{e}_m = j(j+1)\boldsymbol{e}_m \tag{7.36}$$

for any member $\boldsymbol{e}_m$ of the set (7.34). The invariance of the set (7.34) under J_+ now follows using equation (7.32) and (7.35b),

$$\begin{aligned}
\mathrm{J}_+ \boldsymbol{e}_m &= A_m \mathrm{J}_+ \mathrm{J}_- \boldsymbol{e}_{m+1} \\
&= A_m(\mathbf{J}^2 - \mathrm{J}_z^2 + \mathrm{J}_z)\boldsymbol{e}_{m+1} \\
&= A_m\{j(j+1) - (m+1)^2 + (m+1)\}\boldsymbol{e}_{m+1} \\
&= A_m(j+m+1)(j-m)\boldsymbol{e}_{m+1}
\end{aligned} \tag{7.37}$$

This shows that the raising operator cannot produce any vectors outside the set (7.34) which was formed by the lowering operations on $\boldsymbol{e}_j$. (The fact that J_+ is a raising operator does not in itself ensure this result since it could have raised $\boldsymbol{e}_m$ to a new vector $\boldsymbol{e}'_{m+1}$ different from the vector $\boldsymbol{e}_{m+1}$ in the set (7.34).) It follows that $\boldsymbol{e}_j$ must have been the only basis vector with greatest weight in L since, if there were another, the representation would have reduced. Hence the irreducible representations of $\mathscr{R}_3$ must have a basis described by the set (7.34).

Some properties of the irreducible representation D may now be deduced. Using equations (7.35b) and (7.33) we have

$$\begin{aligned}
\mathbf{J}^2 \boldsymbol{e}_{j-t} &= (\mathrm{J}_+ \mathrm{J}_- + \mathrm{J}_z^2 - \mathrm{J}_z)\boldsymbol{e}_{j-t} = \{(j-t)^2 - (j-t)\}\boldsymbol{e}_{j-t} \\
&= (j-t)(j-t-1)\boldsymbol{e}_{j-t}
\end{aligned}$$

and comparing this with equation (7.36) gives $(j-t)(j-t-1) = j(j+1)$, i.e. $(2j-t)(t+1) = 0$. Since t is necessarily a positive integer this implies that $2j = t$ and hence that j may be either an integer or a half-integer. The dimension of D is now $2j+1$. We therefore conclude that the irreducible representations of $\mathscr{R}_3$ may be denoted by $\mathrm{D}^{(j)}$, where the label $j = 0, \frac{1}{2}, 1, \frac{3}{2}, 2, \ldots$, etc., that this representation has dimension $(2j+1)$ and that the basis vectors $\boldsymbol{e}_m$ may be chosen to transform according to the irreducible representation $\mathrm{T}^{(m)}$ of the subgroup $\mathscr{R}_2$, where $m = j, (j-1), (j-2), \ldots, (1-j), -j$. We may write

$$\mathrm{D}^{(j)} = \dot{\sum_{m=-j}^{m=j}} \mathrm{T}^{(m)} \tag{7.38}$$

as a reduction of $\mathrm{D}^{(j)}$ when the group is restricted from $\mathscr{R}_3$ to its subgroup $\mathscr{R}_2$. Notice that the half-integer representations are not periodic over the interval 2π (see subsection 7.3.1). They are encountered only when describing spin in quantum mechanics (see section 8.4).

It now takes only a small step to deduce the matrices of the infinitesimal operators J_q in the representation $\mathrm{D}^{(j)}$. Firstly J_z is diagonal in the basis $\boldsymbol{e}_m$ with matrix elements given by

$$\mathrm{J}_z \boldsymbol{e}_m = m\boldsymbol{e}_m \tag{7.39}$$

The matrix elements of J_+ and J_- are given by the equations (7.32) and (7.37) once the constants A_m are known. To find these constants we use the fact that, since J_q is Hermitian, $J_-^\dagger = J_+$, so that

$$(e_m, J_- e_{m+1}) = (J_+ e_m, e_{m+1}) = (e_{m+1}, J_+ e_m)^*$$

Hence, using equations (7.32) and (7.37), $|A_m|^{-2} = (j+m+1)(j-m)$. The relative phase of the basis vectors is not determined either by the orthogonality or normalisation and so we are free to introduce any complex number of unit modulus into the constants A_m. The usual convention is to take the A_m to be real and positive so that the matrix elements of $J_\pm$ are given by

$$\begin{aligned} J_- e_{m+1} &= [(j+m+1)(j-m)]^{\frac{1}{2}} e_m \\ J_+ e_m &= [(j+m+1)(j-m)]^{\frac{1}{2}} e_{m+1} \end{aligned} \tag{7.40}$$

This choice of phase is called the 'Condon and Shortley convention' and the relative phases of all the $(2j+1)$ basis vectors of the representation $D^{(j)}$ are then uniquely specified.

It may be shown that the set of representations $D^{(j)}$ is complete and indeed, for any single-valued function, only integer j are required

$$\psi = \sum_{j=0}^{\infty} \sum_{m=-j}^{j} \psi_{jm} \tag{7.41}$$

where j is an integer and each term ψ_{jm} in the expansion transforms like the row m of a representation $D^{(j)}$.

The generation of the actual matrices $D^{(j)}_{m'm}(a_x, a_y, a_z)$ for a finite rotation with parameters a_x, a_y, a_z is a rather intricate algebraic process which we leave until section 20.5 of volume 2 and it is usual to present the matrix element $D^{(j)}_{m'm}$ as a function of the three Euler angles rather than the three parameters a_x, a_y and a_z. The reader is warned that a number of different conventions have been used by different authors for the matrices of $D^{(j)}$. The one given here agrees with that used by Brink and Satchler, 1968 (see bibliography), the essential convention lying in equations (4.2), (4.8) and (7.40).

7.4.3 Characters

Since all rotations through the same angle are in the same class of $\mathscr{R}_3$, irrespective of the direction of their axes, it follows that the character of a rotation depends only on the rotation angle. To deduce the character for an irreducible representation $D^{(j)}$ we may therefore choose any axis and since we have used a basis of vectors e_m which transform according to the one-dimensional irreducible representations $T^{(m)}$ of the subgroup $\mathscr{R}_2$ of rotations about the z-axis, it is clearly advantageous to choose a rotation $R_z(a)$ about the z-axis. Then, using the reduction (7.38) of $D^{(j)}$ into irreducible representations of the subgroup $\mathscr{R}_2$ and the formula (7.12) for the characters in $\mathscr{R}_2$, we deduce the character of $D^{(j)}$ for a rotation through angle a,

$$\begin{aligned}\chi_a^{(j)} &= \sum_{m=-j}^{j} \exp(-ima) = \exp(-ija)[1+\exp(ia)+\exp(2ia)+\ldots \\ &\quad +\exp(2jia)] \\ &= \exp(-ija)\{\exp[(2j+1)ia]-1\}/[\exp(ia)-1] \\ &= \{\exp[(j+\tfrac{1}{2})ia]-\exp[-(j+\tfrac{1}{2})ia]\}/\{\exp(\tfrac{1}{2}ia)-\exp(-\tfrac{1}{2}ia)\} \\ &= \frac{\sin(j+\frac{1}{2})a}{\sin\frac{1}{2}a}\end{aligned} \tag{7.42}$$

As a particular case, the character for $j = 1$, the vector representation, is given by

$$\begin{aligned}\chi_a^{(1)} &= \sin\tfrac{3}{2}a/\sin\tfrac{1}{2}a = \cos a + (\cos\tfrac{1}{2}a \sin a/\sin\tfrac{1}{2}a) \\ &= \cos a + 2\cos^2\tfrac{1}{2}a = 2\cos a + 1,\end{aligned}$$

agreeing with equation (4.6). The identity element has $a = 0$ in which limit the character (7.42) reduces to $(2j+1)$ which is just the dimension of $D^{(j)}$.

It may be shown (see appendix 4.3 of volume 2) that the orthogonality relation for the characters of $\mathscr{R}_3$ is given by the integral

$$\frac{1}{2\pi}\int_{a=0}^{2\pi} \chi_a^{(j_1)}\chi_a^{(j_2)}(1-\cos a)\mathrm{d}a = \delta_{j_1 j_2}$$

For the full orthogonal group O_3 which, as explained in section 7.4, is the direct product of $\mathscr{R}_3$ with the inversion group S_2, the irreducible representations are labelled by $D^{(j)+}$ and $D^{(j)-}$. Following section 4.21, the characters for O_3 are then given by

$$\chi^{(j)\pm}(\mathrm{R}(a)) = \chi^{(j)}(\mathrm{R}(a))$$

$$\chi^{(j)\pm}(\mathrm{S}(a)) = \chi^{(j)\pm}(\mathrm{R}(a+\pi)\mathrm{I}) = \pm\chi^{(j)}(\mathrm{R}(a+\pi))$$

where $\mathrm{R}(a)$ is a proper rotation and $\mathrm{S}(a)$ a mirror rotation (see section 9.1), and $\chi^{(j)}(\mathrm{R}(a))$ is given by equation (7.42).

7.4.4 Multiplication of representations

The direct product of two representations, in the sense of section 4.17, will itself be a representation which in general will reduce

$$D^{(j_1)} \times D^{(j_2)} = \dot{\sum_J} c_J D^{(J)}$$

into a sum of irreducible representations. It is the purpose of this section to deduce the coefficients c_J. As with the finite groups, we study the corresponding relation for characters

$$\chi_a^{(j_1)}\chi_a^{(j_2)} = \sum_J c_J \chi_a^{(J)} \tag{7.43}$$

and to be definite we assume $j_1 \geq j_2$. Using equation (7.42), the left-hand side of this equation may be written in the form

$$\frac{\sin(j_1+\frac{1}{2})a\,\sin(j_2+\frac{1}{2})a}{\sin^2\frac{1}{2}a} = \frac{-\cos(j_1+j_2+1)a+\cos(j_1-j_2)a}{2\sin^2\frac{1}{2}a}$$

$$= \frac{2\sin(j_1+j_2+\frac{1}{2})a\,\sin\frac{1}{2}a-\cos(j_1+j_2)a+\cos(j_1-j_2)a}{2\sin^2\frac{1}{2}a}$$

$$= \frac{\sin(j_1+j_2+\frac{1}{2})a}{\sin\frac{1}{2}a}+\frac{\sin(j_1)a\,\sin(j_2)a}{\sin^2\frac{1}{2}a}$$

showing that

$$\chi_a^{(j_1)}\chi_a^{(j_2)} = \chi_a^{(j_1+j_2)}+\chi_a^{(j_1-\frac{1}{2})}\chi_a^{(j_2-\frac{1}{2})}$$

and hence by a repetition of the argument

$$\chi_a^{(j_1)}\chi_a^{(j_2)} = \chi_a^{(j_1+j_2)}+\chi_a^{(j_1+j_2-1)}+\chi_a^{(j_1-1)}\chi_a^{(j_2-1)}$$

and so on. After $2j_2-2$ more steps of this kind and remembering that the character of the identity representation $\mathrm{D}^{(0)}$ is $\chi_a^{(0)}=1$, we have

$$\chi_a^{(j_1)}\chi_a^{(j_2)} = \chi_a^{(j_1+j_2)}+\chi_a^{(j_1+j_2-1)}+\ldots+\chi_a^{(j_1-j_2)}$$

Comparing this result with equation (7.43) shows that the coefficients $c_J=1$ for $(j_1-j_2)\leq J\leq(j_1+j_2)$ and $c_J=0$ otherwise. The condition $j_1 \geq j_2$ clearly caused no loss of generality and we have finally

$$\mathrm{D}^{(j_1)}\times\mathrm{D}^{(j_2)} = \sum_{J=|j_1-j_2|}^{J=(j_1+j_2)} \mathrm{D}^{(J)} \tag{7.44}$$

This result may also be deduced by counting the number of basis vectors in the product representation which transform according to each chosen representation $\mathrm{T}^{(M)}$ of $\mathscr{R}_2$ and using the fact that a representation $\mathrm{D}^{(J)}$ of $\mathscr{R}_3$ contains the $(2J+1)$ vectors with $M=J, J-1, \ldots, -J$ once each. The limitation $|j_1-j_2|\leq J\leq(j_1+j_2)$ is sometimes called a triangular condition because of its similarity to the relations between the lengths j_1, j_2 and J of the three sides of a triangle.

In the language of section 4.17 the group $\mathscr{R}_3$ is 'simply reducible', since in the reduction (7.44) no representation $\mathrm{D}^{(J)}$ occurs more than once. It is possible to deduce a closed formula for the Clebsch–Gordan coefficients defined in section 4.17 using nothing more than the matrix elements (7.40) of the infinitesimal operators. The formula is complicated, however, and we do not reproduce it here. The phase convention (7.40) ensures that these coefficients are real and extensive tables of values have been published in Rotenberg *et al.*, 1959 (see bibliography). An elementary method for calculating the coefficients is explained in problem 7.8. Further details of the Clebsch–Gordan and related coefficients may be found in Brink and Satchler (1968).

The reduction (7.44) leads to explicit relations for products of the matrix elements of the representations $D^{(j)}$. Since it is the Clebsch–Gordan coefficients which actually transform the basis from the product form labelled by $m_1 m_2$ on the left of equation (7.44) to the coupled form labelled by JM on the right, it follows that

$$\sum_{m_1 m'_1 (m_2 m'_2)} C(j_1 j_2 J, m_1 m_2 M) C(j_1 j_2 J', m'_1 m'_2 M') D^{(j_1)}_{m'_1 m_1} D^{(j_2)}_{m'_2 m_2} = \delta_{JJ'} D^{(J)}_{M'M}$$

where the D-matrix elements all refer to the same rotation. (The bracketed variables m_2, m'_2 are determined by m_1, m'_1 through the relations $m_1 + m_2 = M$, $m'_1 + m'_2 = M'$.) Inverting this equation, using the orthogonality of the coefficients, gives

$$D^{(j_1)}_{m'_1 m_1} D^{(j_2)}_{m'_2 m_2} = \sum_J C(j_1 j_2 J, m_1 m_2 M) C(j_1 j_2 J, m'_1 m'_2 M') D^{(J)}_{M'M}$$

These relations must not be confused with the defining relation (4.3) for a representation

$$D^{(j)}_{m'm}(\mathrm{R}(\boldsymbol{c})) = \sum_n D^{(j)}_{m'n}(\mathrm{R}(\boldsymbol{a})) D^{(j)}_{nm}(\mathrm{R}(\boldsymbol{b}))$$

where $\mathrm{R}(\boldsymbol{c}) = \mathrm{R}(\boldsymbol{a})\mathrm{R}(\boldsymbol{b})$.

7.4.5 Examples of basis vectors

As a first example of an irreducible representation $D^{(j)}$ of $\mathscr{R}_3$ we study the three unit vectors $\boldsymbol{e}_x$, $\boldsymbol{e}_y$ and $\boldsymbol{e}_z$ in ordinary three-dimensional space which provide the basis for the representation with $j = 1$. It is clear that these three vectors span an irreducible space of three dimensions and $D^{(1)}$ is the only such representation. This identification may also be made from the character $(2 \cos a + 1)$. To obtain the matrices in the standard form introduced in the previous section it is necessary to take as basis vectors $\boldsymbol{e}_m$ the following combinations

$$\begin{aligned} \boldsymbol{e}_1 &= -(\boldsymbol{e}_x + i\boldsymbol{e}_y)/2^{\frac{1}{2}} \\ \boldsymbol{e}_0 &= \boldsymbol{e}_z \\ \boldsymbol{e}_{-1} &= (\boldsymbol{e}_x - i\boldsymbol{e}_y)/2^{\frac{1}{2}} \end{aligned} \tag{7.45}$$

It is soon verified that $\boldsymbol{e}_m$ transforms according to the representation $T^{(m)}$ of the group $\mathscr{R}_2$ of rotations about the z-axis, see equation (7.13). The three vectors $\boldsymbol{e}_m$ are orthonormal if the scalar product (3.7) is used and the relative phases have been chosen to agree with the Condon and Shortley convention (7.40), as will be seen below. From equation (7.23) we deduce directly that

$$\begin{aligned} &\mathrm{X}_x\boldsymbol{e}_x = 0, &&\mathrm{X}_x\boldsymbol{e}_y = \boldsymbol{e}_z, &&\mathrm{X}_x\boldsymbol{e}_z = -\boldsymbol{e}_y \\ &\mathrm{X}_y\boldsymbol{e}_x = -\boldsymbol{e}_z, &&\mathrm{X}_y\boldsymbol{e}_y = 0, &&\mathrm{X}_y\boldsymbol{e}_z = \boldsymbol{e}_x \\ &\mathrm{X}_z\boldsymbol{e}_x = \boldsymbol{e}_y, &&\mathrm{X}_z\boldsymbol{e}_y = -\boldsymbol{e}_x, &&\mathrm{X}_z\boldsymbol{e}_z = 0 \end{aligned}$$

and hence, using $J_q = iX_q$

$$\begin{aligned} J_z e_1 &= e_1, & J_z e_0 &= 0, & J_z e_{-1} &= -e_{-1} \\ J_+ e_1 &= 0, & J_+ e_0 &= 2^{\frac{1}{2}} e_1, & J_+ e_{-1} &= 2^{\frac{1}{2}} e_0 \\ J_- e_1 &= 2^{\frac{1}{2}} e_0, & J_- e_0 &= 2^{\frac{1}{2}} e_{-1}, & J_- e_{-1} &= 0 \end{aligned}$$

Thus the matrices of the operators J_q in the basis e_m are

$$J_z = \begin{pmatrix} 1 & 0 & 0 \\ 0 & 0 & 0 \\ 0 & 0 & -1 \end{pmatrix}, \quad J_+ = \begin{pmatrix} 0 & \sqrt{2} & 0 \\ 0 & 0 & \sqrt{2} \\ 0 & 0 & 0 \end{pmatrix}, \quad J_- = \begin{pmatrix} 0 & 0 & 0 \\ \sqrt{2} & 0 & 0 \\ 0 & \sqrt{2} & 0 \end{pmatrix} \tag{7.46}$$

which agrees with the general result (7.40) with $j = 1$.

As a second example, the six-dimensional function space of all homogeneous quadratic functions introduced in subsection 3.2.4 is clearly invariant under all rotations, since the degree of any homogeneous polynomial in x, y and z is unchanged by a rotation. It therefore provides the basis for a six-dimensional representation of $\mathscr{R}_3$. The function $x^2 + y^2 + z^2$ is itself invariant and so produces the representation $D^{(0)}$. The remaining five functions in fact form the basis of the representation $D^{(2)}$, as may be shown by deducing the character or explicitly by constructing the functions as follows.

First construct the explicit differential form for the infinitesimal operators using equation (7.21) for X_z and deducing the other two by cyclically permuting the indices,

$$J_z = i\left(y\frac{\partial}{\partial x} - x\frac{\partial}{\partial y} \right) = -i\frac{\partial}{\partial \phi}$$

$$J_y = i\left(x\frac{\partial}{\partial z} - z\frac{\partial}{\partial x} \right)$$

$$J_x = i\left(z\frac{\partial}{\partial y} - y\frac{\partial}{\partial z} \right)$$

and then construct the raising and lowering operators

$$J_+ = J_x + iJ_y = z\left(\frac{\partial}{\partial x} + i\frac{\partial}{\partial y} \right) - (x+iy)\frac{\partial}{\partial z}$$

$$J_- = J_x - iJ_y = -z\left(\frac{\partial}{\partial x} - i\frac{\partial}{\partial y} \right) + (x-iy)\frac{\partial}{\partial z}$$

Using the fact that

$$\left(\frac{\partial}{\partial x} + i\frac{\partial}{\partial y} \right)(x+iy) = 0 \tag{7.47}$$

it follows that the function

$$\psi_2 = (x+iy)^2 = (x^2-y^2)+2ixy = r^2 \exp(2i\phi)$$

has greatest weight with $m = 2$, since $J_+ \psi_2 = 0$ and $J_z \psi_2 = 2\psi_2$. The lowering operator is now used to construct, successively,

$$\psi_1 = \tfrac{1}{2}J_- \psi_2 = -2z(x+iy)$$

$$\psi_0 = (\tfrac{1}{6})^{\frac{1}{2}} J_- \psi_1 = -(\tfrac{2}{3})^{\frac{1}{2}}(x^2+y^2-2z^2)$$

$$\psi_{-1} = 2z(x-iy)$$

$$\psi_{-2} = (x-iy)^2$$

using the relative normalisation constants from equation (7.40). These five functions are linearly independent and together with the invariant $x^2+y^2+z^2$ they span the six-dimensional space of the quadratic functions. Each function ψ_m is a simple multiple of the spherical harmonic $Y_m^{(l)}(\theta, \phi)$ for $l = 2$; in fact $\psi_m = (32\pi/15)^{\frac{1}{2}} r^2 Y_m^{(2)}(\theta, \phi)$.

It is quite convenient to define the spherical harmonics by this procedure. For general values of l the procedure starts from the function $(x+iy)^l$, which has maximum weight because of equation (7.47), and then uses the lowering operator repeatedly. In general one finds, including the correct normalisation and phase,

$$Y_m^{(l)} = (-1)^m \left\{ \frac{(2l+1)(l-m)!}{4\pi(l+m)!} \right\}^{\frac{1}{2}} \exp(im\phi) \sin^m \theta \left(\frac{d}{d\cos\theta} \right)^{l+m} \times (\cos^2\theta - 1)^l / 2^l l! \quad (7.48)$$

for positive m, together with $Y_{-m}^{(l)} = (-1)^m Y_m^{(l)*}$. In particular, $Y_0^{(l)} = \left(\frac{2l+1}{4\pi}\right)^{\frac{1}{2}} P_l(\cos\theta)$, where P_l is a Legendre polynomial. One further property of the spherical harmonics is that the products $\psi(\boldsymbol{r}) = r^l Y_m^{(l)}(\theta, \phi)$ are solutions of Laplace's equation $\nabla^2 \psi = 0$ for any integer l, as we shall show in section 7.5. In fact, this is often taken as the definition of the spherical harmonics.

It is worth recording the explicit transformation properties of the spherical harmonics under rotation. If θ, ϕ are the coordinates of a point $\boldsymbol{r}$ and θ', ϕ' are the coordinates of the point $R^{-1}\boldsymbol{r}$, then from the general definitions (4.2) and (4.8) for the transformation of functions we have

$$D(R)\, Y_m^{(l)}(\theta, \phi) = Y_m^{(l)}(\theta', \phi') = \sum_{m'} D_{m'm}^{(l)}(R)\, Y_{m'}^{(l)}(\theta, \phi)$$

This shows how the spherical harmonic of the rotated coordinates θ', ϕ' is expressed as a sum of spherical harmonics of the original coordinates. The inverse relation is given by

$$Y_m^{(l)}(\theta, \phi) = \sum_{m'} D_{m'm}^{(l)}(\mathrm{R}^{-1})\, Y_{m'}^{(l)}(\theta', \phi') = \sum_{m'} D_{m'm}^{(l)*}(\mathrm{R})\, Y_{m'}^{(l)}(\theta', \phi')$$

making use of the unitarity of the representation matrix.The coordinates θ', ϕ' may be regarded as the coordinates of $\boldsymbol{r}$ with respect to a new set of axes obtained by the rotation R on the original set.

As an example of the expansion (7.41) of a function into irreducible components we may expand any single-valued function of position of a single particle as

$$f(\boldsymbol{r}) = \sum_{l=0}^{\infty} \sum_{m=-l}^{l} f_{lm}(r)\, Y_m^{(l)}(\theta, \phi) \tag{7.49}$$

The orthogonality of the spherical harmonics is ensured by the general result (4.38), since both labels l and m refer to irreducible representations of the groups $\mathscr{R}_3$ and $\mathscr{R}_2$, respectively, i.e.

$$\int_{\phi=0}^{2\pi} \int_{\theta=0}^{\pi} Y_{m'}^{(l)'*}(\theta, \phi)\, Y_m^{(l)}(\theta, \phi) \sin\theta \mathrm{d}\theta \mathrm{d}\phi = \delta_{l'l}\delta_{m'm} \tag{7.50}$$

A normalisation constant is included in the definition of $Y_m^{(l)}$ to achieve the normalisation in equation (7.50). The procedure given above for generating the $Y_m^{(l)}$ ensures that the relative phases of the $Y_m^{(l)}$ for fixed l and different m is consistent with the phase convention in (7.40).

7.4.6 Irreducible sets of operators and the Wigner–Eckart theorem

Following the general definition given in section 4.20, an irreducible set of operators in $\mathscr{R}_3$ is a set $\mathrm{S}_p^{(k)}$ with fixed $k = 0, \frac{1}{2}, 1, \ldots$, etc., and $p = k, k-1, \ldots, -k$ which transform according to the irreducible representation $\mathrm{D}^{(k)}$. The number k is called the degree or rank of the set. In section 7.2 it was shown that such a set must satisfy the commutation relations

$$[\mathrm{J}_q, \mathrm{S}_p^{(k)}] = \sum_r (\mathrm{J}_q)_{rp}^{(k)}\, \mathrm{S}_r^{(k)} \tag{7.51}$$

with the infinitesimal operators J_q of $\mathscr{R}_3$, where $(\mathrm{J}_q)_{rp}^{(k)}$ denotes the matrix elements of J_q in the representation $\mathrm{D}^{(k)}$. Taking the matrix elements from equations (7.39) and (7.40) these defining relations for an irreducible set become

$$\begin{aligned} &[\mathrm{J}_z, \mathrm{S}_p^{(k)}] = p\mathrm{S}_p^{(k)} \\ &[\mathrm{J}_\pm, \mathrm{S}_p^{(k)}] = \{(k \pm p + 1)(k \mp p)\}^{\frac{1}{2}} \mathrm{S}_{p\pm1}^{(k)} \end{aligned} \tag{7.52}$$

The term 'tensor operator', introduced by Racah, is sometimes used for an irreducible set of operators. Irreducible sets may be multiplied together and reduced in the general manner described in subsection 7.4.4 using Clebsch–Gordan coefficients.

As we have remarked before, in subsection 3.8.3, multiplication by a function may be regarded as an operator in a function space and in this sense the spherical harmonics $Y_p^{(k)}$ form an irreducible set with $p = k, k-1, \ldots, -k$.

The infinitesimal operators themselves form an irreducible set with $k = 1$, called a 'vector operator', as may be seen by comparing their commutation relations (7.28) and (7.30) with equation (7.52). In fact, to satisfy the standard equation (7.52), which incorporates the Condon and Shortley phase convention, the irreducible set is given by

$$\mathrm{J}_1^{(1)} = -\mathrm{J}_+/2^{\frac{1}{2}}, \quad \mathrm{J}_0^{(1)} = \mathrm{J}_z, \quad \mathrm{J}_{-1}^{(1)} = \mathrm{J}_-/2^{\frac{1}{2}}$$

as may soon be verified. The overall magnitude of the operators cannot be determined from the relations (7.52) and we have arbitrarily fixed $\mathrm{J}_0^{(1)} = \mathrm{J}_z$. The signs and magnitudes of $\mathrm{J}_{\pm 1}^{(1)}$ are then fixed from (7.52). These 'spherical' components $\mathrm{J}_m^{(1)}$ of $\mathbf{J}$, as they are sometimes called, are analogous to the basis vectors $\boldsymbol{e}_m$ introduced in subsection 7.4.5.

The matrix elements of members of an irreducible set of operators are related through the Wigner–Eckart theorem (4.62). Since the group $\mathscr{R}_3$ is simply reducible, the label t in equation (4.62) is unnecessary and, since the standard Clebsch–Gordan coefficients for $\mathscr{R}_3$ are real, the theorem is conventionally written in the form

$$\langle jm|\mathrm{S}_q^{(k)}|j'm'\rangle = C(j'kj, m'qm)(-1)^{2k}\langle j||\mathrm{S}^{(k)}||j'\rangle \tag{7.53}$$

(Comparison with equation (4.62) shows that the order of some of the labels in the coefficient C has been changed and a factor $(-1)^{2k}$ has been included. These are not significant changes since they only amount to a slightly different definition of the reduced matrix element. We have followed the convention used by Brink and Satchler, 1968 (see bibliography) and note that this differs from that originally used by Racah.)

7.4.7. Equivalent operators

Use of the Wigner–Eckart theorem (7.53) in practical problems demands a knowledge of the vector-coupling (or Clebsch–Gordan) coefficients. Closed formulae, tables and computer programs for these coefficients are available but in a number of simple situations they may be avoided by using what is known as the 'equivalent operator' method. Essentially, what is done in this method is to use the Wigner–Eckart theorem twice, once for the operator of interest and once for an equivalent operator with the same rotational transformation properties but which is easily evaluated. The coupling coefficient then cancels out in the ratio of matrix elements of the two operators. To put this another way, the equivalent operator is really being used to calculate the coupling coefficient.

For an example consider a vector operator V_q, i.e. a set of three operators which transform according to the irreducible representation $D^{(1)}$. The total angular momentum J_q is also a vector operator. Thus from the Wigner–Eckart theorem (7.53) we have

$$\langle JM | V_q | JM' \rangle = C(J1J, M'qM) \langle J \| \mathbf{V} \| J \rangle$$

and

$$\langle JM | J_q | JM' \rangle = C(J1\,J, M'qM) \langle J \| \mathbf{J} \| J \rangle$$

so that

$$\langle JM | V_q | JM' \rangle = \langle JM | J_q | JM' \rangle \left\{ \frac{\langle J \| \mathbf{V} \| J \rangle}{\langle J \| \mathbf{J} \| J \rangle} \right\} \tag{7.54}$$

showing that the dependence of the matrix element of V_q on M, M' and q is the same as that of the matrix element of J_q which is known from equations (7.39) and (7.40). The proportionality factor is the only thing which depends on the detailed form of V_q. This technique may be used in any group and the equivalent operator is usually constructed from the infinitesimal operators of the group as in the example above.

The method of equivalent operators is usually of no help in describing matrix elements between functions from different representations. In the example (7.54) the equivalent operator would vanish between J and $J' \neq J$ although an arbitrary vector operator, like a position vector $\boldsymbol{r}$, will generally be non-zero.

7.5. The Casimir operator

In the group $\mathscr{R}_2$ a function transforming according to an irreducible representation $T^{(m)}$ could be immediately identified by showing it to be an eigenfunction of the single infinitesimal operator J_z with eigenvalue m. In the group $\mathscr{R}_3$ a function $\psi(jm)$ which transforms according to the row m of the representation $D^{(j)}$ may similarly be identified if we introduced the quadratic invariant operator

$$\mathbf{J}^2 = J_x^2 + J_y^2 + J_z^2 \tag{7.55}$$

Since $\mathbf{J}^2$ is an invariant, as may be verified by showing that $[\mathbf{J}^2, J_q] = 0$ for $q = x, y$ and z, it must be a multiple of the unit operator within an irreducible representation. In other words all $\psi(jm)$ for fixed j will be eigenfunctions of $\mathbf{J}^2$ with the same eigenvalue, independent of m. In fact it was shown in equation (7.36) that for any m,

$$\mathbf{J}^2 \psi(jm) = j(j+1)\,\psi(jm) \tag{7.56}$$

In this way a function belonging to $D^{(j)}$ may be identified by showing that it is an eigenfunction of $\mathbf{J}^2$ with eigenvalue $j(j+1)$.

This general result for the operator $\mathbf{J}^2$ may be verified in the two examples introduced in subsection 7.4.5 where explicit forms were deduced for the J_q. In the first example of the three-dimensional vector space we have, using the form (7.23),

$$\mathrm{J}_x^2 \boldsymbol{r} = -\mathrm{X}_x^2 \boldsymbol{r} = -\boldsymbol{e}_x \wedge (\boldsymbol{e}_x \wedge \boldsymbol{r}) = y\boldsymbol{e}_y + z\boldsymbol{e}_z$$

so that $\mathbf{J}^2\boldsymbol{r} = 2\boldsymbol{r}$. In other words $\mathbf{J}^2$ is just twice the unit operator, agreeing with the result $j(j+1)$ for $j = 1$.

In a space of functions of a single particle the form of $\mathbf{J}^2$ follows from the explicit differential forms given in subsection 7.4.5 for the J_q and, after some algebraic manipulation, one finds that

$$\mathbf{J}^2 = -r^2\left\{ \nabla^2 - \left(\frac{\partial^2}{\partial r^2} + \frac{2}{r}\frac{\partial}{\partial r} \right) \right\} \tag{7.57}$$

where $\nabla^2 = \partial^2/\partial x^2 + \partial^2/\partial y^2 + \partial^2/\partial z^2$. We have seen in subsection 7.4.5 that the spherical harmonics $Y_m^{(l)}$ are basis vectors for the representation $\mathrm{D}^{(l)}$ so that $\mathbf{J}^2 Y_m^{(l)} = l(l+1)\, Y_m^{(l)}$ and hence

$$r^2\left\{ \nabla^2 - \left(\frac{\partial^2}{\partial r^2} + \frac{2}{r}\frac{\partial}{\partial r} \right) \right\} Y_m^{(l)}(\theta, \phi) = -l(l+1)\, Y_m^{(l)}(\theta, \phi) \tag{7.58}$$

from which it follows that

$$\nabla^2 r^l Y_m^{(l)}(\theta, \phi) = 0 \tag{7.59}$$

In other words, we have shown that the functions $r^l Y_m^{(l)}(\theta, \phi)$ are solutions of Laplace's equation. Indeed, the spherical harmonics are often defined in this way but we defined them in subsection 7.4.5 as basis vectors of $\mathrm{D}^{(l)}$. The significance of the spherical harmonics in the solution of the Schrodinger equation for a single particle in a spherically symmetric potential $V(r)$ now becomes apparent. The Schrodinger equation (5.4) has the form

$$\left(\frac{-\hbar^2}{2M} \nabla^2 + V(r) - E \right) \psi(\boldsymbol{r}) = 0$$

and if we write $\psi(\boldsymbol{r}) = u_{nl}(r)\, Y_m^{(l)}(\theta, \phi)$ then it follows from equation (7.58) that the radial wave function $u_{nl}(r)$ must satisfy the eigenvalue equation

$$\left\{ \frac{-\hbar^2}{2M}\left(\frac{\partial^2}{\partial r^2} + \frac{2}{r}\frac{\partial}{\partial r} + \frac{l(l+1)}{r^2} \right) + V(r) - E_{nl} \right\} u_{nl}(r) = 0 \tag{7.60}$$

where n is the radial quantum number which distinguishes the different eigenvalues E_{nl} for the same l.

The operator $\mathbf{J}^2$ is called a 'Casimir operator'. For every Lie group it is possible to construct a scalar quadratic in the infinitesimal operators which is called the Casimir operator—see Racah (1950) in the bibliography. In fact it may be shown that this Casimir operator is given by

$$C = \sum_{p,q} (g^{-1})_{pq} X_p X_q \tag{7.61}$$

where g is the matrix with elements $g_{ij} = \sum_{st} c^s_{it} c^t_{js}$ and c^s_{it} is the structure constant from equation (7.7).

In larger groups than $\mathscr{R}_3$ it is necessary to use more than one number like j to label the irreducible representations. The number of numbers needed for this purpose is called the 'rank' of the group and by going to cubic and higher powers in the operators X_q it is possible to form a set of Casimir operators for the group, so that the numbers labelling the representations may all be related to eigenvalues of the Casimir operators.

7.6 Double-valued representations

With finite groups, a representation $T(G_a)$ associated a unique operator $T(G_a)$ with each group element G_a. For a continuous group, the representation $T(\boldsymbol{a})$ was a continuous function of the group parameters $\boldsymbol{a}$. Hence the existence of many-valued functions admits the possibility of many-valued representations of continuous groups. As an example, the choice $m = \frac{1}{2}$ in the representation $T^{(m)}(a) = \exp(-ima)$ of $\mathscr{R}_2$ in subsection 7.3.1 would give $T^{(\frac{1}{2})}(0) = 1$ and $T^{(\frac{1}{2})}(2\pi) = -1$ but since a rotation of 2π is the *same* element of $\mathscr{R}_2$ as a rotation of zero, the representation is double-valued, associating both ± 1 with this element. The choice $m = \frac{1}{3}$ would give a triple-valued representation. In the same way, the expression (7.42) for the character of the representation $D^{(j)}$ of $\mathscr{R}_3$ shows it to be single-valued for integer j and double-valued for half-integer j.

From a physical point of view, the many-valued representations cannot occur in the solution of the Schrodinger equation, see section 7.5, but they are not excluded by the general formulation of quantum mechanics. In fact, the representation $D^{(\frac{1}{2})}$ of $\mathscr{R}_3$ is needed to describe the intrinsic spin of the electron, see section 8.4. Let us consider the general formulation for a moment. If, by definition, the state $T(\boldsymbol{a})\psi$ is to represent the system ψ transformed by $G(\boldsymbol{a})$, then if $G(\boldsymbol{a})\,G(\boldsymbol{b}) = G(\boldsymbol{c})$ it follows that the system transformed by $G(\boldsymbol{c})$ must be represented both by $T(\boldsymbol{c})\psi$ and by $T(\boldsymbol{a})\,T(\boldsymbol{b})\psi$. Since the phase of a state has no physical meaning it follows only that $T(\boldsymbol{a})\,T(\boldsymbol{b}) = \omega(\boldsymbol{a}, \boldsymbol{b})\,T(\boldsymbol{c})$ for some phase ω. This relation is more general than our first definition (4.1) of a representation, being satisfied by the many-valued representations, whereas equation (4.1) is not. Notice that the particular definition (4.8) of the representation operators $T(\boldsymbol{a})$ in a space of single-valued functions led to $\omega = 1$ and so would exclude the many-valued representations.

A convenient mathematical way of describing the many-valued representations is to extend the group by adding new elements so that the representations are single-valued representations of the extended group. (The term 'universal covering group' is used for such an extended group.) In the case of the double-valued representations of the group $\mathscr{R}_3$ the extended group (often called the 'double group') can be described in the following way. A rotation through 2π is denoted by the symbol $\bar{E}$, different from the identity E, but with $\bar{E}^2 = E$. From the relation (2.12) we see that $\bar{E}R(\boldsymbol{a})\bar{E} = R(\boldsymbol{a})$ so that $\bar{E}$ commutes with all rotations and also that $R(\boldsymbol{a})\bar{E}R(\boldsymbol{a})^{-1} = \bar{E}$ so that the new element $\bar{E}$ represents a rotation of 2π about *any* axis. The extended group is obtained by including the new elements $\bar{E}R(\boldsymbol{a})$ with the original elements $R(\boldsymbol{a})$ with $|\boldsymbol{a}| \leq \pi$. In the multiplication of two elements the new element $\bar{E}$ appears wherever the resultant angle of rotation exceeds 2π. This geometrical device for visualising the double group is rather dubious because of the difficulty of distinguishing E from $\bar{E}$ geometrically, but it may be put on a rigorous algebraic footing by identifying the double group with a unitary group in two dimensions. This will be done in subsection 18.13.3 of volume 2. (The use of a double group to avoid double-valuedness is similar to the idea in the theory of complex numbers z, where a double-valued function like $z^{\frac{1}{2}}$ is regarded as single-valued on a Riemann surface with two sheets.)

The fact that $\mathscr{R}_3$ has only single- and double-valued representations while $\mathscr{R}_2$ has n-valued representations for any n could have been predicted from a careful study of the relation between the group elements and the parameter space. We shall give only a very brief discussion of this question which is known as the connectivity property of the group. We are concerned with three objects, the parameter $\boldsymbol{a}$, the group element $G(\boldsymbol{a})$ and the representation function $T(\boldsymbol{a})$. The successive multiplication of infinitesimal group elements corresponds to a path in the space of the parameter $\boldsymbol{a}$. If all paths from the identity to a general point $\boldsymbol{a}$ may be obtained from each other by continuous deformation, then $T(\boldsymbol{a})$ must be single-valued because in a many-valued function the values differ by discrete amounts. Consider first the group $\mathscr{R}_2$ which has a single parameter a in the range $0 \leq a \leq 2\pi$, with the identity corresponding to both end points. In addition to the direct path from 0 to a, there are paths which reach to 2π and reappear at 0, see figure 7.2. Such a path is continuous so far as group elements are concerned. With only a one-dimensional parameter space there is no freedom for deformation of paths so that paths of types (i), (ii), (iii), etc. in the figure cannot be deformed into each other. The existence of these distinct types of path admits the possibility of many-valued representations. For the group $\mathscr{R}_3$ the parameter space is the inside and surface of a sphere of radius π (see section 7.4) with opposite ends of a diameter corresponding to the same rotation, i.e. $R_{\boldsymbol{k}}(\pi) = R_{-\boldsymbol{k}}(\pi)$ for any $\boldsymbol{k}$. Starting from the identity, at the origin $\boldsymbol{a} = 0$ of the sphere, a path to a general $\boldsymbol{a}$ may now go direct, as in figure 7.3(i) or it may strike the sphere and reappear at the other end of the diameter, as in figure 7.3(ii). The interesting difference betweeen $\mathscr{R}_2$ and $\mathscr{R}_3$ is that for $\mathscr{R}_3$ any path which strikes the sphere an even number of times may be continuously deformed into one which goes direct. A

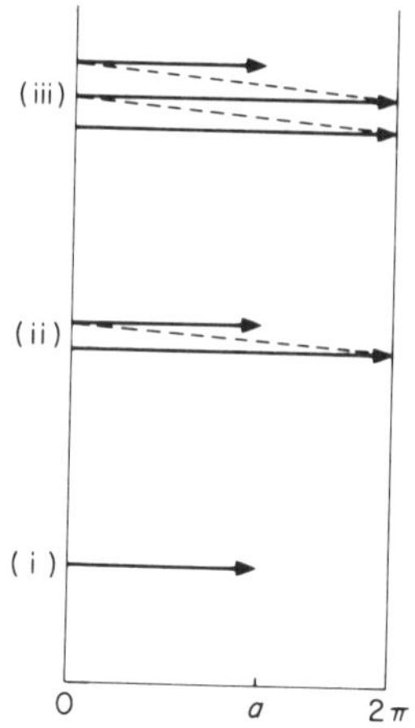

Figure 7.2

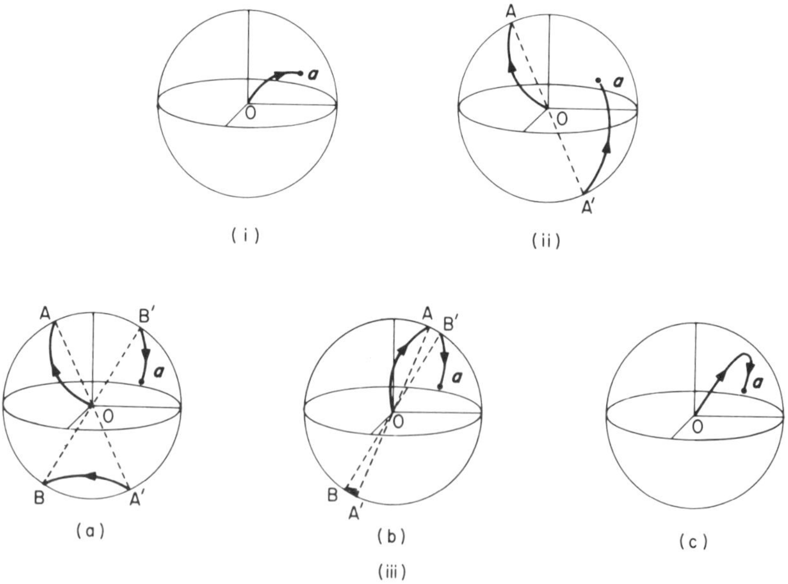

Figure 7.3

path with an odd number of strikes is similarly deformable into one with a single strike, as in figure 7.3(ii). Thus for $\mathscr{R}_3$, only single- and double-valued representations are possible. We illustrate the deformation of a path which strikes the sphere twice in figure 7.3(iii), progressing from figures (a) to (b) to (c). The diameter AA′ is swung around until A reaches B′.

7.7 The complex conjugate representation

In general, if T is a matrix representation then so is the set of complex conjugate matrices T*. However for the group $\mathscr{R}_3$ the representation $D^{(j)*}$ has the same characters as $D^{(j)}$ since the character (7.42) is real and hence these two representations are equivalent. To find the transformation between them we first note that the matrices (7.39) and (7.40) of J_z and $J_\pm$ are real and hence that the matrices of J_z and J_x are real, while that of J_y is pure imaginary. Thus, in passing from $D^{(j)}$ to $D^{(j)*}$ the sign of J_z and J_x must be changed while J_y is unchanged, remembering the factor i in the definition of J_q in subsection 7.4.1. In other words, $J_z \to -J_z$ and $J_\pm \to -J_\mp$. From equations (7.39) and (7.40) we see that this is achieved by using, in place of the basis $\boldsymbol{e}_m$, the basis $(-1)^{j-m}\boldsymbol{e}_{-m}$ (The phase factor is necessary because of the sign change in the raising and lowering operators.) Thus the matrix elements of $D^{(j)*}$ are given by

$$D^{(j)*}_{m'm} = (-1)^{m-m'} D^{(j)}_{-m'-m}$$

This transformation of basis may also be achieved by the transformation matrix $D^{(j)}(R_y(\pi))$ for a rotation of π about the y-axis, since this reverses the x- and z-axes.

Bibliography

For a rigorous treatment of Lie groups one may consult Boerner, 1963 (see chapter 4 bibliography) or

Gilmore, R. (1974). *Lie Groups, Lie Algebras and Some of their Applications* (Wiley, New York)

For details of angular momentum theory and $\mathscr{R}_3$ coupling coefficients see

Brink, D. M. and Satchler, G. R. (1968). *Angular Momentum* (Clarendon Press, Oxford)

Tables of $\mathscr{R}_3$ coupling coefficients have been prepared in

Rotenberg, M., Bivins, R., Metropolis, N. and Wooten, J. K. (1959). *The 3-j and 6-j Symbols* (Technology Press, M.I.T., Cambridge, Mass.)

The following reprint collection contains, among other things, the original papers of Racah on which the algebra of angular momentum is based

Biedenharn, L. C. and van Dam, H. (1965). *Quantum Theory of Angular Momentum* (Academic Press, New York)

The Casimir operator is described in

Racah, G. (1950). *Rend. Accad., Lincei.*, **8,** 108

Problems

7.1 Show from example (2) of subsection 7.3.4 that the functions $x \pm iy$ transform according to the representations $T^{(\pm 1)}$ of $\mathscr{R}_2$. Hence classify the six quadratic functions of x, y and z according to their behaviour under $\mathscr{R}_2$.

7.2 Explain why improper rotations (with determinant equal to -1) cannot form a group by themselves.

7.3 By introducing Cartesian coordinates, write the infinitesimal operator (7.20) for rotations about the z-axis as $X_z = y(\partial/\partial x) - x(\partial/\partial y)$. With corresponding expressions for the other two axes, verify the commutation rules (7.25).

7.4 Show that $[J_+, J_-] = 2J_z$ and $[J^2, J_q] = 0$. (Use equation (7.26).)

7.5 Use equations (7.40) to construct the matrices of J_x, J_y and J_z for $j = \frac{1}{2}$ (the Pauli spin matrices) and the corresponding matrices for $j = 1$—see equation (8.15).

7.6 Find the character of the representation $D^{(2)}$ of $\mathscr{R}_3$ for those rotations which belong to the subgroup D_3 and hence, using the character table (table 4.2), deduce which representations of D_3 occur when the representation $D^{(2)}$ reduces on restriction to the subgroup D_3.

7.7 According to equation (7.44) the product $D^{(\frac{1}{2})} \times D^{(\frac{1}{2})}$ contains $D^{(1)}$ in its reduction. In the notation of equation (4.46) it is trivial that $\Psi_1^{(1)} = \phi_{\frac{1}{2}}^{(\frac{1}{2})} \psi_{\frac{1}{2}}^{(\frac{1}{2})}$ because there is only one product function with $m = 1$. By operating on $\Psi_1^{(1)}$ (using equation (7.40)) with the lowering operator (see equation (7.11)) $J_- = J_-(1) + J_-(2)$ where $J_-(1)$ operates on ϕ and $J_-(2)$ operates on ψ, construct the vector $\Psi_0^{(1)}$ and hence deduce the Clebsch–Gordan coefficients $C(\frac{1}{2}\frac{1}{2}1, \frac{1}{2} - \frac{1}{2}0)$ and $C(\frac{1}{2}\frac{1}{2}1, -\frac{1}{2}\frac{1}{2}0)$ in the notation of equation (4.46).

7.8 Deduce the Clebsch–Gordan coefficients for the construction of the vector with $j = 1$, $m = 0$ from the product $D^{(1)} \times D^{(2)}$ by using the following steps: (i) Write $\Psi_1^{(1)} = \alpha\phi_1^{(1)}\psi_0^{(2)} + \beta\phi_0^{(1)}\psi_1^{(2)} + \gamma\phi_{-1}^{(1)}\psi_2^{(2)}$ and deduce the ratios $\alpha:\beta:\gamma$ by the condition $J_+\Psi_1^{(1)} = 0$. (Note that, from linear dependence, any relation of the kind $a\phi_1^{(1)}\psi_1^{(2)} + b\phi_0^{(1)}\psi_2^{(2)} = 0$ implies that $a = b = 0$.) (ii) Find α, β and γ by imposing the normalisation condition $\alpha^2 + \beta^2 + \gamma^2 = 1$ and the convention that α is positive. (iii) Deduce $\Psi_0^{(1)}$ from $\Psi_1^{(1)}$ by using the lowering operator J_-.

7.9 Use the Clebsch–Gordan coefficients from question 7.7 and the formula at the end of subsection 7.4.4 for coupling D-matrices to deduce the $D^{(1)}$ matrix from the $D^{(\frac{1}{2})}$ matrix given in equation (20.38) (Your result should agree with the general formula (20.40).)

7.10 Use the method of subsection 7.4.5 to construct the spherical harmonics $Y_m^{(3)}$ and check against the general formula (7.48).

7.11 Use the lowering operator to show that the set of functions $-(x+iy)/2^{(\frac{1}{2})}$, z and $(x-iy)/2^{(\frac{1}{2})}$ belong to the representation $D^{(1)}$.

7.12 The functions ψ_m with $m = 2, 1, \ldots, -2$ which were constructed in subsection 7.4.5 to transform according to $D^{(2)}$ form a tensor operator of degree two. Hence their matrix elements in the basis of question 7.11 should all be related to a single reduced matrix element through the Wigner–Eckart theorem (7.53). Verify this for some of the matrix elements by (a) using equation (7.53) with the Clebsch–Gordan coefficients from question 7.8, and (b) carrying out the explicit integrals using the scalar product of subsection 3.2.4, i.e. an integral over the unit sphere.

7.13 Show that $\langle j\|\mathbf{J}\|j\rangle = \{j(j+1)\}^{(\frac{1}{2})}$. (Write $\mathbf{J}^2 = \sum_q \mathrm{J}_q \mathrm{J}_q^{\dagger}$, and evaluate $\langle jm|\mathbf{J}^2|jm\rangle$ by introducing a sum over 'intermediate states' $\sum_{q,m'} \langle jm|\mathrm{J}_q|jm'\rangle\langle jm'|\mathrm{J}_q^{\dagger}|jm\rangle$ and using equation (7.53) together with the normalisation sum over the coupling coefficients.)

7.14 Show that the Clebsch–Gordan coefficients $C(ll0, m-m0)$ have the simple form

$$C(ll0, m-m\,0) = (-1)^{l-m}/(2l+1)^{\frac{1}{2}}$$

(Use the method of question 7.8 by (a) writing $\Psi_0^{(0)} = \sum_m C(ll0, m-m\,0)\,\phi_m^{(l)}\,\psi_{-m}^{(l)}$, (b) making use of $\mathrm{J}_+\Psi_0^{(0)} = 0$ to relate successive coefficients, and (c) using normalisation and the convention that $C(ll0, l-l0) > 0$.) Hence construct an invariant under $\mathscr{R}_3$ from the five spherical harmonics $Y_m^{(2)}$.

7.15 Show that, in the case of the group $\mathscr{R}_3$, the general definition (7.61) of the Casimir operator leads to $\mathrm{G} = -\frac{1}{2}\mathbf{J}^2$.

8

Angular Momentum and the Group $\mathscr{R}_3$ with Illustrations from Atomic Structure

The properties of the group $\mathscr{R}_3$ were investigated in the previous chapter and the relation between the infinitesimal operators and the angular momentum operators was noted. In the present chapter we explore this relation in more detail by studying first one particle and then several particles and by introducing the concept of intrinsic spin. We then move to the realistic problem of atomic structure in which there are many examples of the effects of $\mathscr{R}_3$ symmetry. We consider not only the hydrogen atom but also many-electron atoms.

8.1 Rotational invariance and its consequences

If a Hamiltonian is spherically symmetric, in other words if it has no preferred direction in space, then it has $\mathscr{R}_3$ as its symmetry group. We shall therefore have the various consequences listed generally in chapter 5. Having studied the group $\mathscr{R}_3$ in chapter 7 we are now in a position to explore these consequences.

In the first place, each eigenvalue will carry one of the labels $j = 0, \frac{1}{2}, 1 \ldots$ of the irreducible representations $D^{(j)}$ of $\mathscr{R}_3$. For an energy level E there will be a $(2j+1)$-fold degeneracy with the eigenfunctions forming the basis of the representation $D^{(j)}$. These eigenfunctions may be distinguished from each other by their behaviour with respect to rotations about some chosen axis which we call the z-axis. This corresponds to the reduction (7.38)

$$D^{(j)} = \sum_{m=-j}^{j} T^{(m)} \tag{8.1}$$

of $D^{(j)}$ into irreducible representations $T^{(m)}$ of the subgroup $\mathscr{R}_2$. Thus we may denote the energies by $E(\gamma j)$ and the wave functions by $\psi(\gamma jm)$, where γ denotes all other quantum numbers which have nothing to do with rotational invariance.

As we have seen in the preceding chapter the wave function $\psi(\gamma jm)$ is therefore an eigenfunction both of J_z, the infinitesimal operator for rotations about the z-axis, and of the Casimir operator $\mathbf{J}^2$. From equations (7.39) and (7.36)

$$\begin{aligned} J_z \psi(\gamma jm) &= m\psi(\gamma jm) \\ \mathbf{J}^2 \psi(\gamma jm) &= j(j+1)\psi(\gamma jm) \end{aligned} \tag{8.2}$$

At this stage we have not specified anything about our physical system other than its spherical symmetry so that the explicit form of the operators J_q and their physical significance cannot yet be discussed. Nevertheless, the results (8.2) can be stated.

Because the Hamiltonian is invariant the operators J_x, J_y, J_z and also the Casimir operator $\mathbf{J}^2$ all commute with the Hamiltonian and are therefore conserved in the usual sense, discussed in section 5.5.

If we classify any transition operators according to their behaviour under rotations, then selection rules in j follow immediately from the discussion of section 5.4 using the rule (7.44) for the reduction of product representations in $\mathscr{R}_3$. Thus, for example, if j and j' denote the $\mathscr{R}_3$ labels for initial and final states and if the relevant transition operator transforms like $D^{(k)}$ then given j, the final state j' is restricted by the condition

$$j' = (j+k), (j+k-1), \ldots, |j-k| \tag{8.3}$$

We note from subsection 7.4.5 that a spherical harmonic $Y_q^{(k)}$ transforms like $D^{(k)}$. In particular, the dipole operators described in section 5.4, which transform like the vector representation of $\mathscr{R}_3$, correspond to $k = 1$. The expansion (7.49) of an arbitrary function into spherical harmonics is often called a multipole expansion with the term $Y_q^{(k)}$ being a 2^k pole. (It may be shown that 2^k is the minimum number of point charges needed to produce a potential field which is proportional to a single spherical harmonic $Y_q^{(k)}$ with fixed k.) A transition governed by an operator transforming like $D^{(k)}$ is said to have multipolarity 2^k, i.e. dipole ($k = 1$), quadrupole ($k = 2$), octupole ($k = 3$), etc.

If the group is enlarged from $\mathscr{R}_3$ to O_3 by including the inversion then, since $O_3 = \mathscr{R}_3 \times S_2$, a product group, it follows that the system acquires the properties described above for $\mathscr{R}_3$ together with those of S_2 described in section 5.6. In other words, the states are labelled by jm and the parity $\pm$. In particular, from their manner of construction in subsection 7.4.5, one sees that the spherical harmonics $Y_q^{(k)}$ have parity $(-1)^k$.

8.2 Orbital angular momentum of a system of particles

In quantum mechanics the orbital angular momentum of a single particle is obtained from the classical expression $\mathbf{r} \wedge \mathbf{p}$ by the substitution $\mathbf{p} \rightarrow -i\hbar\nabla$. Thus, in units of $\hbar$,

$$l_z = -i(\mathbf{r} \wedge \nabla)_z = -i\left(x\frac{\partial}{\partial y} - y\frac{\partial}{\partial x}\right) = -i\frac{\partial}{\partial \phi} \tag{8.4}$$

where ϕ is the polar coordinate with respect to the z-axis. But we have seen that the explicit form for the infinitesimal rotation operator in the space of functions of a single particle is, from equation (7.20),

$$\mathrm{J}_z = -i\frac{\partial}{\partial \phi} = l_z \tag{8.5}$$

Thus for a single particle the infinitesimal operators are identical with the angular-momentum operators in units of $\hbar$ (Since the z-axis has no real significance this identification is equally valid for all three components.)

For the reasons discussed in subsection 7.3.1, namely that they are not continuous functions of angle over the range 2π, the wave functions with half-integer angular momentum are not acceptable solutions.

On a question of notation, we have previously denoted the angular momentum operators by J_q and the eigenvalues by the small letters jm. Since we are now about to consider a system of particles it will be convenient to use small letters consistently for a single particle, i.e. for both operators and eigenvalues, and to use capitals for systems of more than one particle. In addition we use L (or l) for the orbital angular momentum and S (or s) for the intrinsic spin, to be described in section 8.4. The symbols J (or j) are reserved for the total angular momentum, orbital plus spin.

Consider now a system of n particles. The total orbital angular momentum of the system, denoted by $\mathbf{L}$ is given by

$$\mathbf{L} = \sum_{i=1}^{n} \mathbf{l}(i) = \sum_i [\mathbf{r}(i) \wedge \mathbf{p}(i)]/\hbar$$

so that in particular

$$L_z = -i \sum_i \frac{\partial}{\partial \phi_i} \tag{8.6}$$

as in equation (8.4), where ϕ_i is the usual polar angle of the ith particle. To connect this operator with rotations we follow the same argument which led to equation (7.20). Denoting any wave function for the system by $\Psi(r_1\theta_1\phi_1, r_2\theta_2\phi_2, \ldots, r_n\theta_n\phi_n)$ the immediate generalisation of equation (7.17) is that a rotation of the entire system through an angle a about the z-axis leaves all r_i and θ_i unchanged and replaces ϕ_i by $\phi_i - a$. Thus, omitting the labels r_i and θ_i for brevity,

$$T(R_z(a))\Psi(\phi_1, \phi_2, \ldots, \phi_n) = \Psi(\phi_1 - a, \phi_2 - a, \ldots, \phi_n - a)$$

For small a we expand the right-hand side of this equation using the Taylor series for a function of n variables. Keeping only the first two terms

$$T(R_z(a))\Psi(\phi_1, \phi_2, \ldots, \phi_n)$$

$$\approx \Psi(\phi_1, \phi_2, \ldots, \phi_n) - \sum_i a\frac{\partial}{\partial \phi_i}\Psi(\phi_1, \phi_2, \ldots, \phi_n)$$

so that we may write

$$R_z(a) \approx 1 - a\sum_i \frac{\partial}{\partial \phi_i} \tag{8.7}$$

Thus the explicit form of the infinitesimal rotation operator for a system of particles is

$$J_z = -i \sum_i \frac{\partial}{\partial \phi_i} = L_z$$

Comparing with equation (8.6) shows that for a system of particles the total angular momentum operator, in units of $\hbar$, is identical with the infinitesimal rotation operator for the system. This is exactly the same conclusion that we reached for a single particle.

8.3 Coupling of angular momenta

Now consider a two-particle wave function which is the product of two single-particle wave functions,

$$\Psi(\boldsymbol{r}_1, \boldsymbol{r}_2) = \psi_{l_1 m_1}(\boldsymbol{r}_1)\psi_{l_2 m_2}(\boldsymbol{r}_2) \tag{8.8}$$

where the suffices denote the single-particle angular momenta and their z-

components. Under rotation R($\boldsymbol{a}$), the induced transformation is

$$\Psi' = \mathrm{T}(\mathrm{R}(\boldsymbol{a}))\Psi = \psi'_{l_1m_1}\psi'_{l_2m_2} = \sum_{m'_1m'_2} D^{(l_1)}_{m'_1m_1}(\boldsymbol{a})D^{(l_2)}_{m'_2m_2}(\boldsymbol{a})\psi_{l_1m'_1}\psi_{l_2m'_2}$$

showing that the set of $(2l_1+1)(2l_2+1)$ product functions (8.8) with fixed l_1 and l_2 span the product representation $\mathrm{D}^{(l_1)}\times\mathrm{D}^{(l_2)}$ of the group $\mathscr{R}_3$. The reduction

$$\mathrm{D}^{(l_1)}\times\mathrm{D}^{(l_2)} = \sum_{|l_1-l_2|}^{(l_1+l_2)} \mathrm{D}^{(L)} \tag{8.9}$$

of the product representation into its irreducible constituents was deduced in subsection 7.4.4. But since the total angular momentum of the pair is associated with overall rotations of the pair it follows that the total angular momentum of the pair may take the values

$$L = (l_1+l_2), (l_1+l_2-1)\ldots|l_1-l_2| \tag{8.10}$$

which is known as the vector-coupling rule for angular momenta. Classically, an angular momentum is a vector and the sum of two angular momenta $\mathbf{l}_1$ and $\mathbf{l}_2$ is given by the usual vector addition $L = \mathbf{l}_1 + \mathbf{l}_2$ so that, depending on the angle between $\mathbf{l}_1$ and $\mathbf{l}_2$, the magnitude of L lies between the sum $|\mathbf{l}_1|+|\mathbf{l}_2|$ and the difference $||\mathbf{l}_1|-|\mathbf{l}_2||$ of the magnitudes of $\mathbf{l}_1$ and $\mathbf{l}_2$. This is the origin of the name 'vector coupling rule' for equation (8.10) which would approach the classical result in the limit of large angular momenta and small $\hbar$.

For rotations about the z-axis the simple additive property of the representations of $\mathscr{R}_2$ (see subsection 7.3.3)

$$\mathrm{T}^{(m_1)}\times\mathrm{T}^{(m_2)} = \mathrm{T}^{(m_1+m_2)}$$

means that the z-component of total angular momentum is just the sum of the z-components for the two particles. The eigenfunctions of total angular momentum will be those combinations of the product functions (8.8) which achieve the reduction (8.9) and the Clebsch–Gordan coefficients were defined (see subsection 7.4.4) to do just this. We write the eigenfunctions as

$$\Psi_{LM}(\boldsymbol{r}_1,\boldsymbol{r}_2) = \sum_{\substack{m_1\\(m_2=M-m_1)}} C(l_1l_2L, m_1m_2M)\psi_{l_1m_1}(\boldsymbol{r}_1)\psi_{l_2m_2}(\boldsymbol{r}_2) \tag{8.11}$$

with the properties

$$\mathbf{L}^2\Psi_{LM} = L(L+1)\Psi_{LM}$$
$$\mathrm{L}_z\Psi_{LM} = M\,\Psi_{LM}$$

By adding the angular momentum of one particle at a time, the rule (8.10) may be used to deduce the total angular momentum of any number of particles. Because of the Pauli principle some of the states of total angular momentum may not be physically acceptable but this point will be discussed in subsection 8.6.4 after the concept of spin has been introduced.

8.4 Intrinsic spin

Physical systems are generally regarded as being built from particles, like electrons, protons, neutrons, etc. We assign to each point particle a finite mass and a finite charge. Experimentally one finds that such a picture is incomplete and in particular it would fail to explain the behaviour of the energy levels of the hydrogen atom in the presence of a weak magnetic field. One is led to the conclusion, as we shall see below, that the point particles should be assigned an intrinsic angular momentum (or spin) and consequently an intrinsic magnetic moment. It is supposed that the intrinsic spin and moment are fundamental properties of the particles like their mass and charge. Even at rest, a particle will still have a spin and moment. Furthermore, a particle, like the neutron, with no charge may (and does) have a magnetic moment which may be imagined to arise from charge currents within the particle which have a net moment but no net charge. Most 'elementary' particles like the electron, proton and neutron are found experimentally to have a spin of $\frac{1}{2}\hbar$ but in contrast to this the ρ-meson has a spin of $\hbar$ and the Ω^- particle has a spin of $\frac{3}{2}\hbar$. The appearance of an angular momentum in half-integer units is something which we did not encounter in describing the orbital motion of particles but it is not completely unexpected. Remember the link established in section 8.2 between the angular momentum and the rotation operators and remember that in section 7.4 we found that the irreducible representations $\mathrm{D}^{(J)}$ of the rotation group were labelled by integer or half-integer values of J. The reason for rejecting the half-integer functions in the description of the orbital motion of a particle does not apply to the spin variable. The orbital motion was described by the Schrodinger differential equation, whose solution must be a continuous function. A half-integer wave function would change by a factor -1 as the angle goes through a full revolution of 2π and so lack continuity. The spin degree of freedom is not described by the Schrodinger equation but, as we shall see in chapter 15 of volume 2, arises naturally from a relativistic treatment. In the solution of these equations the occurrence of half-integer spin brings no conflict. A change of sign in the wave function does not affect the probability $|\Psi|^2$.

To clarify the need, from the point of view of the observed data, for introducing spin let us first consider the effect of an external uniform magnetic field on the energy levels of an electron moving in a spherically symmetric potential. In particular, consider the $(2l+1)$-fold degenerate level E_l with angular momentum l and a set of wave functions $\psi(lm)$ with

$m = l, l-1, \ldots, -l$. The additional term in the Hamiltonian due to a constant magnetic field of strength B is

$$\frac{e\hbar B l_z}{2M_e c} \tag{8.12}$$

where the z-axis has been chosen in the direction of the magnetic field, $-e$ and M_e are the charge and mass of the electron and c is the velocity of light. The operator $-e\hbar l_z/2M_e c$ is called the 'magnetic moment' operator, while the coefficient of Bl_z is called the Bohr magneton and is denoted by

$$\mu_B = e\hbar/2M_e c \tag{8.13}$$

Notice that the contribution (8.12) to the Hamiltonian is proportional to the angular momentum of the particle and the strength of the field. The term (8.12) commutes with the original Hamiltonian $\mathrm{T}+\mathrm{V}(r)$ and its effect on the spectrum is simply to split the degeneracy. The $\psi(lm)$ will be eigenfunctions of the final Hamiltonian with energies $E_{nl} + \mu_B Bm$. This splitting of the levels by a magnetic field is called the Zeeman effect. Group theoretically, the symmetry of the Hamiltonian has been reduced from $\mathscr{R}_3$ to $\mathscr{R}_2$ so that the $(2l+1)$-fold degeneracy splits into $(2l+1)$ levels labelled by the irreducible representations $\mathrm{T}^{(m)}$ of $\mathscr{R}_2$.

The ground state of the hydrogen atom and of almost any attractive spherically symmetric potential has $l = 0$ and is therefore non-degenerate and cannot be split. However it is found experimentally to split into two levels. This immediately suggests that there is another degree of freedom in addition to the position coordinates and that, at least for the ground state, this extra degree of freedom only allows two possible states. Furthermore the interaction with the magnetic field suggests that the splitting is due to an additional source of angular momentum. The fact that the $l = 0$ state splits into two levels suggests a spin of $\frac{1}{2}\hbar$ since $(2J+1) = 2$ for $J = \frac{1}{2}$. It is usual to denote the spin angular momentum of a particle by the symbol s. One soon verifies that the splitting of the excited states with $l \neq 0$ is also consistent with this hypothesis of $s = \frac{1}{2}\hbar$ for the electron. Let us now see how to express these ideas precisely in mathematical form.

To any particle of a given type, e.g. electrons, we assign a spin s and suppose that the possible spin states of the particle are spanned by a set of $(2s+1)$ basis states $\chi^{(s)}_{m_s}$ with $m_s = s, s-1, \ldots, -s$. Under rotations we suppose that the states transform according to the irreducible representation $\mathrm{D}^{(s)}$ of $\mathscr{R}_3$:

$$(\chi^{(s)}_{m_s})' = \mathrm{T}(\mathrm{R}(\boldsymbol{a}))\chi^{(s)}_{m_s} = \sum_{m_s'} D^{(s)}_{m_s' m_s}(\boldsymbol{a})\,\chi^{(s)}_{m_s'} \tag{8.14}$$

where s may be an integer or a half-integer. In other words we suppose that they behave like angular momentum eigenfunctions. Since the space of the spin states is of finite dimension $(2s+1)$, the explicit form for the infinitesimal operators for rotations of the spin states are just the matrices of the infinitesimal operators deduced generally in subsection 7.4.2. Denoting these operators by s_x, s_y and s_z we have from equation (7.40) for particular values of s,

$$s = \tfrac{1}{2} \qquad s_x = \tfrac{1}{2}\begin{pmatrix} 0 & 1 \\ 1 & 0 \end{pmatrix}, \; s_y = \tfrac{1}{2}i\begin{pmatrix} 0 & -1 \\ 1 & 0 \end{pmatrix}, \; s_z = \tfrac{1}{2}\begin{pmatrix} 1 & 0 \\ 0 & -1 \end{pmatrix}$$

$$s = 1 \qquad s_x = \tfrac{1}{2}\begin{pmatrix} 0 & \sqrt{2} & 0 \\ \sqrt{2} & 0 & \sqrt{2} \\ 0 & \sqrt{2} & 0 \end{pmatrix}, \; s_y = \tfrac{1}{2}i\begin{pmatrix} 0 & -\sqrt{2} & 0 \\ \sqrt{2} & 0 & -\sqrt{2} \\ 0 & \sqrt{2} & 0 \end{pmatrix},$$

$$s_z = \begin{pmatrix} 1 & 0 & 0 \\ 0 & 0 & 0 \\ 0 & 0 & -1 \end{pmatrix} \tag{8.15}$$

For $s = \frac{1}{2}$, it is common practice to define $\boldsymbol{\sigma} = 2\mathbf{s}$ and to refer to the three operators $\boldsymbol{\sigma}$ as the Pauli spin matrices. From equations (7.36) and (7.39) the basis states are eigenstates of s_z and $\mathbf{s}^2$,

$$s_z \chi_{m_s}^{(s)} = m_s \chi_{m_s}^{(s)}$$

$$\mathbf{s}^2 \chi_{m_s}^{(s)} = s(s+1)\chi_{m_s}^{(s)}$$

We speak of s_z as the z-component of the spin and $\mathbf{s}^2$ as the square of the total spin. As with angular momentum generally, one may take linear combinations of the basis $\chi_{m_s}^{(s)}$ to form a new basis in which any other chosen component of the spin is diagonal.

For a system of particles, the total spin operators $S_q = \sum_i s_q(i)$ may be defined in a way precisely parallel to the introduction of the total orbital angular momentum L_q in section 8.2. The coupling of spins similarly follows the development in section 8.3.

The complete wave function for a particle with spin s will need to describe both the orbital state and the spin state and may generally be written as

$$\Psi = \sum_{m_s = -s}^{s} \phi_{m_s}(\boldsymbol{r})\chi_{m_s}^{(s)} \tag{8.16}$$

where the coefficients $\phi_{m_s}(\boldsymbol{r})$ are arbitrary functions of position. At every point $\boldsymbol{r}$, the wave function Ψ does not simply have a numerical value; it has a prescribed spin state. The spin state of a particle may be represented as a column vector with $(2s+1)$ components. If we write

$$\chi_s^{(s)} = \begin{pmatrix} 1 \\ 0 \\ 0 \\ \vdots \end{pmatrix}, \; \chi_{s-1}^{(s)} = \begin{pmatrix} 0 \\ 1 \\ 0 \\ \vdots \end{pmatrix}, \text{ etc.}$$

then the general wave function (8.16) may be written

$$\Psi = \begin{pmatrix} \phi_s(\boldsymbol{r}) \\ \phi_{s-1}(\boldsymbol{r}) \\ \cdot \\ \cdot \\ \cdot \end{pmatrix} \tag{8.17}$$

The spin states $\chi^{(s)}_{m_s}$ are taken to be an orthonormal set of basis vectors in the $(2s+1)$-dimensional vector space describing the spin state of the particle. Thus we imagine a scalar product with the property

$$(\chi^{(s)}_{m_s'}, \chi^{(s)}_{m_s}) = \delta_{m_s, m_s'}$$

For two complete wave functions $\tilde{\Psi}$ and Ψ, as defined by equation (8.16), the scalar product is to be imagined in two parts—an integral over position space $\boldsymbol{r}$ and the scalar product in spin space. Thus

$$\begin{aligned}(\tilde{\Psi}, \Psi) &= \sum_{m_s m_s'} \int \tilde{\phi}^*_{m_s'}(\boldsymbol{r})\phi_{m_s}(\boldsymbol{r})\mathrm{d}\boldsymbol{r}(\chi^{(s)}_{m_s'}, \chi^{(s)}_{m_s}) \\ &= \sum_{m_s'} \int \tilde{\phi}^*_{m_s}(\boldsymbol{r})\phi_{m_s}(\boldsymbol{r})\mathrm{d}\boldsymbol{r}\end{aligned}$$

The effect of a rotation on Ψ is given by a rotation of both ϕ and χ,

$$\mathrm{T(R)}\Psi = \sum_{m_s} \phi_{m_s}(\mathrm{R}^{-1}\boldsymbol{r})(\chi^{(s)}_{m_s})'$$

using equations (3.37) and (8.14). For a small rotation about the z-axis through an angle a we have from the definition (7.4) with the introduction of factors i, as explained just before equation (7.26),

$$\begin{aligned}\mathrm{T(R)}\Psi &\approx \sum_{m_s} (1 - ial_z)\,\phi_{m_s}(\boldsymbol{r})(1 - ias_z)\chi^{(s)}_{m_s} \\ &\approx \sum_{m_s} [1 - ia(l_z + s_z)]\,\phi_{m_s}(\boldsymbol{r})\chi^{(s)}_{m_s} \\ &\approx [1 - ia(l_z + s_z)]\,\Psi\end{aligned} \tag{8.18}$$

Thus the infinitesimal operator for the complete wave function Ψ is the sum $(l_z + s_z)$ of orbital and spin angular momenta. We write $j_z = l_z + s_z$ and refer to j as the total angular momentum of the particle, in units of $\hbar$.

The effect of a finite rotation $\mathrm{R}(\boldsymbol{a})$ on Ψ is given by

$$\begin{aligned}\mathrm{T(R)}\Psi &= \sum_{m_s} \phi_{m_s}(\mathrm{R}^{-1}\boldsymbol{r})\,(\chi^{(s)}_{m_s})' \\ &= \sum_{m_s} \phi_{m_s}(\mathrm{R}^{-1}\boldsymbol{r}) \sum_{m_s'} D^{(s)}_{m_s' m_s}(\boldsymbol{a})\,\chi^{(s)}_{m_s'} \\ &= \sum_{m_s'} \left\{ \sum_{m_s} D^{(s)}_{m_s' m_s}(\boldsymbol{a})\phi_{m_s}(\mathrm{R}^{-1}\boldsymbol{r}) \right\} \chi^{(s)}_{m_s'}\end{aligned} \tag{8.19}$$

Hence if we define the transformed component functions $\tilde{\phi}_{m_s'}(\boldsymbol{r})$ by the equation $\mathrm{T(R)}\,\Psi = \sum_{m_s'} \tilde{\phi}_{m_s'}(\boldsymbol{r})\chi_{m_s'}^{(s)}$, which in the form (8.17) would be

$$\mathrm{T(R)}\,\Psi = \begin{pmatrix} \tilde{\phi}_s(\boldsymbol{r}) \\ \tilde{\phi}_{s-1}(\boldsymbol{r}) \\ \cdot \\ \cdot \\ \cdot \end{pmatrix}$$

then

$$\tilde{\phi}_{m_s'}(\boldsymbol{r}) = \sum_{m_s} D_{m_s' m_s}^{(s)}(\boldsymbol{a})\phi_{m_s}(R^{-1}\boldsymbol{r}) \tag{8.20}$$

Note that the summation runs over the second index of the rotation matrix rather than the first index as in equation (8.14). This is appropriate since equation (8.20) represents a transformation of 'coordinates' rather than a transformation of basis vectors.

The effect of a rotation on Ψ given in (8.19) implies a rotation both in the function of position $\boldsymbol{r}$ and in the spin states. In some situations it is convenient to consider separately the effect of rotations on these two parts of the wave function. In fact, the operators s_q may be regarded as infinitesimal operators of the group $\mathscr{R}_3^s$ of rotations in spin while the l_q describe the group $\mathscr{R}_3^l$ of rotations in the position coordinates. The set of six operators s_q, l_q describe the product group $\mathscr{R}_3^l \times \mathscr{R}_3^s$ and the three operators $\mathrm{j}_q = \mathrm{s}_q + l_q$ describe the subgroup $\mathscr{R}_3$ of simultaneous rotations in both spaces.

As a particular case of the wave function (8.16) consider a situation in which the particle is in a particular spin state m_s and also in a particular orbital state with angular momentum l and projection m_l. The appropriate wave function is the simple product

$$\Psi(lsm_lm_s) = \psi_{lm_l}(\boldsymbol{r})\chi_{m_s}^{(s)} \tag{8.21}$$

There is a set of $(2l+1)\,(2s+1)$ such products for fixed l and s corresponding to different values of m_l and m_s. Under rotations the set will transform according to the product representation $\mathrm{D}^{(l)} \times \mathrm{D}^{(s)}$ of $\mathscr{R}_3$ and as in section 8.3 this product will reduce

$$\mathrm{D}^{(l)} \times \mathrm{D}^{(s)} = \dot{\sum_{j=|l-s|}^{(l+s)}} \mathrm{D}^{(j)} \tag{8.22}$$

Thus, as in the coupling of two orbital angular momenta, the total angular momentum is given by the vector coupling rule

$$j = (l+s), (l+s-1), \ldots, |l-s| \tag{8.23}$$

The eigenfunctions of the total angular momentum $j = l+s$ are again built up from the product functions (8.21) with the help of vector-coupling coefficients,

$$\Psi\,(lsjm) = \sum_{\substack{m_l \\ (m_s = m - m_l)}} C(lsj,\; m_l m_s m)\psi_{lm_l}(\mathbf{r})\chi^{(s)}_{m_s} \tag{8.24}$$

The double-valuedness of the half-integer representations does not cause any ambiguity in the coupling coefficients since they follow from the commutation relations of the infinitesimal operators, but care must be taken when considering finite rotations. For example, consider a rotation about the z-axis

$$\mathrm{R}_z(a)\,\psi_{jm} = \exp(-ima)\psi_{jm}$$

so that

$$\mathrm{R}_z(2\pi)\psi_{jm} = \exp(-2\pi mi)\psi_{jm} = -\psi_{jm}$$

if m is a half-integer. This change of sign under an operation which brings the system back into the same physical position from which it started does not cause any contradiction, since no physical significance is attached to the sign of a wave function. Care must be taken, however, to ensure consistency between different parts of any calculation. (In section 18.13 of volume 2 we show how the half-integer representations may be regarded as single-valued representations of a somewhat larger group then $\mathscr{R}_3$.)

For a particle with spin in a magnetic field one now allows the possibility of a contribution to the magnetic moment from the spin of the particle. This is done by including in the Hamiltonian a term

$$Bg_s\mu_B \mathrm{s}_z \tag{8.25}$$

which has the same form as (8.12) but contains a spin g-factor g_s. For the electron it is found experimentally that g_s is very close to 2 and this result can be understood theoretically, see subsection 15.8.2 of volume 2.

In addition to the interaction between the spin magnetic moment and an external field there is also an interaction between the spin magnetic moment and the effective magnetic field due to the orbital motion of the electron (or other charged particle). The derivation of this interaction is best given in a relativistic treatment and here we simply quote that the result is an extra term

$$\xi(r)\mathbf{l}.\mathbf{s} = \xi(r)\,(l_x \mathrm{s}_x + l_y \mathrm{s}_y + l_z \mathrm{s}_z) \tag{8.26}$$

to be added to the Hamiltonian. We stress that this term is present whether there is an external field or not. It is called a spin–orbit interaction. The radial function $\xi(r)$ is related to the central field in which the particle moves.

8.5 The hydrogen atom

In the remainder of this chapter we look at the structure of atoms as an illustration of the results of $\mathscr{R}_3$ invariance and the introduction of spin. First we look at the hydrogen atom treated non-relativistically as an electron moving in the fixed spherically symmetric Coulomb field $V(r) = -e^2/r$ of the proton, the nucleus of the hydrogen atom. Ignoring spin, the wave functions are of the form, see section 7.5,

$$\psi_{nlm_l} = u_{nl}(r)\, Y^{(l)}_{m_l}(\theta, \phi)$$

Solution of the eigenvalue differential equation (7.60) leads (see any book on quantum mechanics) to the formula

$$E_{nl} = -m_e e^4/2\hbar^2 n^2$$

for the energies where $n = 1, 2, \ldots,$ and $l = 0, 1, \ldots, (n-1)$. We shall not need to know the details of the radial wave function $u_{nl}(r)$. The lowest few levels are shown in figure 8.1 (a), where we have used the standard spectroscopic notation in which the symbols s, p, d, f, g, etc., are used for states with $l = 0, 1, 2, 3, 4$, etc., respectively. (The degeneracy between states with the same value of n but different l is not due to the rotational symmetry but is caused by the existence of a rather abstract symmetry group of rotations in a four-dimensional space. A short discussion of this group is given in chapter 19 of

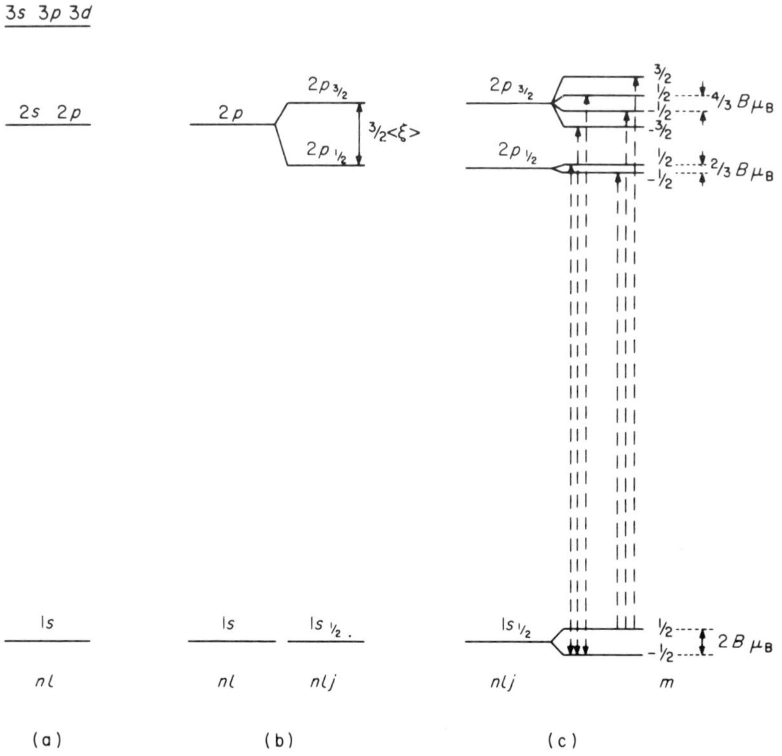

Figure 8.1

volume 2.) To each allowed combination nl there is a $(2l+1)$-fold degeneracy described by the label m_l. If the electron is assigned a spin of $\frac{1}{2}$ but spin–orbit forces are ignored, a further doubling of the degeneracy will result corresponding to the spin projections $m_s = \pm\frac{1}{2}$. Thus for example the $2p$ level is six-fold degenerate with labels $m_l = \pm 1, 0$ and $m_s = \pm\frac{1}{2}$ and has wave functions of the form (8.21).

We now include the spin–orbit term (8.26) in the Hamiltonian but with no external field. Noting that $\mathbf{j}^2 = (\mathbf{l}+\mathbf{s})^2 = \mathbf{l}^2 + \mathbf{s}^2 + 2(\mathbf{l}\,.\,\mathbf{s})$ we may rewrite the spin–orbit force as

$$\xi(r)\mathbf{l}\,.\,\mathbf{s} = \tfrac{1}{2}\xi(r)(\mathbf{j}^2 - \mathbf{l}^2 - \mathbf{s}^2) \tag{8.27}$$

Since the strength of this interaction is small compared with the difference between the energies E_{nl} we need only consider its effect in first-order degenerate perturbation theory (see section 5.8) for each E_{nl}. If, instead of using the product basis $m_l m_s$ for the degeneracy with fixed nl, we choose a coupled basis jm as defined in equation (8.24) then

$$\xi(r)(\mathbf{l}\,.\,\mathbf{s})\Psi(lsjm) = \tfrac{1}{2}[j(j+1) - l(l+1) - \tfrac{3}{4}]\xi(r)\Psi(lsjm)$$

Since $\xi(r)$ is an invariant it will cause no mixing of different l so that the energy shift caused by the spin–orbit interaction is given by

$$\Delta_{nlj} = \tfrac{1}{2}\langle \xi \rangle_{nl}\,\{j(j+1) - l(l+1) - \tfrac{3}{4}\} \tag{8.28}$$

where $\langle \xi \rangle_{nl} = \int_0^\infty u_{nl}^2(r)\,\xi(r)\,r^2\,\mathrm{d}r$ is the radial integral. For example, the six-fold degenerate $2p$ level splits into (1) a four-fold degenerate level with $j = \frac{3}{2}$ and shift $\Delta = \frac{1}{2}\langle \xi \rangle_{2p}$, and (2) a two-fold degenerate level with $j = \frac{1}{2}$ and shift $\Delta = -\langle \xi \rangle_{2p}$.

The splitting of the $2p$ level is shown in figure 8.1 (b). In this way every level with $l \neq 0$ will split into two levels with $j = l \pm \frac{1}{2}$. As a consequence, the absorption and emission lines observed in the spectrum will appear as double lines. For example, the $1s \rightarrow 2p$ transition, the first member of the Lyman series, with an expected energy of $(1 - \frac{1}{4})m_l e^4/2\hbar^2 \approx 82\,000\ \mathrm{cm}^{-1}$ is actually observed as a doublet separated by $\approx 0.4\ \mathrm{cm}^{-1}$ which may be shown to be consistent with the expected splitting $\frac{3}{2}\langle \xi \rangle_{2p}$. In heavier atoms the effect is much greater.

Finally, let us turn on a uniform external magnetic field and choose our z-axis in the direction of the field. The result is to add to the Hamiltonian a term

$$(l_z + 2s_z)\,B\mu_B \tag{8.29}$$

using equations (8.12) and (8.25). The product wave functions with definite m_l and m_s would be eigenfunctions of (8.29) but we have just seen that the spin–orbit interaction (8.27) mixes these product functions. Thus the two terms (8.27) and (8.29) play opposite roles; one tends to couple l and s to a definite resultant j and the other tends to uncouple them, leading to the simple product wave functions with definite m_l and m_s. Here we shall discuss the weak-field case in which B is small enough that the splitting due to (8.29) is small compared with the splitting due to (8.27). In this limit the effect of the magnetic field is to split the $(2j+1)$-fold degeneracy of the levels nlj resulting from the spin–orbit interaction. The operator (8.29) is an invariant with respect to the group $\mathscr{R}_2$ of rotations about the z-axis and hence the degeneracy splits into single levels corresponding to one-dimensional irreducible representations of $\mathscr{R}_2$ and labelled by $m = j, j-1, \ldots, -j$. This is an example of a symmetry-breaking perturbation which we discussed generally in section 5.8. To calculate the energy shift for each m we must evaluate the expectation value

$$\langle jm|l_z + 2s_z|jm\rangle$$

Here we use the notation introduced after equation (5.8) with the further simplification that the symbol Ψ is omitted and only the relevant labels jm are retained.

An argument similar to that given in subsection 7.4.6 shows that both **l** and **s** are vector operators, so that $l_z + 2s_z$ is also a component of a vector operator. Hence the Wigner–Eckart theorem (7.53) tells us that its matrix elements are proportional to those of any other vector operator (see subsection 7.4.7). Hence in particular

$$\langle jm|l_z + 2s_z|jm\rangle = A_j\langle jm|\mathrm{j}_z|jm\rangle = mA_j \tag{8.30}$$

where A_j is a constant independent of m. The simplest way to deduce the constant A_j is first to choose the maximum $m = l+\frac{1}{2}$ which can only occur for $j = l+\frac{1}{2}$ and in which case the wave function is a simple product with $m_l = l$, $m_s = \frac{1}{2}$. Then we have from equation (8.30)

$$A_{l+\frac{1}{2}} = (l+1)/(l+\tfrac{1}{2}) \tag{8.31}$$

To find $A_{l-\frac{1}{2}}$ we consider the two possible states with $m = l-\frac{1}{2}$. In the coupled basis these are labelled by $j = l+\frac{1}{2}$ and $j = l-\frac{1}{2}$ while in the product basis they are labelled by $m_l = l$, $m_s = -\frac{1}{2}$ and $m_l = l-1$, $m_s = +\frac{1}{2}$. Without finding the transformation between these bases we may deduce the required constant by equating the trace of $(l_z + 2s_z)$ in the two bases using equation (8.30),

$$l + 2(-\tfrac{1}{2}) + (l-1) + 2(\tfrac{1}{2}) = (A_{l+\frac{1}{2}} + A_{l-\frac{1}{2}})(l-\tfrac{1}{2})$$

i.e. $A_{l+\frac{1}{2}} + A_{l-\frac{1}{2}} = 2$ and hence using equation (8.31) $A_{l-\frac{1}{2}} = l/(l+\frac{1}{2})$. [A more general method would be to consider

$$\langle jm|(\mathbf{l}+2\mathbf{s}).\mathbf{j}\; jm\rangle = A_j\langle jm|\,\mathbf{j}.\mathbf{j}\,|\,jm\rangle = j(j+1)A_j \tag{8.32}$$

and then to evaluate the left-hand side of the equation by writing

$$(\mathbf{l}+2\mathbf{s}).\mathbf{j} = \mathbf{l}^2 + 2\mathbf{s}^2 + 3\mathbf{l}.\mathbf{s} = \tfrac{3}{2}\mathbf{j}^2 - \tfrac{1}{2}\mathbf{l}^2 + \tfrac{1}{2}\mathbf{s}^2\,]$$

Thus, using the notation nlj the $1s_{\frac{1}{2}}$ level will split into two levels with $m = \pm\frac{1}{2}$ and energy shifts of ± 1 in units of $B\mu_B$. The $2p_{\frac{1}{2}}$ level will split similarly with shifts of $\pm 1/3$. The $2p_{\frac{3}{2}}$ level splits into four levels with $m = \pm\frac{3}{2}$ and shifts of ± 2 and $m = \pm\frac{1}{2}$ with shifts of $\pm 2/3$. These 'Zeeman' splittings are shown in figure 8.1 (c).

As an example of the general arguments given in section 5.6 for the $\mathscr{R}_2$ - selection rules for dipole radiation we see that for radiation polarised in the xy-plane, so that $\Delta m = \pm 1$, the absorption line from $1s$ to $2p$ will split into six lines, as shown in figure 8.1 (c). The relative strengths of the transitions may also be deduced from symmetry arguments, as is shown in appendix 5.3 of volume 2.

8.6 The structure of many-electron atoms

8.6.1 The Hamiltonian

To illustrate further the manifestations of $\mathscr{R}_3$ symmetry in physical systems we go beyond the hydrogen atom to consider the structure of atoms in general. For a neutral atom with a charge Ze on the nucleus and with a number Z of electrons we take the Hamiltonian to be

$$\mathrm{H} = \sum_i \mathrm{T}_i - \sum_i Ze^2/r_i + \sum_{i<j} e^2/r_{ij} + \sum_i \xi(r_i)\mathbf{s}_i \cdot \mathbf{l}_i \tag{8.33}$$

in which the different contributions arise from the kinetic energy, the Coulomb attraction to the 'fixed' nucleus, the Coulomb repulsion between all pairs of electrons and the spin–orbit interaction introduced in section 8.4. It is a good approximation to treat the nucleus as fixed since it has a mass more than 1800 times greater than that of the electrons. We see that all contributions to the Hamiltonian (8.33) except the third are sums of one-particle operators and would therefore lead, as in section 6.3, to a separable solution to the Z-particle Schrodinger equation. Even the third contribution is, in part, equivalent to a sum of one-particle operators since the sum of the repulsions felt by any chosen electron due to the others is similar to a single repulsion directed from the nucleus which lies at the centre of the cloud of electrons. In a sense, the negatively charged electrons screen off the nucleus from an electron at the perimeter of the atom. One may therefore take advantage of this effect by adding an extra contribution $\sum_i U(r_i)$ to the Hamiltonian and then subtracting it. We write

$$\mathrm{H} = \mathrm{H}_0 + \mathrm{H}_1 + \mathrm{H}_2 \tag{8.34}$$

where

$$\mathrm{H}_0 = \sum_i [\mathrm{T}_i - Ze^2/r_i + U(r_i)] \tag{8.35a}$$

$$\mathrm{H}_1 = \sum_{i<j} e^2/r_{ij} - \sum_i U(r_i) \tag{8.35b}$$

and
$$H_2 = \sum_i \xi(r_i)\,\mathbf{s}_i \cdot \mathbf{l}_i \tag{8.35c}$$

Here H_0 is a sum of spin-independent one-body operators, including the screening potential $U(r_i)$, and H_1 is the residual interaction left after the screening potential is subtracted from the Coulomb repulsions. The choice of $U(r_i)$ is then made in such a way that H_1 is small but the details of this choice will not concern us since they have little to do with symmetry.

The solution of the Schrodinger equation now proceeds by perturbation theory. H_0 is taken as the unperturbed Hamiltonian and it has an exact solution in separated form, as we describe in subsection 8.6.2. Of the two remaining parts of H, the residual interaction H_1 is usually the most important and is considered next in subsection 8.6.3 using first-order degenerate perturbation theory (see section 5.8). This leads to what is known as an *LS*-coupling situation. The influence of H_2 which is assumed to be small compared with H_1 is finally included, again in degenerate perturbation theory. In some atoms, the spin–orbit interaction is relatively large, in which case it is necessary to treat $H_1 + H_2$ together in degenerate perturbation theory, a situation known as intermediate coupling. In rare cases H_1 is small compared with H_2, in which case it is simpler to regard $H_0 + H_2$ as the unperturbed Hamiltonian, now spin-dependent but still a sum of one-body operators. The residual interaction H_1 is then treated by degenerate perturbation theory with respect to the new unperturbed Hamiltonian. This limit is referred to as *jj*-coupling but we shall not describe it further.

8.6.2 The Pauli principle and shell filling

From equation (8.35a) the unperturbed Hamiltonian $H_0 = \sum_i H_0(i)$ is spherically symmetric and spin-independent. The eigenfunctions of $H_0(i)$ are therefore of the same form $\psi = u_{nl}(r)\, Y^{(l)}_{m_l}(\theta, \phi)\chi^{(\frac{1}{2})}_{m_s}$ as in the hydrogen atom, except that the radial wave functions $u_{nl}(r)$ and the energies E_{nl} are appropriate to the solution of a radial differential equation which includes the factor Z and the screening term $U(r_i)$. Thus the solutions of the Z-electron Hamiltonian may be written in separable form as products of single-particle wave functions

$$\prod_i \psi_{\gamma_i}(\boldsymbol{r}_i) \tag{8.36}$$

where we have used γ_i to denote the set of single-particle labels nlm_lm_s for particle i. The corresponding energy is $\sum_i E_{\gamma_i}$.

The indistinguishability of identical particles in quantum mechanics, see section 5.9, has led to the postulate that the wave function for a system of particles should be either totally symmetric or totally antisymmetric. By totally symmetric we mean that the wave function should be unchanged by the exchange of any pair of particles. If P_{ij} denotes the exchange operator for particles i and j, then $P_{ij}\ \Psi = \Psi$. A totally antisymmetric wave function is defined to satisfy $P_{ij}\ \Psi = -\Psi$ for all pairs ij. It has been observed that, if the particles have integer spin, like mesons, then their wave functions are symmetric, whereas if the particles have half-integer spin, like electrons, their wave functions are antisymmetric. The implication of this for a Z-electron atom is that products of the kind (8.36) are not acceptable. However, for each set of labels $\gamma_1, \gamma_2, \ldots, \gamma_Z$ in (8.36), assumed all different, there is a total of $Z!$ different products (8.36) obtained by carrying out the $Z!$ different permutations of particle number. Precisely one combination of these products has the required property that it is totally antisymmetric. It may be conveniently written as a determinant, called a 'Slater determinant':

$$\Psi(\gamma_1\gamma_2, \ldots, \gamma_Z) = (Z!)^{-\frac{1}{2}} \begin{vmatrix} \psi_{\gamma_1}(1)\,\psi_{\gamma_2}(1) & \ldots & \psi_{\gamma_Z}(1) \\ \psi_{\gamma_1}(2)\,\psi_{\gamma_2}(2) & \ldots & \psi_{\gamma_Z}(2) \\ \cdot & & \cdot \\ \cdot & & \cdot \\ \cdot & & \cdot \\ \psi_{\gamma_1}(Z)\psi_{\gamma_2}(Z) & \ldots & \psi_{\gamma_Z}(Z) \end{vmatrix} \qquad (8.37)$$

The factor before the determinant ensures normalisation. By multiplying out the determinant it is clear that $\Psi(\gamma_1\gamma_2 \ldots \gamma_Z)$ is a sum of products of the type (8.36). The effect of P_{ij} is to interchange two rows of the determinant and so change its sign. Hence $\Psi(\gamma_1\gamma_2 \ldots \gamma_Z)$ is totally antisymmetric and is still an eigenfunction of H_0 with energy $\sum_i E_{\gamma_i}$. (In group-theoretical language, a totally antisymmetric function transforms according to a one-dimensional irreducible representation of the symmetric group $\mathscr{S}_Z$ of permutations and $\Psi(\gamma_1\gamma_2 \ldots \gamma_Z)$ is the projection on to this representation from the product (8.36). This group is discussed in detail in chapter 17 of volume 2.)

If any two of the labels γ_i are identical then two of the columns of the determinant (8.37) will be identical so that it vanishes for all values of the coordinates $\boldsymbol{r}_1, \boldsymbol{r}_2$ etc. Hence, in the product (8.36) from which $\Psi(\gamma_1\gamma_2 \ldots \gamma_Z)$ is constructed, no two electrons may occupy the same single-particle state γ_i, a result known as the 'Pauli principle'.

The lowest energy levels of the atom will therefore be obtained by choosing those γ_i which correspond to the lowest single-particle energies. Because of the degeneracy with respect to m_l and m_s in the single-particle energies there is an even greater degeneracy in general for the Z-electron wave function (8.37). For example, the six-fold degeneracy of a p-level leads to a fifteen-fold degeneracy of possible determinants (8.37) for two electrons in p-orbits. (Fifteen is the number of ways of choosing two different single particle states $m_s m_l$ from a total of six.) The maximum number of electrons which may be put into orbits with energy E_{nl} and fixed nl is equal to the degeneracy $2(2l+1)$ of that single-particle level and such a system is referred to as a closed shell. A closed shell is clearly non-degenerate since by definition it contains just enough particles that every possible single particle state γ with some chosen values for n and l appears in the determinant (8.37). It can also be argued that the closed shell has zero total spin and zero total orbital angular momentum, $L = S = 0$. To do this we first note that, in a closed shell, the sums $M_s = \Sigma m_s$ and $M_L = \Sigma m_l$ run equally over positive and negative values of m_s and m_l. Hence the sums vanish: $M_S = M_L = 0$. In a similar way, if Ψ denotes the closed-shell wave function, it follows that $\mathrm{L}_+\Psi = \mathrm{S}_+\Psi = 0$ and hence from equation (7.31) that $L = S = 0$. The result $\mathrm{L}_+\Psi = 0$ comes about because $\mathrm{L}_+ = \sum_i l_+(i)$ increases the m_l value of some particle, but since every possible m_l already occurs in the determinant, the function $\mathrm{L}_+\Psi$ will be a determinant with two equal columns and hence vanishes. The result $\mathrm{S}_+\Psi = 0$ follows by the same argument.

Although the effective one-body potential in H_0 is somewhat different from that in the hydrogen atom, nevertheless, the ordering of the single-particle energies E_{nl} for light atoms is the same. It is found that shell closures occur when $Z = 2, 4, 10, 12$ and 18 corresponding to the filling of the shells, $1s$, $2s$, $2p$, $3s$ and $3p$. For an atom with one electron outside a closed shell such as lithium ($Z = 3$), boron ($Z = 5$), sodium ($Z = 11$) and aluminium ($Z = 13$), the spin and angular momentum of the atom will be just that of the last electron, $S = \frac{1}{2}$ with $L = 0, 1, 0$ and 1, in the examples quoted. In many respects the properties of the atom are determined by the electrons outside the closed shells, a feature which underlies the periodic table of the elements. In particular the inert gases are characteristic of the closed shells and the alkalis are characteristic of atoms with one so-called 'valence electron' outside the closed shells.

Since the closed shell ground state is non-degenerate, the inclusion of the perturbations H_1 and H_2 of equation (8.35) produces a small shift in energy without any splitting.

For an atom with one electron outside the closed shells the effect of H_1 is again a simple shift of the ground state energy. Although the ground state has the $2(2l+1)$-fold degeneracy of the single particle level of the valence electron this is not split by H_1 since H_1 is invariant under rotations separately in spin and orbital space, like H_0. The effect of H_2 is qualitatively the same as in the hydrogen atom described earlier and shown in figure 8.1(b).

When inversions are considered, the parity of each single-electron state is $(-1)^l$ and the parity of the Slater determinant is the product of the paritics of the occupicd single-electron states. Thus a closed shell has even parity and all states formed by putting a number t of electrons into a valence state nl have the same parity $(-1)^{tl}$.

8.6.3 Atoms with more than one valence electron—*LS* coupling

In an atom with a number t of electrons in a valence level nl there is a set of determinants (8.37) corresponding to different choices of m_s and m_l for the t occupied orbits. The number of determinants will be equal to the number of ways of selecting t different single-particle states $m_s m_l$ from a total of $2(2l+1)$ and is given by $(4l+2)!/(4l+2-t)!t!$ The perturbation H_1 will split this degeneracy and we now briefly investigate the nature of the splitting. We first simplify the problem by arguing that for an understanding of the splitting it is necessary to consider only the interaction between valence electrons and therefore one need only consider determinants for the t valence electrons. It may be shown that the interaction between pairs of electrons in the closed shells or between an electron in the closed shells and a valence electron contributes the same energy shift in all the states $m_s m_l$ of the valence electrons.

As remarked earlier, the perturbation H_1 is invariant under rotation in position coordinates and spin separately, commuting with both sets of operators L_q and S_q. We may in fact regard the L_q as infinitesimal operators of one $\mathscr{R}_3$ group and the S_q as infinitesimal operators of another. The set of six infinitesimal operators L_q, S_q describe a product group

$$\mathscr{R}_3^L \times \mathscr{R}_3^S$$

whose irreducible representations are labelled by the pair of labels L and S appropriate for each separate factor (see section 4.21). In degenerate perturbation theory (see section 5.8) we must set up the matrix of H_1 in the set of determinants which are degenerate with respect to the unperturbed Hamiltonian H_0. But since H_1 is invariant with respect to this product group it follows that if one uses basis vectors which belong to irreducible representations $D^{(LS)}$ of the product group then the matrix will break up into smaller matrices labelled by LS, with zero matrix elements between states with different labels LS. Furthermore there is no coupling between states with different values for M_L or M_S and there is a $(2L+1)(2S+1)$-fold degeneracy corresponding to the dimension of $D^{(LS)}$. To decide which values of L and S may arise, given values for l and t, is a detailed question which we discuss briefly in subsection 8.6.4. The calculation of the actual matrix elements of H_1 is an even more detailed problem which is described for $t=2$ and 3 in appendix 5 of volume 2. In general, states with definite L and S are rather complicated sums of Slater determinants and one does not normally express them in this way.

For obvious reasons, the situation described above is referred to as LS coupling. The set of $(2L+1)(2S+1)$ states with given L and S is often called a 'term'. Now consider the effect of H_2 when assumed to be small enough that it causes no mixing between different eigenvalues of the matrix of H_1. Its effect is then to split the $(2L+1)(2S+1)$-fold degeneracy and this process may be viewed group-theoretically as a restriction of the symmetry group from $\mathscr{R}_3^L \times \mathscr{R}_3^S$ to the group $\mathscr{R}_3$ of simultaneous rotations of both position and spin which is described by the infinitesimal operators $J_q = L_q + S_q$. The perturbation H_2, while not invariant under the product group, is still invariant under the simultaneous rotations. To prove the invariance we need only show that $[J_q, (\mathbf{l}.\mathbf{s})] = 0$ and this follows directly from the commutation relations for $\mathbf{l}$ and $\mathbf{s}$ remembering that $\mathbf{l}$ commutes with $\mathbf{s}$(see problem 8.4). Thus the

$(2L+1)(2S+1)$-fold degeneracy will split into a number of $(2J+1)$-fold degenerate levels labelled by

$$J = (L+S), (L+S-1), \ldots, |L-S| \tag{8.38}$$

This result for the possible values of J is just the usual vector coupling rule (8.9) because, for simultaneous rotations, the representation $D^{(LS)}$ is simply the product representation $D^{(L)} \times D^{(S)}$ of $\mathscr{R}_3$. A simple expression for the J-dependence of this splitting for fixed LS is now deduced using the Wigner–Eckart theorem. If the symbol α is used to denote all labels of the wave function except LS and J then the splitting due to H_2 is given by

$$\begin{aligned}\Delta_J &= \langle \alpha LSJM | \sum_i \xi(r_i)\mathbf{l}_i.\mathbf{s}_i | \alpha LSJM \rangle \\ &= A_{\alpha LS} \langle \alpha LSJM | \mathbf{L}.\mathbf{S} | \alpha LSJM \rangle \\ &= \tfrac{1}{2} A_{\alpha LS} \{ J(J+1) - L(L+1) - S(S+1) \}\end{aligned} \tag{8.39}$$

using the fact that $\mathbf{L}.\mathbf{S}$, while generally not proportional to the spin orbit force, nevertheless has the same transformation properties under both $\mathscr{R}_3^L$ and $\mathscr{R}_3^S$. To justify this derivation of equation (8.39) one must use the methods of subsection 7.4.7. The constant $A_{\alpha LS}$ cannot be calculated without detailed knowledge of the wave functions but the J-dependence is given clearly by equation (8.39). In particular the levels are ordered with J and the energy difference between adjacent levels is given simply by

$$\Delta_J - \Delta_{J-1} = \tfrac{1}{2} A_{\alpha LS} \{ J(J+1) - (J-1)J \} = J A_{\alpha LS}$$

a result known as the 'Landé interval rule'. In general $A_{\alpha LS}$ is found to be positive, the same sign as ξ, when a shell is less than half full and negative thereafter. This means that the smallest value $|L-S|$ of J is lowest in energy in the first half of a shell and the largest $J = L+S$ is lowest in the second half. An illustration of these results is given in the following section for the cases of two and three valence p-electrons.

The effect of a magnetic field on a many-electron atom is to add to the Hamiltonian a term

$$H_{\text{mag}} = (L_z + 2S_z) B\mu_B$$

obtained by assuming (8.29) over electrons. This will split the $(2J+1)$-fold degeneracy of a level J. Since the operator is the z-component of a vector, the Wigner–Eckart theorem tells us, as in section 8.5, that its matrix elements are proportional to those of J_z for states of a given J-multiplet. Thus we may write

$$H_{\text{mag}} \equiv g_J J_z B\mu_B$$

In LS-coupling the constant g_J, called the Landé g-factor, may be calculated for given L and S following equation (8.32):

$$g_J = \tfrac{3}{2} + \{ S(S+1) - L(L+1) \}/2J(J+1) \tag{8.40}$$

Thus the splitting of the J-multiplet is proportional to $\langle J_z \rangle = M$ and the eigenstates are labelled by M. The magnitude of the splitting, when measured in an experiment, gives information about S and L through equation (8.40).

8.6.4 Classification of terms

Atoms with two valence electrons

If an atom consists of closed shells and just two valence electrons in orbits with angular momentum l it is clear from the usual vector-coupling rules that the only possible values of L and S are $L = 0, 1, 2, \ldots 2l$ and $S = 0, 1$. However, not all combinations of these L and S values are allowed when one imposes the necessary condition that the wave function must be antisymmetric. We show below that the states with even L are symmetric under permutation of the orbital co-ordinates while those with odd L are antisymmetric. We also show that for permutation of the spin co-ordinates of the two particles, states with $S = 1$ are symmetric and those with $S = 0$ are antisymmetric. For a totally antisymmetric state the allowed combinations are therefore $S = 1$ with odd L and $S = 0$ with even L.

It is quite common to use the notation ${}^{2S+1}L$ to denote states with given L and S, sometimes also referred to as a term. The index $2S+1$ is called the multiplicity, since it determines the number of levels J into which the term splits when H_2 is included, see equation (8.38). Thus, for two valence electrons with $l = 1$, as in the carbon atom ($Z = 6$), we shall expect to find terms ${}^3P$, ${}^1S$ and ${}^1D$, again using the spectroscopic notation of $S, P, D \ldots$ for $L = 0, 1, 2, \ldots$. Inclusion of the perturbation H_2 cannot split the 'singlet' terms ${}^1S$ and ${}^1D$, since with $S = 0$, $J = L$ is the only possibility. However, the 'triplet' term splits into three levels with $J = 2, 1$ and 0 with energies given by equation (8.39) and illustrated in figure 8.2.

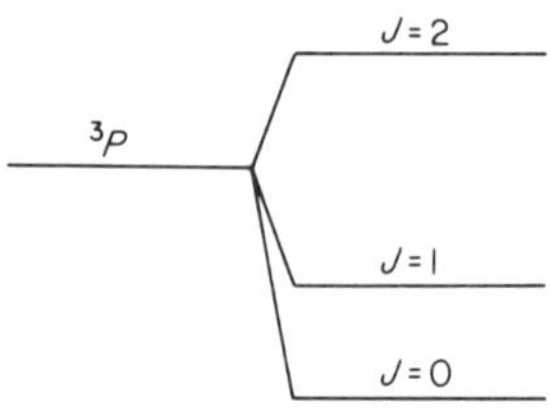

Figure 8.2

Derivation of permutation symmetry of two-particle wave functions

In this short section we derive the result quoted above which relates the L

and S values of a pair of electrons to the permutation symmetry in the position and spin variables separately. Consider first the possible wave functions of the position variables. If the two particles have the same single-particle labels nl then the two-particle wave functions may be labelled either by $m_1 m_2$ in a simple product basis or by LM, where $L = 2l, 2l-1, \ldots, 1, 0$ in a coupled basis. We first note that the state with maximum value of $M = L = 2l$ is a simple product with $m_1 = m_2 = l$ and is therefore symmetric, each particle being in the same state. Since the lowering operator $\mathrm{L}_- = l_-(1) + l_-(2)$ is also symmetric it follows that states with maximum $L = 2l$ and any M are also symmetric. Next consider the states with $M = 2l-1$. In the product basis there are two such states $m_1 = l$, $m_2 = l-1$ and $m_1 = l-1$, $m_2 = l$, while in the coupled scheme the basis states would be labelled by $L = 2l$ and $L = 2l-1$. In the product basis the permutation P_{12} simply interchanges the two states and so has a matrix $\left(\begin{smallmatrix} 0 & 1 \\ 1 & 0 \end{smallmatrix}\right)$ with character zero. Hence from the coupled basis we deduce that since $L = 2l$ is symmetric, $L = 2l-1$ must be antisymmetric to give the same character of zero. Proceeding in this way we see that for $M = 2l-2$ there is a set of three states with character equal to 1 so that $L = 2l-2$ is symmetric. Generalising, we deduce that even L are symmetric and odd L antisymmetric.

For a half-integer spin s a parallel argument shows that the maximum spin $S = 2s$ is again symmetric, $S = 2s-1$ is antisymmetric and so on. However since s is half-integer this means that odd S are symmetric and even S antisymmetric. In particular for an electron with $s = \frac{1}{2}$, the two-particle state with $S = 1$ is symmetric and $S = 0$ is antisymmetric.

Atoms with three-valence electrons

The listing of terms for two valence electrons was done by studying the permutation symmetry of the orbital and spin parts of the wave function. With three particles we must in the same way use the group $\mathscr{S}_3$ of all permutations of three objects. This group was introduced in section 2.2 (10) and was shown to be isomorphic with the point group D_3, see section 2.3 and subsection 2.7.3. It therefore has a character table which is the same as that given in table 4.2 for D_3 and repeated in table 8.1. We have used the labels s, a and m to distinguish the irreducible representations, being abbreviations for symmetric, antisymmetric and mixed. $\mathrm{T}^{(s)}$ is the identity representation and is therefore symmetric under any permutation. $\mathrm{T}^{(a)}$ is antisymmetric, since any interchange P_{ij} induces a change of sign. Notice that the elements in the second class like $\left(\begin{smallmatrix} 1 & 2 & 3 \\ 2 & 3 & 1 \end{smallmatrix}\right) = \mathrm{P}_{13}\mathrm{P}_{12}$ are the products of two interchanges and so have an entry $+1$ in the table for $\mathrm{T}^{(a)}$. The two-dimensional representation $\mathrm{T}^{(m)}$ has basis functions which are neither symmetric nor antisymmetric and is simply called the mixed representation. (For the group $\mathscr{S}_n$ with $n > 3$ there is more than one type of mixed representation and a more precise label has to be used, see chapter 17 of volume 2.)

Using the character table we may study the reduction of product repre-

Table 8.1

$\mathscr{S}_3$	E	$\binom{1\,2\,3}{2\,3\,1}, \binom{1\,2\,3}{3\,1\,2}$	P_{12}, P_{23}, P_{31}
$T^{(s)}$	1	1	1
$T^{(a)}$	1	1	-1
$T^{(m)}$	2	-1	0

sentations as in section 4.17 and deduce that $T^{(a)}$ occurs in the following products only

$$T^{(s)} \times T^{(a)},\ T^{(a)} \times T^{(s)} \text{ and } T^{(m)} \times T^{(m)} \tag{8.41}$$

Consequently a totally antisymmetric function for three electrons may be constructed from separate orbital and spin parts only if the parts transform as shown in (8.41). We now proceed to find which values of L and S are associated with which permutation symmetry as we did for the two-electron case. (In general, since the operators $L_{\pm}$ (and $S_{\pm}$) are symmetric, it follows that, for given L (or S), states with different projections M_L (or M_S) have the same permutation symmetry.)

By vector coupling particles 1 and 2 to $S = 1$ and then coupling in particle 3 we can produce $S = \frac{3}{2}$ or $S = \frac{1}{2}$, while coupling the first two particles to $S = 0$ gives another $S = \frac{1}{2}$ state. Taking note of the multiplicities $2S + 1$, this gives the correct total of $2^3 = 8$ states. Using the same argument as for two electrons, the four states with $S = \frac{3}{2}$ must be symmetric. In fact, this exhausts the symmetric states since in general the number of symmetric triple products that can be formed from n different single particle states is $\frac{1}{6}n(n+1)(n+2)$ and this number is four when $n = 2$. The number of antisymmetric states is similarly given by $\frac{1}{6}n(n-1)(n-2)$, which is zero for $n = 2$—a result to be expected with only two single-particle states since two of the three particles must then be in the same state so that antisymmetry is impossible. Hence the $S = \frac{1}{2}$ states have mixed symmetry and the fact that there are two of them, in addition to the $2S+1$-fold multiplicity, is to be expected since $T^{(m)}$ is two-dimensional.

For orbital states, a similar argument shows that, when $l = 1$, there is one antisymmetric state, which must therefore have $L = 0$ and is simply the determinant formed from the three single-particle states with $m = 1, 0$ and -1. From vector coupling, the $3^3 = 27$ states break up into L-values as follows: $S + 3P + 2D + F$ and again the F-state, with maximum L, must be symmetric. However, the number of symmetric states is now $\frac{1}{6} \times 3 \times 4 \times 5 = 10$, showing that one of the P-states must also be symmetric. The remaining pair of P- and D-states must therefore have mixed symmetry.

Putting together these results with the help of equation (8.41) we deduce that the possible terms for an atom like nitrogen ($Z = 7$) with three valence p-electrons are ^4S, ^2P and ^2D. To construct in detail the wave functions for the latter two, one would need to use the Clebsch–Gordan coefficients for the group $\mathscr{S}_3$ in the reduction of the product $T^{(m)} \times T^{(m)}$ to form an antisymmetric function.

The results which we have deduced above in a rather *ad hoc* fashion may be obtained also by using the characters of symmetrised products described in appendix 3.1 of volume 2. See also section 18.10 of volume 2. The extension to a larger number of valence particles is straightforward but laborious.

8.6.5 Ordering of terms

To deduce the relative energies of the different possible terms listed by the methods given above, and in particular to deduce the values of S and L for the ground state, one must evaluate the matrix of the perturbation H_1 in these states. Generally this involves setting up and diagonalising a small matrix for each combination of S and L. In the simple examples quoted above there is never more than one term with given S and L so that the matrix, set up in an LS coupling basis, will be already diagonal. The calculation of matrix elements is generally quite an intricate process and we shall not go into detail. However, there are several general features of the calculation which result from symmetry considerations and have simple explanations.

There is an intimate connection between the value S of the total spin and the symmetry of the wave function under permutation of the spin or positional coordinates separately. In particular we have seen that the wave functions for a system of t electrons with maximum spin $S = \frac{1}{2}t$ must be totally symmetric with respect to permutations of spin and therefore totally antisymmetric with respect to permutations in position coordinates. For general values of S we may obtain a measure of the symmetry by noting that for any two particles i and j there is a relation between the total spin operator $\mathbf{S}(ij) = \mathbf{s}(i) + \mathbf{s}(j)$ and the spin exchange operator P_{ij}^S, which interchanges the spin coordinates of particles i and j. The relation is

$$\mathbf{S}^2(ij) = 1 + P_{ij}^S \tag{8.42}$$

which is proved by showing that, in the complete set of four spin states of two spin half-particles, both sides of this equation have the same matrix (see problem 8.8). Turning now to the system of t electrons and summing the left-hand side of equation (8.42) over all pairs of particles we have

$$\sum_{i<j} [\mathbf{s}(i) + \mathbf{s}(j)]^2 = \sum_{i<j} [\mathbf{s}^2(i) + \mathbf{s}^2(j)] + 2 \sum_{i<j} \mathbf{s}(i) . \mathbf{s}(j)$$

while the square of the total spin is given by

$$\mathbf{S}^2 = \left(\sum_i \mathbf{s}(i) \right)^2 = \sum_i \mathbf{s}^2(i) + 2 \sum_{i<j} \mathbf{s}(i) . \mathbf{s}(j)$$

Combining these two expressions with equation (8.42) gives

$$\mathbf{S}^2 = \sum_{i<j} \mathrm{P}_{ij}^S + \sum_i \mathbf{s}^2(i) + \sum_{i<j} [1 - \mathbf{s}^2(i) - \mathbf{s}^2(j)]$$

$$= \sum_{i<j} P_{ij}^S + t - \tfrac{1}{4}t^2 \qquad (8.43)$$

using the result that $\mathbf{s}^2(i) \equiv \frac{3}{4}$ for any particle with $s = \frac{1}{2}$. The operator $\mathscr{M}^S = \sum_{i<j} \mathrm{P}_{ij}^S$ gives a direct measure of the permutation symmetry, since any symmetric pair contributes $+1$ while an antisymmetric pair contributes -1. Equation (8.43) shows how it is related to the total spin S. In particular one sees that states with maximum $S = \frac{1}{2}t$ are also eigenfunctions of $\mathscr{M}^S$ with eigenvalue $\frac{1}{2}t(t-1)$, confirming that all pairs are symmetric. For the $S = \frac{1}{2}$ states of $t = 3$ this equation shows that $\mathscr{M}^S$ has the value zero, showing that the states of mixed symmetry $\mathrm{T}^{(m)}$ for three particles have equal proportions of symmetric and antisymmetric pairs. In general, from equation (8.43), one sees that as S decreases so does the value of $\mathscr{M}^S$ implying a gradual reduction in the number of symmetric spin pairs.

Every totally antisymmetric wave function Ψ must, by definition, satisfy $\mathrm{P}_{ij}\Psi = -\Psi$ where P_{ij} exchanges all coordinates of particles i and j. But $\mathrm{P}_{ij} = \mathrm{P}_{ij}^r \mathrm{P}_{ij}^S$, where P_{ij}^r denotes the exchange of position coordinates so that $\mathrm{P}_{ij}^r \Psi = -\mathrm{P}_{ij}^S \Psi$. Hence, on an antisymmetric wave function, we have the equivalence $\mathrm{P}_{ij}^r \equiv -\mathrm{P}_{ij}^S$. Thus defining $\mathscr{M}^r = \sum_{i<j} \mathrm{P}_{ij}^r$ we have $\mathscr{M}^r \equiv -\mathscr{M}^S$ and hence from equation (8.43)

$$\langle \alpha LS | \mathscr{M}^r | \alpha LS \rangle = t - \tfrac{1}{4}t^2 - S(S+1) \qquad (8.44)$$

For a fixed number t of electrons a decrease in the total spin S therefore implies an increase in the value of $\mathscr{M}^r$ and hence an increase in the number of pairs which are symmetric in their position coordinates. For maximum spin $S = \frac{1}{2}t$ the value of $\mathscr{M}^r$ from equation (8.44) is $-\frac{1}{2}t(t-1)$, corresponding to the antisymmetry of every pair in their position coordinates.

Let us now return to the study of atomic energy levels and consider the consequences of this relation between spin and permutation symmetry. Since it is the repulsion between electrons which determines the ordering of terms, the ground state will be the state in which this interaction is smallest. Generally, the repulsion is smallest when the electrons have least symmetry in their position coordinates. From equation (8.44) this corresponds to the highest $S = \frac{1}{2}t$, when all pairs are antisymmetric in position. Thus the ground state is expected to have maximum spin $S = \frac{1}{2}t$ and the energies for other values of S are expected to increase as S decreases. It is interesting to see this result expressed so clearly in terms of the spin, since H_1 is spin independent. The point is that S determines the spin symmetry which through the constraint of antisymmetry determines the positional symmetry of the electrons.

The value of $\frac{1}{2}t$ for the maximum spin in a system of t valence electrons with the same quantum numbers nl requires modification beyond the middle of the shell. There is a maximum number of $2(2l+1)$ electrons in the shell but only half of these will have $m_s = +\frac{1}{2}$. Hence the value $S = \frac{1}{2}t$ can only be achieved for the first half of the shell $t \leq (2l+1)$. Beyond this point the maximum value of S decreases as t increases. In fact the terms for $\{2(2l+1)-t\}$ electrons are the same as for t electrons.

The ordering of terms with the same t and S is more difficult to understand but again there is a simple general rule. The ground states are found experimentally to have the highest possible value of L. This result can be understood in general terms by arguing that the probability of two particles being close together tends to be less in states of high orbital angular momentum, so that bearing in mind the shape of the Coulomb potential one tends to get a smaller repulsion in states of high L.

The fact that atomic ground states have maximum S and the maximum L consistent with that value of S was observed empirically by Hund in 1927 (Hund's rule). Some detailed calculations of the ordering of levels is described in appendix 5.1 of volume 2.

Bibliography

Although many of its techniques have been superseded by those described in this book through the work of Racah, we nevertheless recommend the classic

Condon, E. U. and Shortley, G. H. (1935). *The Theory of Atomic Spectra* (Cambridge University Press, London)

For a more recent text see

Slater, J. C. (1960). *Quantum Theory of Atomic Structure* (McGraw-Hill, New York)

For details of the use of symmetry in atomic structure see

Judd, B. R. (1963). *Operator Techniques in Atomic Spectroscopy* (McGraw-Hill, New York)

Problems

8.1 A state with $j = 4$ decays to a state with $j' = 2$. What are the possible multipolarities of the transition?

8.2 Use equation (8.3) to deduce the minimum value j of the angular momentum of a state in which the quadrupole moment ($k = 2$) is non-zero.

8.3 By calculating the Clebsch–Gordan coefficients as in question 7.8 (or by looking them up in Rotenberg *et al.*, 1959 in the bibliography to chapter 7), write down the wave function (8.24) for the case $l = 2$, $s = \frac{1}{2}$, $j = \frac{3}{2}$, $m = \frac{3}{2}$.

8.4 Use equations (7.26) to prove the commutation relation

$$[\mathrm{J}_q, (\mathbf{l}.\mathbf{s})] = 0, \text{ where } \mathrm{J}_q = \mathrm{L}_q + \mathrm{S}_q.$$

8.5 Use the method of equation (8.32) to derive the results given in section 8.5 for the Zeeman splitting amplitudes A_j.

8.6 Show that the nitrogen atom ($Z = 7$) has the electron configuration $(1s)^2(2s)^2(2p)^3$.

8.7 Use the result of question 7.7 to show that the $S = 0$ state of two particles, each with spin $s = \frac{1}{2}$, is antisymmetric under permutation of the particles.

8.8 Justify equation (8.42) by calculating matrix elements of both sides of this equation in the four-dimensional space of spin states for two particles with spin $\frac{1}{2}$. (Use a coupled basis with labels SM_S.)

8.9 Find the possible combinations of L and S for three d-electrons.

9

Point Groups with an Application to Crystal Fields

We have already met several examples of point groups in earlier chapters and seen their application in the study of molecular vibrations in chapter 6. In this chapter we first develop the formal properties of all the point groups, which are the finite subgroups of the group O_3 of orthogonal transformations in three-dimensional space. They are important physically because they describe the symmetry of molecules and geometrical figures such as the regular solids. Some of them, the 32 crystallographic point groups, are of particular importance since they describe the symmetry of regular crystal lattices and are therefore much used in solid state physics.

In the last section (section 9.9) we investigate the effect of placing an atom in a potential which has a point group symmetry, a situation which occurs naturally in a crystal where the surrounding ions are arranged with the crystalline symmetry. This example shows very clearly the power of symmetry arguments and illustrates most of the techniques which are used in systems having a point group symmetry.

9.1 Point-group operations and notation

There is unfortunately more than one conventional notation for the elements of the point groups. In this book we use the Schoenflies notation which is also used in most published physical applications of point-group theory. An alternative notation is the 'international' system. A table for comparison of notation is provided in appendix 1.

The elements of the point groups are proper and improper rotations (i.e., rotations with or without inversion). The origin remains fixed under all these groups elements; in other words all rotation axes and reflection planes contain the origin. If a rotation through an angle a is an element of a point group then this element raised to any finite integer power will also be an element and for some power must be the identity E since the group is finite. Hence we must be able to write $[R(a)]^n = R(na) = E = R(2\pi m)$ so that the angle a must be of the form $2\pi m/n$, where m and n are integers. If the smallest angle of rotation for a particular symmetry axis is $2\pi/n$ then the axis is said to be n-fold. The rotation through $2\pi/n$ is denoted by C_n in the Schoenflies notation.

The axis of greatest symmetry is conventionally called the vertical axis and, naturally, the perpendicular plane is called horizontal. A reflection in a horizontal plane is denoted by σ_h and a reflection in a vertical plane by σ_v. The product of a rotation C_n with the reflection σ in the plane perpendicular to the axis of C_n is denoted S_n and called a mirror rotation. A mirror rotation can be written in terms of the inversion I and proper rotations since the product $\sigma\, C_2$ has the effect of inverting all the axes, i.e. $\sigma\, C_2 = I$ so that $\sigma = I\, C_2$ and hence

$$S_n = \sigma\, C_n = IC_2\, C_n$$

The point groups can now be constructed by starting with the simplest and adding symmetry elements. The addition of a new symmetry element to an existing group implies the existence of others. From simple geometrical consideration it is possible to deduce the following rules:

(1) The addition of the inversion (which commutes with all other elements) doubles the size of a group.

(2) The addition of a two-fold horizontal axis to an n-fold vertical axis implies another $n-1$ horizontal two-fold axes.

(3) The addition of a vertical reflection plane σ_v to an n-fold axis implies another $n-1$ vertical reflection planes.

9.2 The stereogram

In order to describe the symmetries of a point group of rotations one could take a sphere centred on the origin and mark on its surface an arbitrary point and all the positions to which this point would move under the group rotations. To present the results in two dimensions, the resulting points are

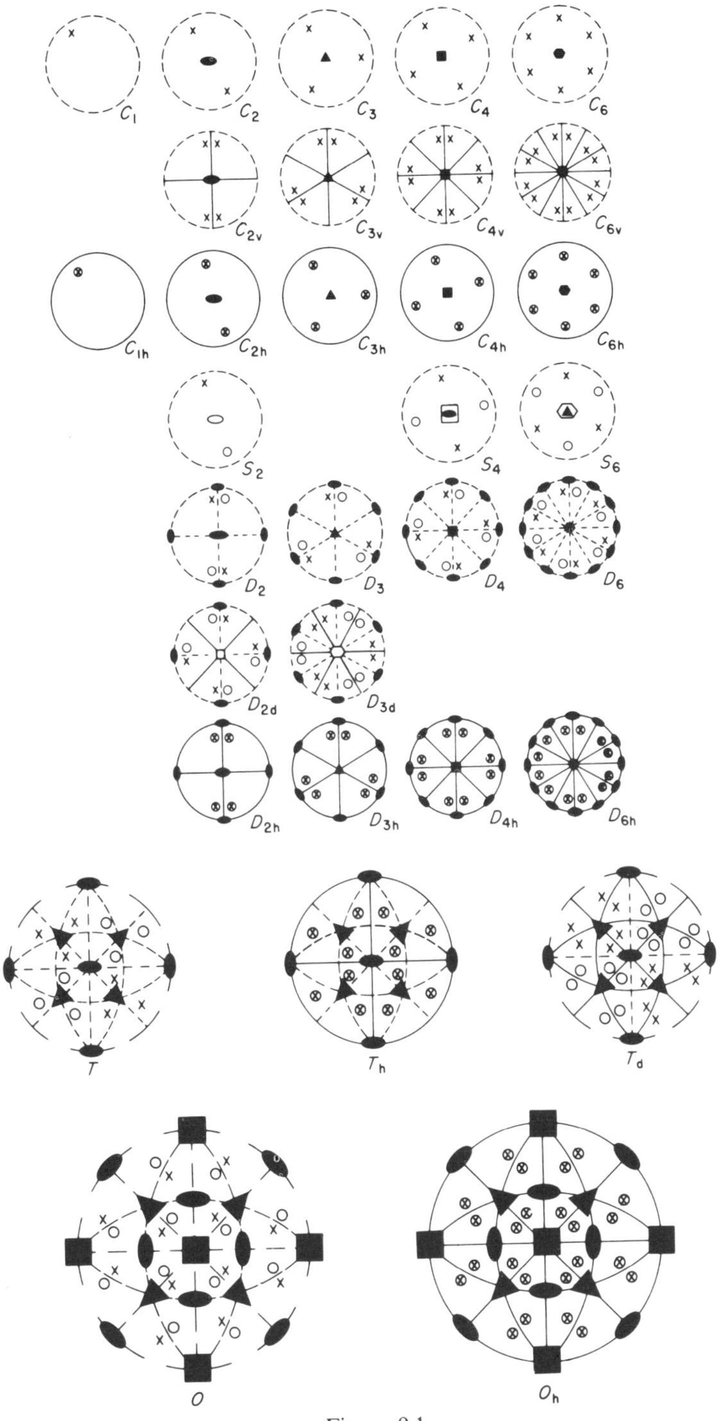
C1
C2
C3
C4
C6
C2v
C3v
C4v
C6v
C1h
C2h
C3h
C4h
C6h
S2
S4
S6
D2
D3
D4
D6
D2d
D3d
D2h
D3h
D4h
D6h
T
Th
Td
O
Oh

Figure 9.1

mapped on to a plane as follows. Every point in the 'north' hemisphere is projected on to the equatorial plane by straight line projection through the 'south' pole and marked by a cross. Every point on the 'south' hemisphere is projected via the 'north' pole and marked by a circle. This particular mapping has the property that points lying on a circle in one of the hemispheres map on to a circle in the plane – but their centres do not map on to each other. A set of 'stereograms' is shown in figure 9.1. The notation used in this figure is as follows. A rotation axis is marked by a characteristic symbol which itself has n-fold symmetry; thus we use ⬬, ▲, ■ and ⬢ for C_2, C_3, C_4 and C_6, respectively, and ⬭, △, □ and ⬡ for S_2, S_3, S_4 and S_6, respectively. Mirror planes are denoted by solid lines while other construction lines and rotation axes are dotted. This representation is particularly useful for the simpler groups having only one n-fold axis with $n > 2$. The other groups are more easily thought of as symmetry operations of regular solids such as the tetrahedron and cube.

9.3 Enumeration of the point groups

We do little more than describe geometrically the operations in each group. To fully appreciate any particular group the reader should refer to the stereograms in figure 9.1 and identify geometrically the product of any pair of group operations to build up the multiplication table. In section 9.4 we describe the class structure of the elements. We first consider groups which have only proper rotations.

9.3.1 Proper groups

C_n groups

The simplest point groups we can construct have a single n-fold axis for proper rotations. They are Abelian groups of order n, since rotations about the same axis commute. They are also cyclic groups generated from a single element C_n and its powers $C_n^2, C_n^3, \ldots, C_n^n = E$.

D_n (dihedral) groups

The only way that we can add another symmetry axis to the group C_n without generating further n-fold axes (from the C_n axis) is to add a horizontal 2-fold axis. From rule (2) in section 9.1 this implies $n - 1$ other horizontal two-fold axes. The group D_2 which has just three mutually perpendicular two-fold axes is sometimes called the Vierergruppe and labelled V. The group D_1 is not a new group since it is simply C_2 with its axis horizontal.

We noted above that any other proper point group must have two or more n-fold axes and we now see which of these are possible by considering a group with several n-fold axes intersecting at the centre of a sphere. If we mark on the

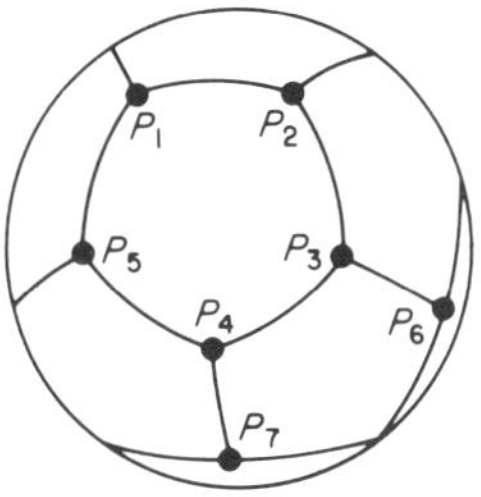

Figure 9.2

sphere the points P_i where the n-fold axes cross the surface then, because of the n-fold rotation symmetry about each axis, the points P_i are the vertices of a regular polyhedron. If we now join up nearest neighbour points P_i by segments of great circles (edges) on the sphere we shall have marked out a network on the surface of the sphere. Figure 9.2 illustrates this construction for the case of ten three-fold axes. Now there is a remarkable geometrical theorem due to Euler which relates the number V of vertices, the number E of edges and the number F of faces in such a network. It states that

$$V - E + F = 2 \tag{9.1}$$

Each edge, however, has a vertex at its two ends and n edges meet at each vertex so that $E = nV/2$. Furthermore, if we denote the number of edges of each face by s, then since each edge is shared by two faces, we also have $E = Fs/2$ and hence $nV = Fs$. We can now rewrite equation (9.1) as

$$\frac{Fs}{n} - \frac{Fs}{2} + F = 2 \tag{9.2}$$

i.e.

$$\frac{1}{n} + \frac{1}{s} = \tfrac{1}{2} + \frac{2}{Fs}$$

and since Fs is necessarily positive, this implies the restriction

$$\frac{1}{n} + \frac{1}{s} > \tfrac{1}{2} \tag{9.3}$$

For $n = 2$ equation (9.2) implies $F = 2$ and a little thought shows that this procedure will simply duplicate the dihedral groups discussed above. For $n > 2$ we have the possibilities listed in table 9.1 together with the corresponding values of F, V and E. The final column of the table indicates the form of the regular polyhedron marked out by the points P_i in each case. For $n \geq 6$ we get $s = 2$ and this gives only the dihedral groups again.

Table 9.1

n	s	F	V	E	*Polyhedron*
3	3	4	4	6	tetrahedron
3	4	6	8	12	cube
3	5	12	20	30	dodecahedron
4	3	8	6	12	octahedron
5	3	20	12	30	icosahedron

The complete symmetry groups for each of these cases will involve other symmetry elements besides the n-fold rotations and we describe them below. Their character tables are given in the appendix.

The tetrahedral group T

From the first row of table 9.1 we have four three-fold axes passing through the vertices of a tetrahedron as in figure 9.3. Products of the three-fold rotations can be seen to generate two-fold rotations about axes through the centres of opposite edges of the tetrahedron. For example a rotation through $2\pi/3$ about the axis C in figure 9.3 followed by a rotation through $-2\pi/3$ about the 'vertical' axis A, interchanges the axes A and B and also C and D. It is thus a two-fold rotation about the axis E. No other rotations are generated. The group is in fact the set of proper covering operations of the regular tetrahedron.

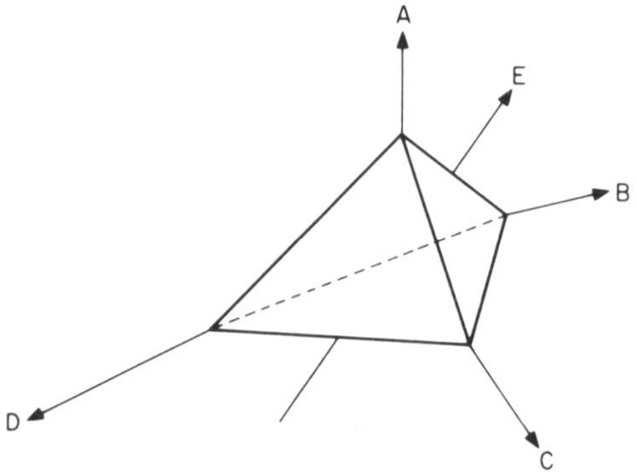

Figure 9.3

The octahedral group O

If we consider the second row of table 9.1 the axes for three-fold rotations pass through the corners of a cube as in figure 9.4. The complete symmetry group will again contain all the proper covering operations of the cube since

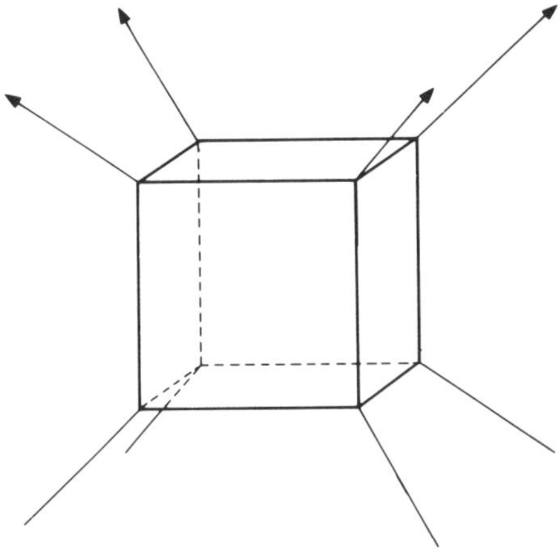

Figure 9.4

they rotate the four-fold axes into each other. The new elements are four-fold and two-fold rotations about axes through the centres of opposite faces of the cube and two-fold rotations about axes passing through the midpoints of opposite edges.

The same group of symmetry elements is generated by starting from the octahedron of row four of the table so no new symmetry group arises there. Figure 9.5 shows an octahedron inscribed in a cube to demonstrate the equivalent symmetries. It is for this reason that the group is called 'octahedral'.

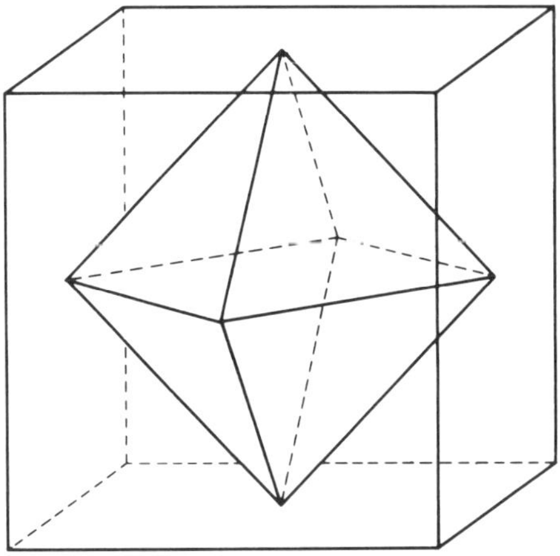

Figure 9.5

The icosahedral group *Y*

The group corresponding to the third row of the table has three-fold rotation axes passing through the vertices of a dodecahedron as in figure 9.6. Again the complete symmetry group will contain all the proper covering operations of the dodecahedron. The new elements are five-fold rotations about axes through opposite face centres and two-fold rotations about axes through the midpoints of opposite edges of the dodecahedron.

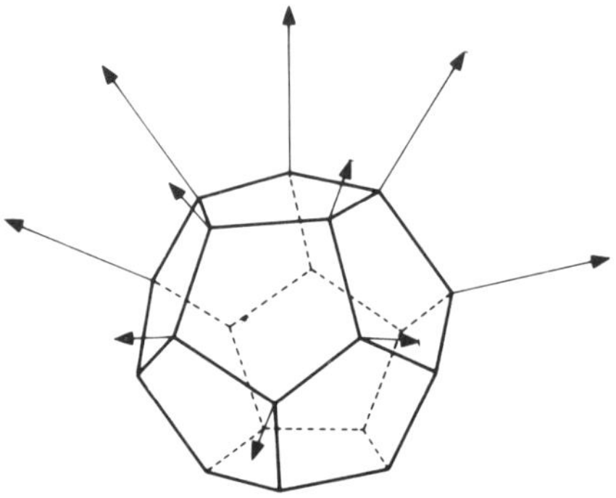

Figure 9.6

Again the same group of symmetry elements is generated by starting from the icosahedron of row five of table 9.1 and figure 9.7 shows a dodecahedron inscribed in an icosahedron to demonstrate this.

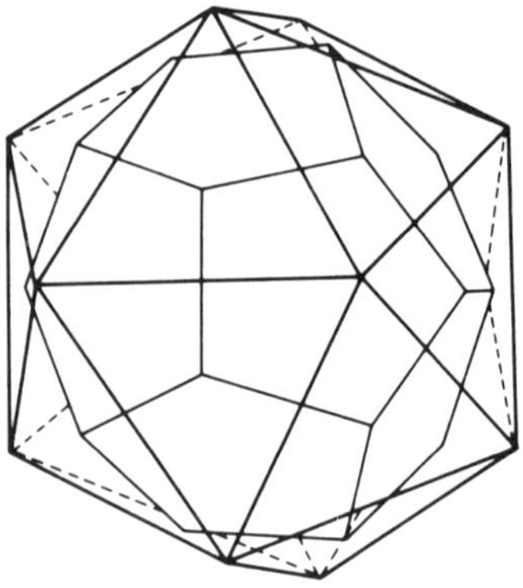

Figure 9.7

9.3.2 Improper groups

The improper groups will be formed by adding an improper element S_n to the proper groups discussed above. This need be done only in such a way that no new proper rotations are produced since otherwise there would be duplication. In the case of the C_n groups, with a single axis for proper rotations, we can add either a horizontal or vertical reflection plane without generating further proper rotations. In this way two new sets of point groups C_{nh} and C_{nv} are produced.

C_{nh} groups

Obtained by adding a horizontal reflection σ_h to the group C_n. They will contain the inversion as an element if n is even, since then $\sigma_h C_n^{\frac{1}{2}n} = I$.

C_{nv} groups

Obtained by adding a vertical reflection σ_v to the group C_n implying the existence of a total of n vertical mirror planes. There is of course no difference between C_{1h} and C_{1v} which each contain the identity and a reflection only.

S_{2n} groups

In the particular case of C_1 (which is simply the identity element) we can add any of the elements S_p to form a group S_p with a single p-fold axis for mirror rotations. However the groups S_p with p odd are identical with C_{ph} since they contain C_p and σ_h as elements. This follows because $S_p^{p+1} = (C_p \sigma_h)^{p+1} = C_p^{p+1} \sigma_h^{p+1} = C_p$ for p odd, so that C_p is contained. Also $S_p^p = (C_p \sigma_h)^p = C_p^p \sigma_h^p = \sigma_h$ for p odd, so that σ_h is also contained. When p is even, a new group S_{2n} is obtained where we have written $p = 2n$ with n any integer. When n is odd the group S_{2n} will also contain the inversion as an element since $S_{2n}^n = C_{2n}^n \sigma_h^n = C_2 \sigma_h = I$, and in particular the group S_2 consists of the identity E and the inversion I only. The groups S_{2n} are cyclic groups of order $2n$ consisting of the elements $S_{2n}, S_{2n}^2, \ldots, S_{2n}^{2n} = E$.

In the case of the groups D_n we can again add horizontal or vertical reflection planes.

D_{nh} groups

Obtained by adding a horizontal reflection σ_h to the groups D_n. From rule (3) of section 9.1 there must also be vertical reflection planes containing each of the n two-fold horizontal axes.

D_{nd} groups

Obtained by adding to D_n a vertical reflection σ_v whose plane bisects the angle between a pair of adjacent horizontal two-fold axes. $n-1$ additional planes of this type are of course generated by the n-fold rotation, and these relections are usually denoted σ_d where the d, as in D_{nd}, stands for diagonal. The groups can also be thought of as arising from the addition of a 'horizontal' two-fold axis to the groups S_{2n}.

The improper groups generated from T, O and Y by the addition of improper elements are:

The full tetrahedral group T_d

This is the complete group of proper and improper covering operations of the tetrahedron and is obtained from T by adding the reflection planes which pass through two corners of the tetrahedron and the midpoint of the opposite edge. The existence of these six mirror planes at angles $\pi/4$ to the two-fold axes implies six S_4 mirror rotations about these axes.

The group T_h

Obtained by adding the inversion to the group T. The inversion is not a symmetry element of the tetrahedron and therefore not contained in T_d. The group T_h is the direct product group formed from T and S_2: $T_h = T \times S_2$.

The full octahedral group O_h

Obtained by adding the inversion to the group O. The group is thus the direct product group formed from O and S_2: $O_h = O \times S_2$. O_h is the complete group of proper and improper covering operations of the cube or octahedron.

The group Y_h

Obtained by adding the inversion to the group Y and is thus the direct product group formed from Y and S_2: $Y_h = Y \times S_2$. Y_h is the complete group of proper and improper covering operations of an icosahedron or dodecahedron.

9.4 The class structure of the point groups

In order to facilitate the construction of the irreducible representations of the point groups by the methods of chapter 4 we first divide up the elements into classes using some of the results of sections 2.6 and 2.7. The classes are labelled by a typical element in the class and, since all elements in a given class are rotations through the same angle, this will be a rotation C_n or a mirror rotation S_n. This label is preceded by the number of elements in the class. If the elements of a class consist of the mth power $(C_n)^m$ of a rotation C_n about an n-fold axis, then the symbol C_n^m is used, prefixed by the number of elements in the class.

σ and I are usually used instead of S_1 and S_2, respectively, and the σ carries a label to indicate whether it is vertical or horizontal. If there are several inequivalent classes of rotations C_n then distinct classes will be distinguished by primes or occasionally by an index denoting the axis of rotation where, conventionally, the z-axis is taken as the axis of highest symmetry.

9.4.1 Proper point groups

C_n groups

As these groups are Abelian, every element is in a class by itself.

D_n groups

The additional two-fold axes now make the elements C_n^p and C_n^{-p} in the same class and the axis is said to be bilateral. (They are not distinct if $p = n/2$ with n even, since C_n^p is then a two-fold rotation.) When n is odd all the horizontal two-fold rotations are in the same class but when n is even, alternate two-fold axes fall into two non-equivalent classes. For example the classes of D_4 are E, $2C_4$, C_4^2, $2C_2$, $2C_2'$, and the classes of D_5 are E, $2C_5$, $2C_5^2$, $5C_2$.

The tetrahedral group T

Here the two-fold rotations are in one class since the three-fold rotations take the two-fold axes into each other. The eight three-fold rotations are not all in the same class but separate into two classes of four elements each of which may be taken as the clockwise and counterclockwise rotations about the four three-fold axes. There are no rotations which make these axes bilateral. The class structure is thus E, $3C_2$, $4C_3$, $4C_3'$.

The octahedral group O

The eight three-fold rotations are in the same class since there is a two-fold rotation perpendicular to each three-fold axis making it bilateral. These six two-fold rotations are all in one class but are distinct from the three two-fold rotations about the four-fold axes. The class structure is thus E, $8C_3$, $3C_4^2$, $6C_2$, $6C_4$.

The icosahedral group Y

Here there are six bilateral five-fold axes, fifteen two-fold axes, and ten bilateral three-fold axes. The class structure, see Murnaghan (1938) in the bibliography, is E, $12C_5$, $12C_5^2$, $15C_2$, $20C_3$.

9.4.2 Improper point groups

Before proceeding to the improper point groups in detail it is worth considering their general structure. The first point to notice is that the product of two improper elements is a proper element. This follows because improper rotations have determinant -1 (see section 7.4) and the product of two improper rotations must therefore have determinant $+1$ and be a proper rotation. Similarly the product of a proper and an improper element is an improper element. Consider now an improper group $\mathscr{G}$ with the elements divided in two sets, the proper elements $\mathscr{H}$ and the improper elements $\mathscr{K}$, so that the set $\mathscr{G}$ may be written $\mathscr{G} = \mathscr{H} + \mathscr{K}$. Taking the product of each side of this equation with a particular improper element $\mathscr{K}_i$ we have, using the group rearrangement theorem of section 2.9,

$$\mathscr{G} = \mathscr{K}_i\, \mathscr{G} = \mathscr{K}_i\, \mathscr{H} + \mathscr{K}_i\, \mathscr{K} \tag{9.4}$$

Since the elements $\mathscr{K}_i\, \mathscr{H}$ are improper and $\mathscr{K}_i \mathscr{K}$ are proper it follows that $\mathscr{K}_i\, \mathscr{H} = \mathscr{K}$ and $\mathscr{K}_i\, \mathscr{K} = \mathscr{H}$ enabling us to write

$$\mathscr{G} = \mathscr{H} + \mathscr{K}_i\, \mathscr{H} \tag{9.5}$$

where $\mathscr{K}_i$ is any one of the improper elements. If the inversion exists in the group as one of the improper elements then we can choose $\mathscr{K}_i = \mathrm{I}$ and, since I commutes with all other elements, the group is a direct product group: $\mathscr{G} = \mathscr{H} \times \{\mathrm{E}+\mathrm{I}\} = \mathscr{H} \times S_2$. If the group does not contain the inversion it may still be possible to write it as a direct product $\mathscr{G} = \mathscr{H} \times \{\mathrm{E}+\mathscr{K}_i\}$ if there exists an improper element $\mathscr{K}_i$ which commutes with all other elements. Also, when the inversion is not present the group $\mathscr{G}$ is isomorphic with the group $\mathscr{G}' = \mathscr{H} + \mathrm{I}\mathscr{K}$ which is a proper point group since $\mathrm{I}\mathscr{K}$ is a proper rotation. To see this isomorphism we first note that since I is not contained in $\mathscr{G}$ the group elements $\mathrm{I}\mathscr{K}_j$ cannot coincide with any of the elements $\mathscr{H}$. It is then trivial to show the isomorphism by associating the element $\mathrm{I}\mathscr{K}_j$ of $\mathscr{G}'$ with the element $\mathscr{K}_j$ of $\mathscr{G}$, while the elements $\mathscr{H}$ are common to both groups—for example if $\mathscr{H}_e\, \mathscr{K}_j = \mathscr{K}_m$ then it follows that $\mathscr{H}_e\,(\mathrm{I}\mathscr{K}_j) = \mathrm{I}\mathscr{H}_e\, \mathscr{K}_j = (\mathrm{I}\mathscr{K}_m)$. We now use these properties to derive the class structure for the improper point groups.

S_{2n} groups

These are cyclic groups of order $2n$ with each element in a class on its own. When n is odd they may be written as direct product groups since they contain the inversion. In this case $S_{4m+2} = C_{2m+1} \times S_2$.

C_{nh} groups

The horizontal reflection commutes with the n-fold rotations so that the groups are Abelian with each element in a class on its own. When n is even the inversion is present and the groups may be written as a direct product $C_{2mh} = C_{2m} \times S_2$. When n is odd they may also be written as a direct product of C_n with the groups S_1 consisting of the identity and σ_h only $C_{2m+1h} = C_{2m+1} \times S_1$.

C_{nv} groups

In these groups the vertical reflection does not commute with the n-fold rotations so that the groups are not Abelian, e.g. $C_n^k \sigma_v = \sigma_v C_n^{-k}$. This also shows that C_n^k and C_n^{-k} are in the same class. The reflections are all in the same class if n is odd but if n is even they split into two classes. From the general results above, these groups will be isomorphic with the groups obtained by multiplying the improper elements (reflections) by the inversion. This yields two-fold rotation axes perpendicular to the n-fold axes and consequently C_{nv} will be isomorphic with D_n, and have the same class structure. For example the classes of C_{4v} are E, $2C_4$, C_4^2, $2\sigma_v$, $2\sigma_v'$ and the classes of C_{5v} are E, $2C_5$, $2C_5^2$, $5\sigma_v$.

D_{nh} groups

Here the reflection plane σ_h commutes with all the elements of D_n so that the groups can be written as direct products $D_{nh} = D_n \times S_1$. When n is even the inversion is present and they may alternatively be written $D_{2mh} = D_{2m} \times S_2$. In either case the class structure is simply related to that of D_n, there being two classes for each class in D_n. For example the classes of D_{3h} are E, σ_h, $2C_3$, $2S_3$, $3C_2$, $3\sigma_v (= C_2\sigma_h)$.

D_{nd} groups

When n is odd there is a two-fold rotation axis perpendicular to each of the vertical reflection planes so that the group contains the inversion and can be written $D_{2m+1\,d} = D_{2m+1} \times S_2$ and its class structure follows from D_{2m+1}. When n is even the inversion is not a member of the group but multiplying the improper elements by the inversion yields further two-fold axes containing the vertical mirror planes and converts the group to D_{2n}. Thus for n even D_{nd} is isomorphic with D_{2n} and has the same class structure. As examples the classes of D_{2d} are E, $2S_4$, C_2, $2C_2'$, $2\sigma_d$, and the classes of D_{3d} are E, I, $2C_3$, $2S_3$, $3C_2$, $3\sigma_d$.

The group T_d

From the description of T_d above, the classes are

$$E,\ 3C_2,\ 8C_3,\ 6\sigma,\ 6S_4$$

since the three-fold rotation axes of T are now bilateral.

The groups T_h, O_h and Y_h

These all contain the inversion and are therefore direct product groups with S_2. Their class structure follows directly from that of T, O and Y.

9.5 The crystallographic point groups

One of the main applications of the theory of point groups occurs in the study of crystalline solids where the arrangement of atoms may be invariant under a point group. Only a few of the point groups listed in section 9.3 can occur in this context and the purpose of the present section is to find these so-called 'crystallographic point groups'. The crystal consists of an atom or arrangement of atoms repeated periodically throughout space. To describe this mathematically, we define a space lattice to be the set of points

$$\boldsymbol{n} = n_1 \boldsymbol{a}_1 + n_2 \boldsymbol{a}_2 + n_3 \boldsymbol{a}_3 \tag{9.6}$$

where n_1, n_2, n_3 are integers and $\boldsymbol{a}_1, \boldsymbol{a}_2, \boldsymbol{a}_3$, the 'primitive translation vectors' from one lattice point to the next, are fixed. The crystal then has an arrangement of atoms with specific relative orientation (sometimes called a basis) associated with each lattice point. We first consider the possible symmetry elements of the space lattice and will find that only certain point groups can occur. The relation of these symmetry groups of the lattice to the crystallographic point groups is then discussed.

To leave the lattice invariant a rotation R must take each lattice point $\boldsymbol{n}$ into another $\boldsymbol{m}$ and we now show that this implies a restriction on the possible angles of rotation. Let us write

$$\mathrm{R}\boldsymbol{n} = \boldsymbol{m} = m_1 \boldsymbol{a}_1 + m_2 \boldsymbol{a}_2 + m_3 \boldsymbol{a}_3$$

Clearly we can form a three-dimensional matrix representation for this rotation operation R which carries $\boldsymbol{n}$ into $\boldsymbol{m}$ by writing

$$\begin{pmatrix} m_1 \\ m_2 \\ m_3 \end{pmatrix} = \begin{pmatrix} R_{11} & R_{12} & R_{13} \\ R_{21} & R_{22} & R_{23} \\ R_{31} & R_{32} & R_{33} \end{pmatrix} \begin{pmatrix} n_1 \\ n_2 \\ n_3 \end{pmatrix}$$

By considering the case $n_2 = n_3 = 0$, $n_1 = 1$, we deduce that $m_p = R_{p1}$ and hence that R_{p1} is an integer. Similarly by putting $n_1 = n_3 = 0$, $n_2 = 1$, etc., we find that R_{p2} and R_{p3} are also integers. Hence the trace of R must also be an integer. If we now make a similarity transformation to a Cartesian set of basis vectors then the trace remains invariant and must still be an integer. But, in the Cartesian basis, a rotation of a vector through an angle ϕ has a trace of $(1 + 2\cos\phi)$, see equation (4.6). Thus the only allowed values for ϕ are $\phi = 0°$, $60°, 90°, 120°$ and $180°$, and hence five-fold axes and axes of order greater than six are excluded. Similarly for an improper rotation $\mathrm{S}(\phi)$ the trace is $(2\cos\phi - 1)$ and must also be an integer, so that the angle ϕ takes the same set of values. The space lattice defined by equation (9.6) clearly has inversion symmetry and, if it contains an n-fold rotation axis with $n > 2$, it will also have n 'vertical' mirror planes. These conditions, when put together, are seen to limit the possible point groups for space lattices to the following seven: S_2, C_{2h}, D_{2h},

D_{3d}, D_{4h}, D_{6h}, O. These seven different lattice symmetries or syngonies are called triclinic, monoclinic, orthorhombic, rhombohedral, tetragonal, hexagonal and cubic, respectively.

For simple monatomic crystals with one atom per unit cell these seven are the only possible crystallographic point groups. For more complicated crystals with a molecule or an arrangement of atoms associated with each lattice point, the symmetry will generally be reduced to the subgroup which leaves the molecule or arrangement of atoms invariant. The complete list of all possible crystallographic point groups will therefore be given by the above seven together with all their subgroups. There are 32 of them: $C_1, C_{1h}, C_n, C_{nv}, C_{nh}$, D_n, D_{nh}, with $n = 2, 3, 4$ or 6, $S_2, S_4, S_6, D_{2d}, D_{3d}, T, T_d, T_h, O$ and O_h. Groups which might at first sight appear to be missing from the list are C_{1v}, D_1, D_{1h}, S_1 and S_3 but these are the same as C_{1h}, C_2, C_{2v}, C_{1h} and C_{3h}, respectively.

9.6 Irreducible representations for the point groups

Having found the class structure of the point groups, irreducible representations can be generated as in chapter 4. In particular we can find the characters by the methods introduced there. Since the 32 crystallographic point groups occur frequently in problems in solid state physics we have listed, in appendix 1, the character tables for 11 proper point groups and the groups isomorphic to them. All the others follow, as indicated above, by taking direct products with the groups S_1 or S_2. These isomorphisms and the results of taking the direct products are listed in table 9.2. The 32 crystallographic point groups all appear in the first three rows of the table and the remaining two rows exhibit some other useful relationships. In constructing the character tables we need consider only the first row of table 9.2. The first five groups in the first row are cyclic Abelian groups with trivial one-dimensional representations, see section 4.8. We have already met the irreducible representations of D_3 in section 4.10. D_2 is isomorphic with $C_{2h} = C_2 \times S_2$ and D_6 is

Table 9.2

Proper point group $\mathcal{G}$	C_1	C_2	C_3	C_4	C_6	D_2	D_3	D_4	D_6	T	O
Direct product group $\mathcal{G} \times S_2$	S_2	C_{2h}	S_6	C_{4h}	C_{6h}	D_{2h}	D_{3d}	D_{4h}	D_{6h}	T_h	O_h
Improper group not containing I and isomorphic with $\mathcal{G}$		S_1		S_4	C_{3h}	C_{2v}	C_{3v}	C_{4v} D_{2d}	C_{6v} D_{3h}		T_d
Other improper group isomorphic with $\mathcal{G}$		S_2			S_6	C_{2h}			D_{3d}		
Direct product groups $\mathcal{G} \times S_1$	S_1	C_{2h}	C_{3h}	C_{4h}	C_{6h}	D_{2h}	D_{3h}	D_{4h}	D_{6h}		

isomorphic with $D_{3h} = D_3 \times S_1$ and may itself be written $D_6 = D_3 \times C_2$, see problem 2.7. Thus the only tables which need to be constructed *ab initio* are those for D_4, T and O.

In the limit when $n \to \infty$, the group C_n approaches the continuous group $\mathscr{R}_2$ discussed in section 7.3. Similar limits exist for the groups C_{nv}, C_{nh}, D_n and D_{nh} as $n \to \infty$, and they are usually denoted by $C_{\infty v}$, $C_{\infty h}$, D_∞ and $D_{\infty h}$. The group $C_{\infty v}$ is obtained from $\mathscr{R}_2$ by adding any vertical reflection plane and this generates all other vertical reflection planes. It is soon verified that this infinity of reflection planes is in the same class. As with finite n, the rotations through angles a and $-a$ form a class of two elements. The relation $\sigma_v \mathrm{R}(a) = \mathrm{R}(-a)\sigma_v$ for any vertical reflection means that the product of a rotation and a reflection is just a reflection in a plane rotated through half the angle,

$$\mathrm{R}(a)\sigma_v = \mathrm{R}(\tfrac{1}{2}a)\mathrm{R}(\tfrac{1}{2}a)\sigma_v = \mathrm{R}(\tfrac{1}{2}a)\sigma_v\mathrm{R}(-\tfrac{1}{2}a) = \sigma_v'$$

It also means that the reflections and rotations do not commute, so that the group $C_{\infty v}$ is not a product group. The group $C_{\infty v}$ is, however, isomorphic with the group O_2 of all orthogonal matrices in two dimensions since, in the basis formed by the unit vectors $\boldsymbol{e}_x$ and $\boldsymbol{e}_y$ in the horizontal plane, the rotations provide all orthogonal matrices with determinant $+1$ while the reflections provide those with determinant -1. (Note that the relation between O_2 and $\mathscr{R}_2$ differs from that between O_3 and $\mathscr{R}_3$ where there is a direct product, see section 7.4.) The other groups mentioned above are simply related to $C_{\infty v}$, since $C_{\infty h} = \mathscr{R}_2 \times S_2$, D_∞ is isomorphic with $C_{\infty v}$ and $D_{\infty h} = C_{\infty v} \times S_2$. From a practical point of view, the group $C_{\infty v}$ is the symmetry group of a linear molecule with no centre of symmetry while, if the molecule is symmetrical about the midpoint, as in a homonuclear diatomic molecule, the symmetry group is $D_{\infty h}$.

The irreducible representations of $C_{\infty v}$ may be built up from those of the subgroup $\mathscr{R}_2$. If $\boldsymbol{e}_m$ is a basis vector belonging to the representation $\mathrm{T}^{(m)}$ of $\mathscr{R}_2$, for example the function $\exp(im\phi)$, the vertical reflection transforms it into a vector $\boldsymbol{e}_{-m}$ which belongs to the representation $\mathrm{T}^{(-m)}$. Thus for $m \neq 0$, the irreducible representations of $C_{\infty v}$ are two-dimensional with character given by

$$\chi^{(m)}(a) = \exp(ima) + \exp(-ima) = 2\cos ma$$

for a rotation and $\chi^{(m)} = 0$ for a reflection, since there is then no diagonal matrix element. For the special case $m = 0$, a reflection transforms $\boldsymbol{e}_0$ into another vector $\boldsymbol{e}_0'$ which also has $m = 0$. If $\boldsymbol{e}_0'$ is a multiple of $\boldsymbol{e}_0$ then, since $\sigma_v^2 = 1$, the character of the reflection must be ± 1 and the same conclusion is reached if $\boldsymbol{e}_0$ and $\boldsymbol{e}_0'$ are independent because we may then form the two combinations $\boldsymbol{e}_0 \pm \boldsymbol{e}_0'$ which have characters ± 1 for reflections. Thus, for the case $m = 0$, there are two one-dimensional irreducible representations. These results are summarised in table 9.3.

Table 9.3

$C_{\infty v}$	E	2R(a)	σ_v
A_+	1	1	1
A_-	1	1	-1
E_m $(m \geq 1)$	2	$2 \cos ma$	0

9.7 Double-valued representations of the point groups

We have seen the appearance of double-valued representations of the group $\mathscr{R}_3$ in chapter 7 and their significance has been discussed in section 7.6 and in section 8.4 in connection with spin. Since the point groups are subgroups of $\mathscr{R}_3$ we shall expect to find the need for double-valued representations of the point groups also. A device for interpreting the double-valued representations as single-valued representations of a double group was explained in section 7.6 for $\mathscr{R}_3$. We now use the same device for the point groups. Recall that the double group consisted of the usual rotations R together with the new elements $\bar{\mathrm{E}}\mathrm{R}$, where $\bar{\mathrm{E}}$ was a rotation of 2π about any axis. The identity E is now associated with a rotation of 4π, in other words $\bar{\mathrm{E}}^2 = \mathrm{E}$. The new element $\bar{\mathrm{E}}$ commutes with all rotations.

Given a point group $\mathscr{G}$ with elements G_a the corresponding double group $\bar{\mathscr{G}}$ is thus generated by introducing the new elements $\bar{\mathrm{E}}\mathrm{G}_a$ which we denote by $\bar{\mathrm{G}}_a$. If the group contains a rotation C_n then

$$\mathrm{C}_n^n = \bar{\mathrm{E}} \text{ and } \mathrm{C}_n^{2n} = \mathrm{E}, \text{ with } \bar{\mathrm{C}}_n = \bar{\mathrm{E}}\,\mathrm{C}_n$$

In contrast with the group $\mathscr{R}_3$, the techniques used in deducing the irreducible representations of finite groups in chapter 4 made use of the relation (4.1) and hence did not produce any double-valued representations. We now use the device of the double group to find them. The representations of the double group will be of two kinds depending on the sign in the relation $\chi(\bar{\mathrm{G}}_a) = \pm\chi(\mathrm{G}_a)$. (To justify this relation we argue that since $\bar{\mathrm{E}}$ commutes with all group elements it must, from Schur's lemma, be represented by a multiple of the unit matrix in any irreducible representation. But since $\bar{\mathrm{E}}^2 = \mathrm{E}$ this multiple must be ± 1.) Those representations with the plus sign are clearly single valued as representations of $\mathscr{G}$, since the new element $\bar{\mathrm{E}}$ has the same matrix as the identity. The double-valued representations of $\mathscr{G}$ will be given by those representations of $\bar{\mathscr{G}}$ with the minus sign. To find how many such representations there are and to deduce the character table we follow the general methods of chapter 4, applied to the group $\bar{\mathscr{G}}$. First we must count the number of classes which tells us the number of irreducible representations. Knowing already the character table of the single-valued representations it is then straightforward to complete the character table using orthogonality and other techniques as in chapter 4.

The class structure of the proper double groups is easily deduced from that of the original point groups. Since $\overline{E}$ commutes with all elements it forms a class by itself. To each class of n-fold rotations ($n \neq 2$) there are two classes of the double group which we denote by C_n and $\overline{C}_n$. In the special case of two-fold rotations the elements C_2 and $\overline{E}C_2 = \overline{C}_2$ will be conjugate (i.e. in the same class) if the axis is bilateral, i.e. there is a group element which reverses the direction of the two-fold axis. To see this we note that rotations $R(\pm\theta)$ about a bilateral axis are conjugate so that, in particular, rotations $R(\pm\pi)$ are conjugate. But in the double group, although these are not identical, $R(-\pi) \equiv R(3\pi)$ and so $R(\pi)$ is conjugate with $R(3\pi)$. In other words, C_2 and $\overline{E}C_2$ are in the same class.

Table 9.4

$\overline{D}_4$	E	$\overline{E}$	$2C_4^2$	$2C_4$	$2\overline{C}_4$	$4C_2$	$4C_2'$
A_1	1	1	1	1	1	1	1
A_2	1	1	1	1	1	-1	-1
B_1	1	1	1	-1	-1	1	-1
B_2	1	1	1	-1	-1	-1	1
E	2	2	-2	0	0	0	0
E_1	2	-2	0	$\sqrt{2}$	$-\sqrt{2}$	0	0
E_2	2	-2	0	$-\sqrt{2}$	$\sqrt{2}$	0	0

As an example we consider the proper group $\overline{D}_4$. The group D_4 has classes E, C_4^2, $2C_4$, $2C_2$, $2C_2'$. As shown above, the new elements $\overline{E}C_2$ lie in the same class as C_2 so that the classes of $\overline{D}_4$ are E, $\overline{E}$, $2C_4^2$, $2C_4$, $2\overline{C}_4$, $4C_2$, $4C_2'$. The addition of two new classes means two more irreducible representations, with characters satisfying $\chi(\overline{G}_a) = -\chi(G_a)$. This implies that, for the bilateral two-fold rotations, the character must be zero. We can construct the complete character table displayed in table 9.4 by noting that we already have five irreducible representations of the single group D_4 which will form representations of the double group with $\chi(\overline{G}_a) = \chi(G_a)$ and have dimensions 1, 1, 1, 1, and 2. If the two new irreducible representations have dimensions s_6 and s_7, then from equation (4.33),

$$16 = 1^2 + 1^2 + 1^2 + 1^2 + 2^2 + s_6^2 + s_7^2$$

i.e.

$$8 = s_6^2 + s_7^2$$

and the only possible solution is $s_6 = s_7 = 2$. This gives us immediately the character for E and $\bar{E}$ and the remaining entries for $2C_4$ and $2\bar{C}_4$ are easily found from the orthogonality relations (4.25) and (4.36).

The class structure and irreducible representations of the double improper point groups can be deduced from the double proper point groups by the same methods which were used in subsection 9.4.2 and section 9.6. Again, groups containing the inversion may be expressed as direct products with S_2 and groups not containing the inversion are isomorphic with certain double proper groups. In fact, the relationships given in the first three rows of table 9.2 remain valid for the corresponding double groups. We note that the inversion, when present, commutes with all elements and satisfies $I^2 = E$ as before but a reflection σ now satisfies

$$\sigma^2 = (IC_2)^2 = EC_2^2 = \bar{E}$$

and hence $\sigma^4 = E$.

A complete list of character tables for the crystallographic double point groups is provided in appendix 1.

9.8 Time-reversal and magnetic point groups

In section 5.10 we introduced the time-reversal operator Υ which is a symmetry operator for the Hamiltonians of many physical systems. For a crystalline system with time-reversal symmetry the full point group symmetry will then be the product of the ordinary point group with the group of the identity and the time-reversal, since the latter commutes with the point group operations. However, we note that this can only be true for non-magnetic crystals since Υ reverses the direction of currents and spins and consequently the direction of magnetisation in magnetically ordered crystals. One would therefore expect that magnetic crystals would have only an ordinary point group symmetry but this is not always the case since even magnetic crystals may be invariant under an operation which is a product of Υ with a rotation, even though it is not invariant under Υ itself. For instance, in a ferromagnetic crystal with magnetisation along the z-axis, the operation ΥR, where R is a two-fold rotation about the x-axis, may be a symmetry operator since R reverses the direction of magnetisation and Υ restores it. Symmetry groups of this kind which contain time-reversal only in combination with a rotation or reflection, are called magnetic point groups and there are altogether 58 of them—see Bradley and Cracknell (1972) in the bibliography. (They are also called colour groups or Shubnikov groups because Shubnikov first studied them by considering the symmetry of regular solids with faces painted black or white. He included also an operation which reversed the colours and is analogous with time-reversal.)

As an example, we consider a ferromagnetic crystal which, disregarding time-reversal, has symmetry group D_3. In the magnetically disordered state, i.e.

above the transition temperature T_N at which magnetic ordering occurs, the inclusion of time-reversal leads to the larger group $D_3 \times \{E, \Upsilon\}$, since there is no magnetic moment. Below T_N, the crystal becomes ferromagnetic with moment along the three-fold axis. Since the three two-fold rotations C_2 about axes in the xy-plane reverse the direction of magnetisation they are no longer symmetry operations and neither is Υ. However, the products ΥC_2 remain as symmetry operations and the new symmetry group contains E, C_3, C_3^2 and the three operators ΥC_2 which involve the time-reversal. Notice that the magnetic point group here contains a subgroup C_3 of the original group D_3 and that, in the language of section 20.3 of volume 2, it is a normal subgroup of index 2. It may be shown that this feature, the existence of a normal subgroup of index 2, is present in all the magnetic point groups and may be used to construct all 58 of them. The notation for these groups is illustrated by the above example, for which one would write $D_3(C_3)$.

We shall not discuss the irreducible representations of either the product groups with time-reversal, or the magnetic point groups, since this is rather intricate (see, however, Bradley and Cracknell, 1972). In the former case, one must distinguish between point group representations which (a) are equivalent to a real representation, (b) are not equivalent to their complex conjugate representations and (c) are equivalent to their complex conjugate representations but not to a real representation. At all times one must take note of the anti-linear and anti-unitary nature of the operator Υ, see subsection 15.7.4 of volume 2.

9.9 Crystal field splitting of atomic energy levels

We now leave the mathematical discussion of the point groups to see how the point group symmetry helps us to understand the properties of atoms in a crystal field. A further application, to the theory of electron states in a molecule, is given in chapter 13 of volume 2. The regular arrangement of atoms in an infinite crystal can lead to symmetry with respect to a point group of operations centred on any one of the atoms. Furthermore, the complete symmetry of the crystal involves not only this local point group at a particular lattice site but also the translations which displace the atoms through an integer number of cells, leaving the same set of lattice points. To a certain extent, the properties which arise from the translational symmetry can be separated out and will be discussed in chapter 14 of volume 2. Here, we shall be concerned only with the effect of the local point group on the properties of one of the atoms of which the crystal is composed.

9.9.1 Definition of the physical problem

In chapter 8 we discussed the atomic states of a free atom using the invariance of the Hamiltonian under the group O_3 of rotations and inversion in three dimensions to classify the energy levels. When an atom is placed in a crystal the presence of the surrounding atoms produces electric and magnetic fields which lower the symmetry to one of the crystallographic point groups. As shown in section 5.8 this reduction of symmetry will cause a splitting of some of the degeneracies present in the energy levels of the free atom and the residual degeneracy will be determined by the point group. The details of the splitting and the exact energies of the states will, however, depend on the details of the crystal field potential and the wave functions, but by use of the Wigner–Eckart theorem some ratios of splittings may be deduced by group theory alone. Experimentally it is possible to investigate the low-lying energy states by a variety of techniques. At first the splittings were studied by means of heat capacity and magnetic susceptibility measurements, then more detail was obtained from optical absorption experiments. Finally, paramagnetic resonance spectroscopy has given a wealth of high precision data for the transition and rare-earth metal ions in crystalline salts.

When atoms combine to form a crystal there is usually some rearrangement of the outer electrons with a tendency towards closed-shell structures and ionisation. From the point of view of symmetry this would be uninteresting since a closed shell is non-degenerate and gives no scope for the expected splitting of degeneracies due to the restriction in symmetry from O_3 to a point group. However, certain elements from the lanthanide, actinide or iron transition series have an incomplete inner electron shell surrounded by closed shells. When these atoms combine into crystals the inner shells remain incomplete, being shielded by the outer electrons. Such crystals then provide a means of studying the effect of the crystal field on the atomic energy levels due to the incomplete inner shell. The detailed comparison between the theoretical predictions and the experimental results for these systems has not only yielded information about the strength of crystalline fields but also shed some light on the basic theory of atomic states.

The Hamiltonian for an atom or ion in a crystal field is given by $\mathrm{H} = \mathrm{H}_0 + \mathrm{H}_1 + \mathrm{H}_2 + \mathrm{H}_3$ in the notation of subsection 8.6.1, where H_0 is the central field, H_1 is the Coulomb repulsion between electrons, H_2 is the spin–orbit force and H_3 is the crystal field. In considering the effect of the crystal field H_3 it is instructive to study three simple limiting cases:

(1) *Weak field* in which $\mathrm{H}_3 \ll \mathrm{H}_2 \ll \mathrm{H}_1 \ll \mathrm{H}_0$;

(2) *Intermediate field* in which $\mathrm{H}_2 \ll \mathrm{H}_3 \ll \mathrm{H}_1 \ll \mathrm{H}_0$;

(3) *Strong field* in which $\mathrm{H}_2 \ll \mathrm{H}_1 \ll \mathrm{H}_3 \ll \mathrm{H}_0$.

The lanthanide or rare-earth salts, such as cerium ethyl sulphate, provide examples of the weak-field case. The electron configuration of the lanthanide ion has full shells up to and including the 4*d*-shell; there is then an incomplete 4*f*-shell containing 1 to 13 electrons as we run through the series and finally two full shells $5s^2\,5p^6$. The two outer shells of electrons surround the 4*f*-electrons and shield them from the crystal field so that the effective field they see is very weak. Here the $(2J+1)$-fold multiplicity of a spin–orbit multiplet J, described in subsection 8.6.3, will be split by the smaller perturbation H_3.

The iron-group salts, such as vanadium bromate, have fields in the intermediate to strong region. Here, the electron configuration has full shells up to and including the $3p$-shell. The incomplete $3d$-shell is then outermost and feels the full effect of the crystal field. Depending on the particular salt chosen the field here will vary from intermediate to strong. In the intermediate case H_3 is treated as a perturbation to the Coulomb interaction and has the effect of splitting the $(2L+1)$-fold degeneracy of an LS-coupled multiplet ^{2S+1}L. The spin–orbit interaction is then imposed as a final perturbation. In the strong field case the crystal field H_3 is second in importance only to the central field H_0 so that it has the effect of perturbing the central field in which the electrons move.

9.9.2 Deduction of the manner of splitting from symmetry considerations

Rather than take three different salts corresponding to the three cases (1), (2) and (3) we will illustrate the methods by using a single salt, vanadium bromate, and treating the crystal field in all three ways. In fact, the field for this salt belongs to case (2) but in practice one is not always sure *a priori* which of the three cases is appropriate for a particular salt. It may be necessary to do the calculation several ways and comparison with experiment will then enable us to decide which treatment is best.

Vanadium bromate has a chemical formula $Vd(BrO_3)_2, 6H_2O$. The reason for choosing what at first sight seems a most complicated hydrated salt is that we are interested in the behaviour of a single vanadium ion in a crystal field. If we were to choose a molecule with a high proportion of vanadium then there would be complicated interactions between the vanadium ions themselves. In the bromate the vanadium ions are widely spaced so that their interactions can be ignored. In this salt the six water molecules surround the vanadium ion in a very nearly regular octahedron so that the symmetry group can be taken to be O. (In reality there is very small trigonal distortion which we can ignore.) The vanadium ion is doubly ionised with the electron configuration $1s^2 \ldots 3p^6 3d^3$, i.e. it has three d-electrons in the incomplete shell. The ground-state multiplet of this free ion is ^4F with $L = 3$ and $S = 3/2$. (We simply quote this result here but it could be deduced following the method of subsection 8.6.4.) In the absence of the crystal field, the spin–orbit interaction would split the ^4F multiplet into four levels with $J = 3/2, 5/2, 7/2$ and $9/2$, as shown in the second column of figure 9.8, using equation (8.39).

Weak field

If the crystal field is weak,then the $(2J+1)$-fold degeneracy of each level J will be split by a small amount and the nature of the splitting may be deduced by studying the reduction of the $\mathscr{R}_3$ representation

$$D^{(J)} = \dot{\sum_{\alpha}} m_\alpha T^{(\alpha)} \tag{9.7}$$

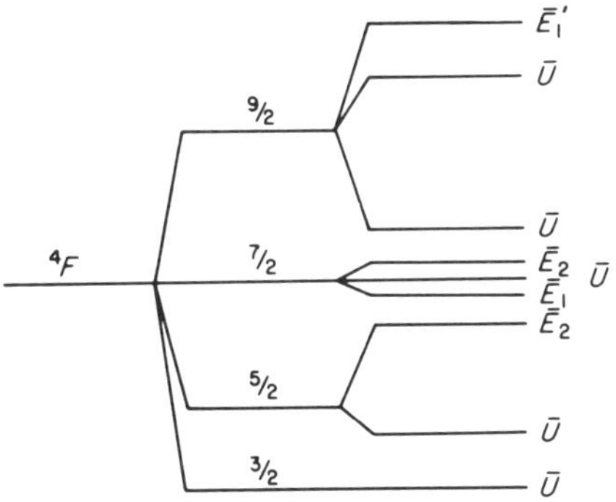

Figure 9.8

Table 9.5

	E	$\overline{E}$	$8C_3$	$8\overline{C}_3$	$3\overline{C}_2$ $3C_2$	$6C_4$	$6\overline{C}_4$	$6\overline{C}_2$ $6C_2$	
A_1	1	1	1	1	1	1	1	1	
A_2	1	1	1	1	1	-1	-1	-1	
E	2	2	-1	-1	2	0	0	0	
T_1	3	3	0	0	-1	1	1	-1	
T_2	3	3	0	0	-1	-1	-1	1	
$\overline{E}_1$	2	-2	1	-1	0	$\sqrt{2}$	$-\sqrt{2}$	0	
$\overline{E}_2$	2	-2	1	-1	0	$-\sqrt{2}$	$\sqrt{2}$	0	
$\overline{U}$	4	-4	-1	1	0	0	0	0	
$D^{(\frac{3}{2})}$	4	-4	-1	1	0	0	0	0	$= \overline{U}$
$D^{(\frac{5}{2})}$	6	-6	0	0	0	$-\sqrt{2}$	$\sqrt{2}$	0	$= \overline{E}_2 \dot{+} \overline{U}$
$D^{(\frac{7}{2})}$	8	-8	1	-1	0	0	0	0	$= \overline{E}_1 \dot{+} \overline{E}_2 \dot{+} \overline{U}$
$D^{(\frac{9}{2})}$	10	-10	-1	1	0	$\sqrt{2}$	$-\sqrt{2}$	0	$= \overline{E}_1 \dot{+} 2\overline{U}$
$D^{(3)}$	7	7	1	1	-1	-1	-1	-1	$= A_2 \dot{+} T_1 \dot{+} T_2$
$\overline{U} \times A_2$	4	-4	-1	1	0	0	0	0	$= \overline{U}$
$\overline{U} \times T_1$	12	-12	0	0	0	0	0	0	$= \overline{E}_1 \dot{+} \overline{E}_2 \dot{+} 2\overline{U}$
$\overline{U} \times T_2$	12	-12	0	0	0	0	0	0	$= \overline{E}_1 \dot{+} \overline{E}_2 \dot{+} 2\overline{U}$
$D^{(2)}$	5	5	-1	-1	1	-1	-1	1	$= E \dot{+} T_2$
$\{T_2 \times T_2 \times T_2\}_a$	1	1	1	1	1	-1	-1	-1	$= A_2$
$\{T_2 \times T_2\}_a$	3	3	0	0	-1	1	1	-1	$= T_1$
$T_1 \times E$	6	6	0	0	-2	0	0	0	$= T_1 \dot{+} T_2$
$\{E \times E\}_a$	1	1	1	1	1	-1	-1	-1	$= A_2$
$A_2 \times T_2$	3	3	0	0	-1	1	1	-1	$= T_1$
$D^{(1)}$	3	3	0	0	-1	1	1	-1	$= T_1$
$D^{(4)}$	9	9	0	0	1	1	1	1	$= A_1 \dot{+} E \dot{+} T_1 \dot{+} T_2$
$D^{(5)}$	11	11	-1	-1	-1	1	1	-1	$= E \dot{+} 2T_1 \dot{+} T_2$

where $T^{(\alpha)}$ denotes the irreducible representations of the point group O. We reproduce (from appendix 1) the character table for O in the first eight rows of table 9.5, including the double-valued representations. We also give the characters for the representations $D^{(J)}$, calculated from the formula (7.42). The coefficients m_α are then quickly deduced. The resulting splitting is shown in the third column of figure 9.8. We defer a discussion of the ordering of the levels $T^{(\alpha)}$ within each multiplet until appendix 5.4 of volume 2.

Intermediate field

Having illustrated the effect of a weak field in this example we hasten to add that the *actual* field in vanadium bromate is found to be sufficiently strong that it must be treated in the intermediate field approximation, case (2). Let us then move on to consider this case in which the crystal field is large compared with the spin–orbit force. Starting again from the single multiplet ^4F we must now consider the crystal field effect before putting on the small spin–orbit force. Thus we may first ignore the spin and study the reduction of the orbital state $L=3$ into irreducible representations of the point group O. For this we simply need the character of $D^{(3)}$, again calculated from formula (7.42) and shown in table 9.5, together with its reduction into $A_2 \dot{+} T_1 \dot{+} T_2$. Thus we have column two of figure 9.9 with the ^4F level splitting into three levels, denoted by $^4A_2$, $^4T_1$ and $^4T_2$ with degeneracies, 4, 12 and 12, including the four-fold spin degeneracy.

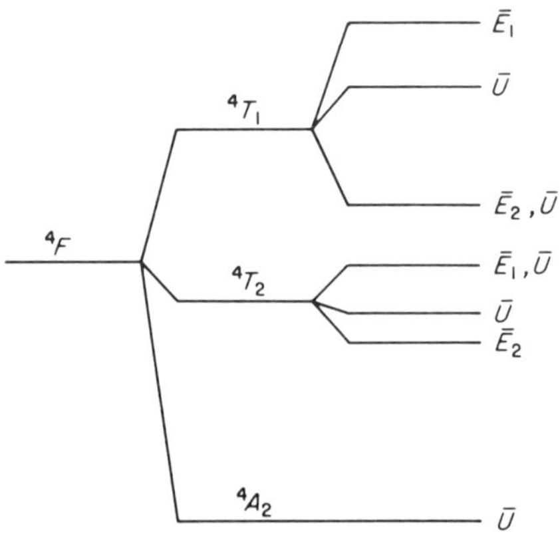

Figure 9.9

When the small spin–orbit interaction is taken into account a further splitting will take place. To find this splitting we note that the spin state $S=3/2$ transforms like $\bar{U}$ under the group O, see table 9.5 for $D^{(3/2)}$, so that, for example, the ${}^4A_2$ level transforms like the product $\bar{U} \times A_2$. But, products like this reduce into irreducible representations of O as follows,

$$\begin{aligned} \bar{U} \times A_2 &= \bar{U} \\ \bar{U} \times T_2 &= 2\bar{U} \dotplus \bar{E}_2 \dotplus \bar{E}_1 \\ \bar{U} \times T_1 &= 2\bar{U} \dotplus \bar{E}_2 \dotplus \bar{E}_1 \end{aligned}$$

deduced from the characters. This step is analogous to the reduction $D^{(3/2)} \times D^{(3)} = D^{(9/2)} \dotplus D^{(7/2)} \dotplus D^{(5/2)} \dotplus D^{(3/2)}$ for $\mathscr{R}_3$ which was used in deducing column two of figure 9.8. Thus the spin–orbit splitting leads to the picture given in the third column of figure 9.9. We shall not discuss the ordering and magnitudes of this spin–orbit splitting.

Strong field

Finally we study the nature of the splitting to be expected in the strong field limit, again using the same salt, vanadium bromate. In this case the crystal field is stronger than the Coulomb interaction e^2/r_{ij} between electrons so that it must be taken into account first. But this is simple to do since the crystal field is a one-body operator and its effect is to modify the central field in which the electrons move. Thus, instead of the three valence electrons moving in d orbits of a spherical field with the usual five-fold degeneracy due to $m = 2, 1, \ldots, -2$, the field will now have symmetry group O and the single electron orbit will be labelled by irreducible representations of O. Reference to table 9.5 shows that $D^{(2)} = E \dotplus T_2$ so that the five d-orbits split up into a doublet which we call e and a triplet t_2. In practice, the crystal field of vanadium bromate puts t_2 at a lower energy then e.

With three electrons there is a choice of which of the orbits e or t_2 to use but clearly the ground state will have the configuration of t_2^3 with all three particles in the orbit t_2 with lower energy. We also expect to find states corresponding to configurations t_2^2e, t_2e^2 and e^3 at increasing excitation energy. The total spin S will still be a good quantum number even though L is not and Hund's rule, favouring high S, still applies. We therefore restrict our discussion here to the maximum $S = 3/2$, as in cases (1) and (2).

The Coulomb interaction between the electrons now splits the orbital degeneracies of each configuration like t_2^3. The Hamiltonian at this stage with H_0, H_1 and H_3 is, of course, still invariant under the group O and the resulting states will be classified by those irreducible representations of O which appear on reduction of the product representations such as $T_2 \times T_2 \times T_2$ for the configuration t_2^3. However, since the spin part of the wave function corresponding to $S = 3/2$ is symmetric under all permutations, the Pauli principle demands that the orbital state be totally antisymmetric. We shall therefore need to know the character for the antisymmetrised products. However, in section 6.6 we deduced a formula (6.33) for the character for the symmetrised product of two factors so that, by subtraction from the character $\{\chi^{(\alpha)}(G_a)\}^2$ for the full product space, we have the result

$$\chi^{(\alpha\times\alpha)}_{\text{antisym}}(G_a) = \tfrac{1}{2}\{\chi^{(\alpha)}(G_a)\}^2 - \tfrac{1}{2}\chi^{(\alpha)}(G_a^2)$$

in the notation of section 6.6. For three factors we quote, from appendix 3.1 of volume 2, a similar formula

$$\chi^{(\alpha\times\alpha\times\alpha)}_{\text{antisym}}(G_a) = \tfrac{1}{6}[\{\chi^{(\alpha)}(G_a)\}^3 - 3\chi^{(\alpha)}(G_a^2)\chi^{(\alpha)}(G_a) + 2\chi^{(\alpha)}(G_a^3)]$$

For the lowest configuration t_2^3, the result for the product $T_2 \times T_2 \times T_2$ shown in table 9.5 indicates that only the A_2 state has the correct symmetry so that the ground state in the strong field is again A_2. A similar calculation for e^3 would give zero character, showing that no antisymmetric states can be found for this configuration, a result to be expected since e is only a doublet and two of the three electrons must inevitably be in the same state. For the other configurations t_2^2e and t_2e^2 we need to find the character of the antisymmetrised products $\{T_2 \times T_2\}_a$ and $\{E \times E\}_a$ and then multiply these by E and T_2, respectively. As shown in table 9.5 this procedure leads to the classification

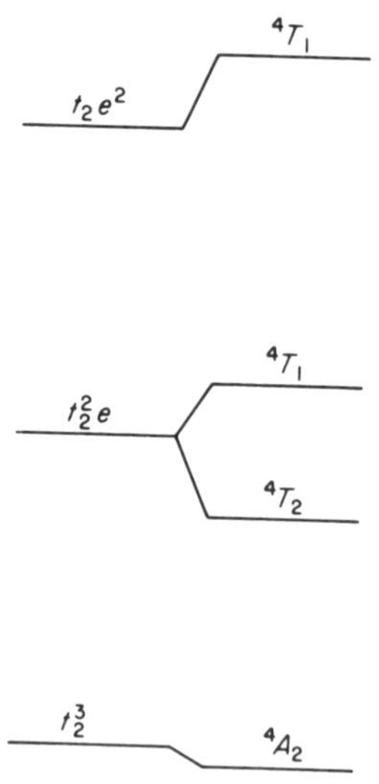

Figure 9.10

shown in the second column of figure 9.10. (To find the magnitude of the splitting between the new levels in the second column of figure 9.10 would involve a calculation with the two-body Coulomb interaction somewhat like that described in appendix 5 of volume 2.) The reason why there are more levels shown in figure 9.10 is that all states with $S = 3/2$ have been considered. The corresponding treatment for cases (1) and (2) would have included also the excited ^4P multiplet. The pattern of spin–orbit splitting in the strong field case is the same as that for an intermediate case, although the magnitudes may be different.

For accurate results or for a field whose strength does not lie close to one of the three limiting cases discussed above, allowance must be made for mixing between states with the same O-label. This is a detailed numerical problem involving a matrix diagonalisation. The passage from weak field through intermediate to strong field is of course continuous.

Selection rules

The discussion of selection rules in sections 5.4 and 5.6 finds immediate application in this crystal field problem. In section 5.4 we saw that the operator for dipole transitions transformed according to the vector representations $D^{(1)}$ of $\mathcal{R}_3$ and hence, from table 9.5, according to the representation T_1 of the group O. Since $T_1 \times \bar{U} = \bar{E}_1 \dot{+} \bar{E}_2 \dot{+} 2\bar{U}$, from the table, one would expect all states shown in figures 9.8 to 9.10 to be excited from the ground state $\bar{U}$. However, a closer look at the properties of the states shows that selection rules do occur. In the weak-coupling case of figure 9.8 the J-selection rule of equation (8.3) forbids excitation to the 7/2 and 9/2 states, so that only the first two states $\bar{U}$ and $\bar{E}_2$ may be excited. In the intermediate and strong-field cases illustrated in figures 9.9 and 9.10 the reduction $T_1 \times A_2 = T_2$ deduced from table 9.5 shows that the T_1 states cannot be excited by the dipole process. Thus the first four states $\bar{U}$, $\bar{E}_2$, $\bar{E}_1$ and $\bar{U}$ would be expected to be excited.

However, we must remember that all the states considered in this example originated from the d^3 configuration and thus have even parity, see subsection 8.6.2. Electric dipole transitions between these states are therefore forbidden, see the discussion of the group S_2 in section 5.6. Only the much slower magnetic dipole transitions are allowed. In practice, there is often a small admixture of states of the opposite parity due to a lack of inversion symmetry in the crystal, and weak electric dipole transitions may then occur.

9.9.3 Effect of a magnetic field

Experimental measurements of the magnetic susceptibility and the paramagnetic resonance spectrum at low temperatures give details of the energy levels of the low-lying states in the presence of a weak magnetic field.

A measurement of the magnetic susceptibility yields the energy gaps indirectly since it involves the population of the levels as a function of temperature. Paramagnetic resonance spectroscopy measures the energy gaps directly by radio frequency transitions between the levels. Many refinements (see Abragam and Bleaney, 1970, in the bibliography) are necessary in a complete treatment of these small splittings (e.g. spin–spin interaction and hyperfine interaction which have a comparable magnitude in many cases) but in this section we investigate the effect of the magnetic field alone for the vanadium bromate salt discussed in previous sections.

As in section 8.6 the magnetic field adds a further term

$$H_{mag} = (\mathbf{L} + 2\mathbf{S}) . \boldsymbol{B} \mu_B \tag{9.8}$$

to the Hamiltonian H of subsection 9.9.1, and it will be treated as a perturbation to H.

Weak field

Here the ground state $\overline{U}$ was a quartet corresponding to states $J = \frac{3}{2}$, $M_J = \frac{3}{2}, \frac{1}{2}, -\frac{1}{2}, -\frac{3}{2}$.

Since in this case there was no crystal field splitting and the ground state still has definite J, the splitting due to the magnetic field is exactly as described in section 8.6 with a Lande g-factor of $g_J = 0.4$ from equation (8.40) and a splitting proportional to M.

Intermediate and strong fields

The calculations illustrated in figure 9.9 show that when the crystal field is stronger than the spin–orbit coupling the ground state, although still a quartet $\overline{U}$, is now an orbital singlet A_2 so that, in the absence of a spin moment, the state could not be split by a magnetic field. This follows from the general selection rule arguments. The operator $\mathbf{L}$ transforms like a vector $D^{(1)}$ under rotations and thus, using table 9.5, like the representation T_1 of the group O. But such an operator must have zero expectation value in A_2 through the usual selection rule argument because the product $T_1 \times A_2$ does not contain A_2. The energy shift due to the magnetic field is thus simply given from equation (9.8) by

$$\begin{aligned} \Delta E_{M_S} &= 2\mu_B \langle \boldsymbol{B} . \mathbf{S} \rangle \\ &= 2\mu_B B M_S \end{aligned}$$

with a g-factor of 2. This result that the magnetic properties of the ground state are due to the spin only is often referred to as the 'quenching' of the orbital motion by the strong crystal field.

Notice that although the ground state has the same symmetry property $\overline{U}$ under the group O in all three cases, the g-factor is 0.4 in case (1) and 2.0 in the others. From experiment the g-value is measured to be 1.96, consistent with an intermediate or strong field.

Bibliography

The first use of point groups in describing crystal field splittings was made by

Bethe, H. A. (1929). *Annln Phys.*, **3**, 133

Very complete sets of tables relating to the crystallographic point groups have been given by

Koster, G. F., Dimmock, J. O., Wheeler, R. G. and Statz, H. (1963). *Properties of the Thirty-two Point Groups* (Technology Press, M.I.T., Cambridge, Mass.)

The icosahedral group is described by

Murnaghan, F. D. (1938). *Theory of Group Representations* (Johns Hopkins Press, Baltimore)

and for a detailed discussion of the double-valued representations see

Opechowski, W. (1940). *Physica*, **7**, 552
Bradley, C. J. and Cracknell, A. P. (1972). *The Mathematical Theory of Symmetry in Solids* (Oxford University Press)

For further reading in the theory of crystal fields we suggest either Judd (1963), see bibliography to chapter 8, or

Abragam, A. and Bleaney, B. (1970). *Electron Paramagnetic Resonance of Transition Ions* (Oxford University Press)

Problems

9.1 Use the stereogram to generate all the elements of the group O from the four three-fold rotations.

9.2 A cubic crystal has symmetry group O. (a) If the crystal is then distorted by stretching along the (111) direction, a three-fold axis, what is the resulting symmetry group? (b) If, alternatively, it is stretched along a four-fold axis, what symmetry group results?

9.3 A cubic crystal with symmetry group O has eigenstates labelled by the irreducible representations of O. If the symmetry is lowered as in problem 9.2 deduce, for each case, which states are split and label the resulting states by irreducible representations of the new symmetry groups.

9.4 Show that the function $f(\theta, \phi) = Y^{(l)}_m(\theta, \phi) + Y^{(l)}_{-m}(\theta, \phi)$, where l and m are even integers, is invariant under the symmetry group D_{2h}.

9.5 Using the basis of functions x, y, z, construct the matrices for one element in each class of the group O. By comparing the character given by these matrices with the character table in the appendix, show that the functions x, y, and z belong to the representation T_1. This implies that an atomic p-state is not split by a cubic perturbation.

9.6 Show that the space of the second-order polynomials in x, y and z reduces to the sum $A_1 \dot{+} E \dot{+} T_2$ of representations of the group O. Use the symmetrised product (see subsection 9.9.2) of T_1 with itself. By projection or otherwise, construct the polynomials which transform according to each of these representations.

9.7 Use equation (7.42) to determine, for the group O, the character of the representation generated by the set of seven spherical harmonics $Y_m^{(3)}$. Reduce this representation into irreducible representations of O. Hence deduce the nature of the splitting of an atomic f-state in the presence of a cubic perturbing field.

9.8 An electron moves in a potential field with symmetry group C_{4v} and its wave function has symmetry A_1. (See appendix 1 for the character table.) Does the energy of the state change (in a first-order approximation) when a weak constant electric field is applied (a) along the four-fold axis and (b) in a plane perpendicular to the four-fold axis. (For a field along the z-axis the relevant operator is proportional to z.)

9.9 An electron moves in a potential with symmetry groups D_{2h} so that its eigenstates are labelled by the irreducible representations $A_1^+, B_1^+, B_2^+, B_3^+, A_1^-, B_1^-, B_2^-, B_3^-$ (see appendix 1). Deduce selection rules for electric and magnetic dipole transitions between these states.

9.10 Use the character table for D_3 to deduce the character table for the double group $\bar{D}_3$. (Check your result against table A2 of appendix 1.)

9.11 For each of the following molecules, deduce the symmetry group and classify the vibrational modes, using the methods described in chapter 6: (a) The phosphorus molecule P_4, with the atoms at the vertices of a regular tetrahedron. (b) The methane molecule CH_4, which has the four H-atoms at the vertices of a regular tetrahedron and the C-atom at the centre. (c) The ethylene molecule C_2H_4, which is coplanar with two H-atoms attached to each C-atom and the CH-bonds making an angle of about 120° with the CC-bond. (Take the molecule to lie in the xz-plane with the CC-bond as z-axis.)

9.12 An atom has a single valence electron in a d-state. The atom is placed in a crystalline environment with point group symmetry O_h. Assuming the crystal field to be weak and neglecting the effect of spin, calculate the number of levels into which you would expect the d-state to split.

9.13 If, in the atom of the previous problem, we include the effect of spin and assume the spin–orbit coupling to be large compared with the crystal field, what would be the nature of the resulting state?

9.14 Deduce selection rules for dipole transitions between the states formed in problem 9.3.

10

Isospin and the Group SU_2

The symmetries which have been used in all previous applications have been connected with the position coordinates of the particles which make up the system. In particular we have studied the physical consequences of full rotational symmetry and of symmetry under a finite group of rotations. Inversion of axes and reflections were also studied. The intrinsic spin of a particle, which we described in section 8.4, is not connected with the position coordinates but rather describes an intrinsic property of the particle. Nevertheless it is transformed under rotations of the system and so forms a part of the properties of the system under rotation. In this chapter and the next we study symmetries which have nothing to do with position coordinates and in fact nothing to do with ordinary three-dimensional space. Instead, the new symmetries are concerned entirely with intrinsic properties of the particles such as their electric charge.

The forces of nature are believed to be of four distinct types: gravitational, electromagnetic, strong nuclear and weak nuclear. The first two of these are familiar in the everyday world. The strong nuclear force is that which holds the atomic nucleus together and is responsible for the enormous energy release in atomic bombs and nuclear reactors. In the present chapter we study a symmetry of the strong nuclear force which is known as 'isospin'. (The words 'isotopic spin' and 'isobaric spin' are sometimes used in place of isospin. All three terms have precisely the same meaning.) In chapter 11 this symmetry is extended in the realm of the very short-lived elementary particles into what is known as SU_3 symmetry. We shall not, in this chapter, discuss the weak nuclear force which is responsible for β-decay and similar processes in nuclei.

The mathematical description of isospin will be completely analogous to that of the spin in a system of particles each with intrinsic spin $s = \frac{1}{2}$ and we shall be able to use all the results derived for the group $\mathscr{R}_3$ in chapter 7. The reason for this is that the group underlying isospin symmetry is a group SU_2 which is homomorphic with $\mathscr{R}_3$. We defer a detailed mathematical description of the relation between these two groups until section 18.13 of volume 2.

10.1 Isospin in nuclei

The most elementary example of isospin symmetry occurs in the structure of atomic nuclei. These nuclei, which carry a positive charge Ze equal in magnitude to the charge on the Z electrons of the atom, are known to be composed of a number Z of protons and N of neutrons held together by the strong attractive nuclear forces in a small volume of radius about $(10)^{-12}$ cm. A proton has positive charge equal to that on an electron and a neutron has no charge. The masses of the proton and neutron are very similar, $M_p c^2 = 938.26$ MeV and $M_n c^2 = 939.55$ MeV, and they both have a spin of $\frac{1}{2}$. We use p and n as abbreviations for proton and neutron. Furthermore, experiment shows that, apart from the contribution of the electromagnetic interaction, the forces between pp, nn and np are the same within about 1%. Thus the strong nuclear force in nuclei is found experimentally to be independent of whether the particles involved are neutrons or protons. This clearly indicates a symmetry in the strong nuclear force and our main interest will be in studying the consequences of this symmetry in complex nuclei, where many neutrons and protons are interacting together. Because of the similarities between neutron and proton, it is sometimes convenient to use the word 'nucleon' to refer to either a neutron or a proton and to introduce the 'nucleon number' $A = N + Z$. (Because of the near-equality of masses of neutron and proton the number A is sometimes called the mass number of the nucleus.) One may in fact regard the neutron and proton as two different states of the same particle. There is actually no physics in this last sentence. It is always possible to regard any two particles as different states of the same particle but usually this description would lead to unnecessary complication. For the neutron and proton the near-equality of masses and interactions brings a considerable simplification as we shall soon see. It means that the Hamiltonian is invariant with respect to transformations between proton and neutron states and to mixtures of these two states. We first explore the mathematical nature of these transformations.

The two states, proton and neutron, of a nucleon may be taken to define a two-dimensional abstract vector space with $|p\rangle \equiv \binom{1}{0}$ and $|n\rangle \equiv \binom{0}{1}$.

Consider the group U_2 of unitary 2×2 matrices in this space (any more general transformation would not conserve normalisation of the states). As shown in section 7.2 the infinitesimal matrices must be Hermitian, apart from a factor i, and conversely, any Hermitian matrix H leads to a unitary matrix U through exponentiation $\mathrm{U} = \exp i\mathrm{H}$. But any Hermitian 2×2 matrix may be written as a linear combination of the four matrices

$$1 = \begin{pmatrix} 1 & 0 \\ 0 & 1 \end{pmatrix}, \ \tau_x = \begin{pmatrix} 0 & 1 \\ 1 & 0 \end{pmatrix}, \ \tau_y = \begin{pmatrix} 0 & -i \\ i & 0 \end{pmatrix}, \ \tau_z = \begin{pmatrix} 1 & 0 \\ 0 & -1 \end{pmatrix} \quad (10.1)$$

Hence these four matrices may be taken as the infinitesimal operators of the group U_2. If the unitary matrices are restricted to having a determinant equal to $+1$ then the group is called SU_2. This restriction implies that the trace of the infinitesimal matrices must be zero and so the unit matrix 1 is excluded, leaving the three matrices τ_q as the infinitesimal operators of SU_2. (The restriction from U_2 to SU_2 simply removes the freedom to change the phase of both states of the single nucleon simultaneously.) One sees immediately by comparison with equation (8.15) that the matrices τ_q are the same as the spin matrices for a particle with $s = \frac{1}{2}$, apart from a factor $\frac{1}{2}$. Thus if we define $\mathrm{t}_q = \frac{1}{2}\tau_q$ then the matrices t_q will be identical with the spin matrices s_q and will satisfy the same commutation relations (7.26) as the infinitesimal operators of the group $\mathscr{R}_3$. Since all the properties of the irreducible representations of $\mathscr{R}_3$ were deduced in section 7.4 from these commutation relations, it follows that the same irreducible representations will occur for the group SU_2. We shall use the notation $\mathrm{D}^{(T)}$ for an irreducible representation where, as in R_3, the label T may take the values $T = 0, \frac{1}{2}, 1, \ldots$. We shall call T the 'isospin' and, remembering the link established in chapter 8 between the group $\mathscr{R}_3$ and angular momentum, we shall expect that a Hamiltonian which is invariant under SU_2 will lead to an angular momentum-like structure in the eigenfunctions. (We use a capital T to denote isospin in general for any system and reserve a small t for the isospin of a single nucleon, a convention which is consistent with the use of L and l in section 8.2.) In particular, the two states $|p\rangle$ and $|n\rangle$ of a single nucleon will have $t = \frac{1}{2}$ and conventionally, we choose $|p\rangle$ to have $m_t = \frac{1}{2}$ and $|n\rangle$ to have $m_t = -\frac{1}{2}$, where m_t denotes the eigenvalue of t_z. The operator corresponding to the charge on a nucleon is then given by $\mathrm{Q} = e(\frac{1}{2} + t_z)$ which has the required property $\mathrm{Q}|p\rangle = e|p\rangle, \mathrm{Q}|n\rangle = 0$. The choice made here, by which we associate, respectively, p and n with eigenvalues $\pm\frac{1}{2}$ of t_z, is the convention usually followed in elementary-particle physics. It is the custom in nuclear physics to use the opposite convention since this leads to positive values of T_z in most nuclei where, because of the repulsive Coulomb forces, there are more neutrons than protons.

10.1.1 Isospin labelling and degeneracies

The interest in isospin lies in its use when more than one nucleon is present and one considers simultaneous isospin transformations in all nucleons. The vector space V_A, describing the possible charge states of a system of A nucleons, has total dimension 2^A and we are considering the transformations induced in this space by simultaneous SU_2 transformations of each nucleon. The unitary transformation in V_A would formally be written as a product $\mathrm{U} = \prod_{i=1}^{A} \mathrm{U}(i)$, where $\mathrm{U}(i)$ is the SU_2 transformation in the 2-dimensional space of nucleon i. For the infinitesimal operators T this product leads to the sum $\mathrm{T} = \sum_{i=1}^{A} \mathrm{t}(i)$ and one sees immediately that the T satisfy the same commutation relations as the 2×2 matrices $\mathrm{t}(i)$. The space V_A will clearly provide a representation of SU_2 which in general will not be irreducible. However it may always be reduced, leading to a classification of states by the label T. This extension to many nucleons is similar to the extension from the angular momentum of a single particle to the total angular momentum L of a system of particles, as described in section 8.2. Isospins may be coupled together in exactly the same way as angular momenta so that for a nucleus with A nucleons it is possible to construct a total isospin T which may have any value up to $\frac{1}{2}A$. Since the strong nuclear force is charge-independent it will not depend on the charge states of the nucleons—it will be a unit operator in the abstract two-dimensional isospace of each nucleon. Thus the Hamiltonian commutes with the operators of the SU_2 group and so, by the general arguments of section 5.3, the eigenfunctions of the Hamiltonian may be labelled by an irreducible representation $\mathrm{D}^{(T)}$ of SU_2 or in other words by the isospin T.

For a system of N neutrons and Z protons, with $N+Z=A$, the total isospin operators are written as $\mathrm{T}_q = \sum_i \mathrm{t}_q(i)$ for $q = x, y, z$, where i runs over all nucleons. Thus in particular, the total charge is given by $\mathrm{Q} = \sum_i \mathrm{Q}(i) = e\{\frac{1}{2}(N+Z) + \mathrm{T}_z\}$ but since clearly, in a given nucleus, the charge operator is diagonal with value $\langle \mathrm{Q} \rangle = eZ$ it follows that T_z must be diagonal with value $M_T = \frac{1}{2}(Z-N)$. Since the representation $\mathrm{D}^{(T)}$ has dimension $(2T+1)$ it follows that an eigenvalue labelled by T will have a $(2T+1)$-fold degeneracy. In the group $\mathscr{R}_3$ it was convenient to label these degenerate states by the value of J_z, namely $M = J, J-1, \ldots, -J$. In precisely the same way the $(2T+1)$-fold degeneracy is labelled by the value of T_z which may take the values $M_T = T, T-1, \ldots, -T$. Since, as shown above, $M_T = \frac{1}{2}(Z-N)$, it follows that the $(2T+1)$ components of a T-multiplet correspond to different nuclei, i.e. with different values of $(Z-N)$ but of course with the same number of nucleons $A = (N+Z)$.

As a first example, consider two nucleons only and we should expect to form states with $T=1$ and $T=0$. Calling the two particles i and j there are clearly four possible charge states

$$\psi_1 = |p_i\, p_j\rangle,\ \psi_2 = |p_i\, n_j\rangle,\ \psi_3 = |n_i\, p_j\rangle,\ \psi_4 = |n_i\, n_j\rangle$$

whose values of M_T are 1, 0, 0 and -1, respectively. It is soon verified that the symmetric combination $\psi'_2 = (\psi_2 + \psi_3)/2^{\frac{1}{2}}$ has $T = 1$, together with ψ_1 and ψ_4, while $\psi'_3 = (\psi_2 - \psi_3)/2^{\frac{1}{2}}$ has $T = 0$. To see this we first note that since ψ_1 has $M_T = +1$ it must have $T = 1$. Then we may construct the state with $T = 1$, $M_T = 0$ by using the isospin lowering operator T_- defined in complete analogy with the angular momentum operator J_- of equation (7.40). Explicitly, for a single particle, t_- is given from equation (10.1) as the 2×2 matrix

$$\mathrm{t}_- = \tfrac{1}{2}(\tau_x - i\tau_y) = \begin{pmatrix} 0 & 0 \\ 1 & 0 \end{pmatrix}$$

with the property $\mathrm{t}_-|p\rangle = |n\rangle$, $\mathrm{t}_-|n\rangle = 0$. For two particles $\mathrm{T}_- = \mathrm{t}_-(i) + \mathrm{t}_-(j)$ so that

$$\mathrm{T}_-|p_i p_j\rangle = |n_i p_j\rangle + |p_i n_j\rangle = \psi_2 + \psi_3 = 2^{\frac{1}{2}}\psi'_2$$

The factor $2^{\frac{1}{2}}$ is expected from equation (7.40). Similarly $\mathrm{T}_-\psi'_3 = 0$ so that since ψ'_3 has $M_T = 0$ it must have $T = 0$. It is worth commenting here that although charge independence implies a degeneracy between the three members ψ_1, ψ'_2 and ψ_4 of the $T = 1$ multiplet it does *not* imply degeneracy also with the $T = 0$ state ψ'_3. The reason for this is that the $T = 1$ states are even under permutation of the particle labels i and j while the $T = 0$ state is odd. Consequently, to satisfy the Pauli principle, the $T = 1$ states must be combined with odd states in the spin and orbital coordinates, while the $T = 0$ state must be combined with an even state of those remaining coordinates. In reality, the strong nuclear force leads to a lower energy for the even state of spin and orbital coordinates. The ground state of the *np* system (the deuteron) therefore has $T = 0$. In the $T = 1$ states the nuclear force is insufficiently strong to give any bound state which explains why, in spite of charge-independence, the *np* system has a bound state while the *nn* system does not.

In the structure of atoms it is found that the ground state always has the maximum value of the total spin S, a result known as Hund's rule and in section 8.6.5 we gave a reason for this. In the structure of nuclei, one finds that the ground state has the *minimum* value of the total isospin T and very similar reasons can be used to explain the result. The reason why it is the minimum T rather than the maximum is that in nuclei the forces between nucleons is attractive whereas the forces between electrons in the atom are repulsive. For a given nucleus, N and Z are fixed and so therefore is M_T. Thus, since necessarily $T \geq |M_T|$, the minimum value of T for a given nucleus is $|M_T| = \frac{1}{2}|Z - N|$. States with greater values of T will occur as excited states.

As a further example of the isospin labelling and degeneracy, consider the nuclei with mass number $A = 13$. The only stable nucleus of this kind is carbon 13 with $Z = 6$, $N = 7$. However, the nucleus nitrogen 13 ($Z = 7$, $N = 6$) has been produced artificially in nuclear reactions and its properties studied. The nuclei boron 13 ($Z = 5$, $N = 8$) and oxygen 13 ($Z = 8$, $N = 5$) have also been produced but they are very unstable and rather little is known about them.

From the arguments given above we would expect ^{13}C and ^{13}N to have $T = \frac{1}{2}$ and to correspond to the two possible projections $M_T = \mp\frac{1}{2}$. On the other hand, ^{13}B and ^{13}O would have $T = \frac{3}{2}$ and $M_T = \mp\frac{3}{2}$, respectively. The two remaining projections $M_T = \mp\frac{1}{2}$ of the $T = \frac{3}{2}$ system would correspond to excited states of the nuclei ^{13}C and ^{13}N. Thus we expect the lowest states of each isospin to be as illustrated in figure 10.1. We use broken lines for $T = \frac{3}{2}$ states.

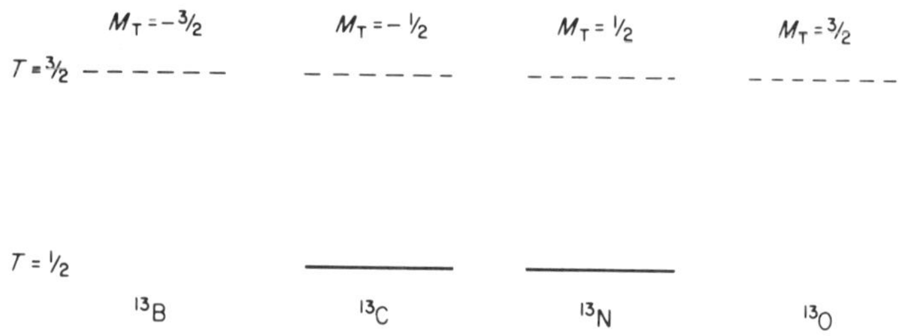

Figure 10.1

In reality, the situation is complicated by the presence of the electromagnetic Coulomb force between the protons. This force is clearly not charge-independent and so tends to break the isospin symmetry but fortunately it is small compared with the nuclear forces and its contribution may be estimated quite simply. If this Coulomb energy is calculated and subtracted from the experimental data, one finds the picture given in figure 10.2 for the low-lying energy levels. The figure records the spin J and parity of each level and also its excitation energy. It is seen that the ground states do in fact exhibit the isospin degeneracies indicated in the previous figure. Furthermore, apart from small shifts due to the Coulomb force, there is degeneracy for each excited state. That is, to each excited state of ^{13}C with some J and parity, there is a corresponding state of ^{13}N with the same J and parity and differing only in its M_T-value. Thus although chemically ^{13}N and ^{13}C are different species, the structure of their nuclei is very similar.

Because the isospin relates nuclei with the same mass number A, it is sometimes called isobaric spin. The word 'isotopic' was first used but this is not very appropriate since isotopes are nuclei with the same chemical properties, i.e. the same Z, but different N and therefore different A. Isospin does not give any relation between isotopes such as ^{12}C, ^{13}C and ^{14}C. The term 'isobaric analogues' is used to describe states like those in figure 10.1 which differ only in the projection M_T.

10.1.2 Splitting of an isospin multiplet

As remarked above, the Coulomb force breaks the isospin symmetry, but

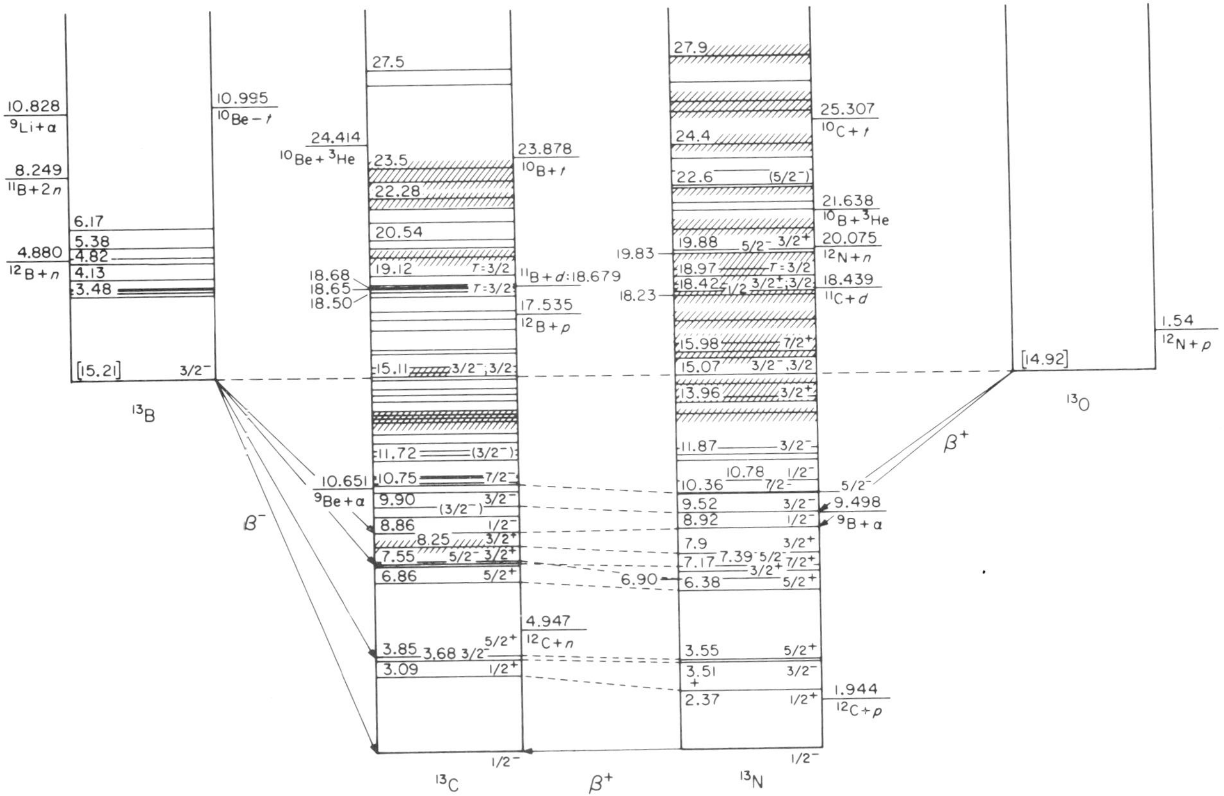

Figure 10.2 Isobar diagram, $A = 13$. The diagrams for individual isobars have been shifted vertically to eliminate the neutron–proton mass difference and the Coulomb energy, taken as $E_c = 0.60Z(Z-1)/A^{\frac{1}{3}}$. Energies in square brackets represent the (approximate) nuclear energy, $E_n = M(Z, A) - ZM(\mathrm{H}) - NM(n) - E_c$, minus the corresponding quantity for ^{13}C: here M represents the atomic mass excess in MeV. Levels which are presumed to be isospin multiplets are connected by dashed lines (from *Nuclear Physics*, **A152,** 32, 1970)

being weak, this breaking may be treated in perturbation theory. In the previous section we quoted experimental data in figure 10.2 from which the Coulomb energy had been subtracted. This was done in order to exhibit the isospin degeneracies. We now take a closer look at the Coulomb energies—see Wilkinson (1969) in the bibliography. Although the Coulomb force is not an isoscalar it certainly conserves the charge so that M_T remains a good quantum number. Thus the splitting of an isospin multiplet, with fixed T and $M_T = T, T-1, \ldots, -T$ is very like the Zeeman splitting of an angular momentum multiplet, with fixed J and $M_J = J, J-1, \ldots, -J$, by a magnetic field in the z-direction, see section 8.5. In the case of the Zeeman splitting it was shown from a symmetry argument that the splitting was proportional to M_J, because the relevant operator was a vector operator. In the case of the Coulomb energy splitting of an isospin multiplet we must study the nature of the Coulomb operator and this will enable us to deduce a formula for the M_T dependence of the splitting without making detailed calculations for any particular nucleus. The Coulomb force may be written as a sum over protons

$$V_c = \sum_{\substack{i<j \\ \text{(protons)}}} e^2/r_{ij}$$

or as a sum over all nucleons

$$V_c = \sum_{i<j} [\tfrac{1}{2} + t_z(i)]\,[\tfrac{1}{2} + t_z(j)]\,e^2/r_{ij} \qquad (10.2)$$

using the isospin operators. (Since $\langle t_z(i) \rangle = -\frac{1}{2}$ for a neutron the only contribution to the sum occurs when i and j are both protons.) The isospin dependence of V_c is contained in the two factors

$$[\tfrac{1}{2} + t_z(i)]\,[\tfrac{1}{2} + t_z(j)] = \tfrac{1}{4} + \tfrac{1}{2}[t_z(i) + t_z(j)] + t_z(i)\,t_z(j)$$

As in subsection 7.4.6 for the group $\mathscr{R}_3$ we may define tensor operators $S_q^{(k)}$ in isospin. One has simply to replace the operators J in equation (7.52) by the infinitesimal operators **T** for isospin. The infinitesimal operators themselves will form an isovector, analogous to the angular momentum **J** in $\mathscr{R}_3$. Hence it follows from the usual vector coupling rule (7.44) that, since V_c contains at most a product of two such operators t_z, it may be expressed as a sum of three parts which are isoscalar, isovector and isotensor of rank 2 in the language of subsection 7.4.6. Each part must have $q = 0$ because V_c is diagonal in the total charge. The isoscalar can produce no splitting and, like the Zeeman splitting, the isovector gives a contribution proportional to M_T. To find the M_T dependence of the isotensor part we could use the Wigner–Eckart theorem (7.53) and conclude that it is given by the vector-coupling coefficient $C(T2T, M_T0M_T)$ which has the explicit form $\{3M_T^2 - T(T+1)\}/\{T(T+1)(2T-1)(2T+3)\}^{\frac{1}{2}}$. Alternatively we could construct an equivalent operator in the sense described in section 7.4.7. Starting from the vector operator **T** it is easy to show that the operator $2T_z^2 - T_x^2 - T_y^2$ is a second rank tensor $T_q^{(k)}$ with $k = 2$, $q = 0$ (see problem 10.2). But this operator may be written as $3T_z^2 - \mathbf{T}.\mathbf{T}$ which has the value $3M_T^2 - T(T+1)$ in a state of definite T and M_T.

In conclusion therefore we have shown that for fixed T the splitting due to V_c is given by a quadratic in M_T

$$\Delta E(M_T) = a + b\,M_T + cM_T^2 \tag{10.3}$$

To find the coefficients one must go beyond these symmetry arguments and make very elaborate calculations with the detailed nuclear wave functions. However, the form (10.3) is independent of such details. In fact it is only a simple extension of the Zeeman splitting formula applied to a different symmetry. When $T > 1$ so that there are more than three values of M_T this formula may be tested since it contains only three constants, a, b and c. Thus for the quartet of $T = \frac{3}{2}$ states shown in figures 10.1 and 10.2 one has a test of the formula. Knowing the binding energies of ^{13}B, ^{13}C and ^{13}N and the excitation energies of the $T = \frac{3}{2}$ states in ^{13}C and ^{13}N we can predict the binding energy of ^{13}O. (The binding energy is defined as the difference between the mass of the nucleus and the sum of the masses of its constituent nucleons and is a measure of the interaction energy due to the strong nuclear force.) The measured binding energies of these four states are given in the first four columns of table 10.1 in MeV, and the last column contains the value for ^{13}O calculated from the first three entries using the quadratic (10.3). It can be seen that the calculated and measured values agree within experimental error.

Table 10.1

^{13}B	^{13}C	^{13}N	^{13}O	^{13}O(calculated)
84.45	82.00	79.04	75.56	75.57

10.1.3 Selection rules

In transition processes there will be selection rules in isospin in exact analogy with the rule (8.3) for angular momentum. If the transition operator is a tensor of rank k in isospin then, if T_i denotes the isospin of the initial state, the rule is given by

$$|T_i - k| \leq T_f \leq (T_i + k) \tag{10.4}$$

where T_f refers to the final state. The common decay processes, like β and γ decay, are governed by one-body operators of the type $S = \sum_{i(\text{proton})} S_p(i) + \sum_{i(\text{neutron})} S_n(i)$, where the sums run over protons and neutrons, respectively, and S_p may be different from S_n. By noting that the operator $\frac{1}{2} + t_z$ has the value $+1$ for a proton and 0 for a neutron we may rewrite the operator S as

$$S = \tfrac{1}{2}\sum_i [S_p(i) + S_n(i)] + \sum_i t_z(i)[S_p(i) - S_n(i)]$$

where the sum runs over all nucleons. In this form the first term is an isoscalar ($k = 0$) while the second is an isovector ($k = 1$). This puts strong limits on the possible values of T_f through equation (10.4) for a given T_i. One finds this rule to be well satisfied and there is considerable interest in looking for very weak transitions in situations where the rule forbids them—see Wilkinson (1969) in the bibliography. This gives a measure of the small mixing of isospin due to the Coulomb force.

10.2 Isospin in elementary particles

Although neutrons and protons are the only strongly interacting particles which are stable, or exist in stable nuclei, it is possible to produce a variety of new strongly interacting particles in high energy collisions. They decay rapidly, in times of the order 10^{-8} sec or less. Strongly interacting particles are generally called 'hadrons' and they include the π-mesons and Λ and Σ-particles, about which more will be said in chapter 11. For the present we simply notice that π-mesons are found to be of three kinds π^+, π^0 and π^- with charges of 1, 0 and -1, respectively, in units of e. In the same way the Σ-particles are found to have charges of ± 1 or zero, while the Λ-particle has zero charge only. The masses of the three different charge states of the π-meson are only slightly different from each other and the same is true of the Σ-particles. This situation is very similar to that of the neutron and proton and suggests that the π-meson and Σ-particles have isospin $T = 1$ while the Λ-particle has $T = 0$. It is found quite generally that all hadrons occur in isospin multiplets and so may be assigned an isospin T. This is then another intrinsic property of the particle, like its mass, spin, charge and parity. Notice, however, that whereas for the nucleon the charge was related to the isospin projection by $Q = e(\frac{1}{2} + T_z)$, the corresponding relation for the π-mesons, Σ- and Λ-particles is $Q = eT_z$. These two relations are brought together in chapter 11 within the broader symmetry of SU_3.

The validity of these isospin assignments and the invariance of the strong interactions under isospin transformations may now be tested by studying various nuclear collisions involving nucleons, π-mesons, Σ- and Λ-particles. By assigning isospins to different types of particle we are assuming some abstract transformation which acts simultaneously on the isospin labels of all particles. This is completely analogous to the idea that particles have different spins but a rotation of axes affects them all simultaneously. One therefore vector-couples isospins together and as we shall see, experiments are consistent with an invariance of the strong nuclear force with respect to these overall isospin transformations.

10.2.1 Collisions of π-mesons with nucleons

Since a π-meson has $T = 1$ and a nucleon has $T = \frac{1}{2}$, the combined system may be coupled into states of $T = \frac{3}{2}$ or $T = \frac{1}{2}$ by the usual vector coupling rule. Using the properties (7.40) of the raising and lowering operators it is soon shown that the transformation between uncoupled and coupled states is given by

$$\begin{aligned}
&|\tfrac{3}{2}\tfrac{3}{2}\rangle = |\pi^+ p\rangle \\
&|\tfrac{3}{2}\tfrac{1}{2}\rangle = \sqrt{\tfrac{1}{3}}|\pi^+ n\rangle + \sqrt{\tfrac{2}{3}}|\pi^0 p\rangle, \qquad &&|\tfrac{1}{2}\tfrac{1}{2}\rangle = \sqrt{\tfrac{2}{3}}|\pi^+ n\rangle - \sqrt{\tfrac{1}{3}}|\pi^0 p\rangle \\
&|\tfrac{3}{2}-\tfrac{1}{2}\rangle = \sqrt{\tfrac{2}{3}}|\pi^0 n\rangle + \sqrt{\tfrac{1}{3}}|\pi^- p\rangle, \qquad &&|\tfrac{1}{2}-\tfrac{1}{2}\rangle = \sqrt{\tfrac{1}{3}}|\pi^0 n\rangle - \sqrt{\tfrac{2}{3}}|\pi^- p\rangle \\
&|\tfrac{3}{2}-\tfrac{3}{2}\rangle = |\pi^- n\rangle
\end{aligned} \tag{10.5}$$

where the notation $|TM_T\rangle$ has been used for the coupled states. These equations may of course be inverted to give for example,

$$|\pi^- p\rangle = \sqrt{\tfrac{1}{3}}|\tfrac{3}{2}-\tfrac{1}{2}\rangle - \sqrt{\tfrac{2}{3}}|\tfrac{1}{2}-\tfrac{1}{2}\rangle \tag{10.6}$$

Now if the interaction is assumed to be isospin invariant then we should expect many relations between scattering cross sections for π-mesons on nucleons. For example the two elastic scattering processes $\pi^+ + p \rightarrow \pi^+ + p$ and $\pi^- + n \rightarrow \pi^- + n$ should have the same cross section since they represent different M_T projections of the same $T = \frac{3}{2}$ isospin state. On the other hand equation (10.6) shows that the $\pi^- + p \rightarrow \pi^- + p$ process involves both the $T = \frac{3}{2}$ and $T = \frac{1}{2}$ states of the system. The isospin invariance says nothing about the relation between the scattering in $T = \frac{3}{2}$ and $T = \frac{1}{2}$ states so that, unless there was some further symmetry, we should not expect the $\pi^- p$ scattering to be the same as the $\pi^+ p$. Experiment confirms the results expected from the isospin invariance but does not indicate any further symmetry. In other words the scattering from $T = \frac{3}{2}$ and $T = \frac{1}{2}$ states is different. In fact, at low energy the scattering in $T = \frac{3}{2}$ dominates so that from equations (10.6) and (10.5) we should expect the transition matrix elements for $\pi^+ p$ and $\pi^- p$ scattering to be in the ratio 3 : 1 and thus the cross sections to be in the ratio 9 : 1. This result is confirmed by experiment—see Bohr and Mottelson (1969) in the bibliography.

10.3 Isospin symmetry and charge-independence

When we introduced isospin for nuclei in section 10.1 it was seen that isospin invariance of the Hamiltonian was equivalent to charge-independence. The concept of isospin invariance was then extended to other particles and experiment shows that the isospin invariance of the strong interactions persists among all the hadrons, apart from the electromagnetic corrections. However, we emphasise that, in this extension, isospin invariance is *not* equivalent to charge-independence. Strict charge-independence in the example of section

10.2 would imply the equality of $\pi^+ p$ and $\pi^- p$ scattering, in disagreement with experiment.

In the same way, in the interaction between two π-mesons we have isospin invariance but *not* charge-independence. In an *s*-state of relative angular momentum two π-mesons, each with $T = 1$, can exist in states of total $T = 2$ or $T = 0$, taking account of the fact that they are bosons (see section 5.9). Experiments on the $\pi\pi$-system confirm the validity of isospin but show that the interactions in $T = 2$ and $T = 0$ states are not equal. Thus there is again no charge-independence.

We therefore have the remarkable situation that, although it was the charge-independence of neutron–proton interactions which first suggested isospin symmetry, we find that the isospin invariance of the strong interactions has general validity, whereas charge-independence does not. It is probably correct to say that the full significance of this result is not yet understood. In fact, the strong interaction shows an approximate symmetry with respect to a larger group SU_3 which contains the isospin SU_2 as a subgroup. We describe this larger group SU_3 in chapter 11.

Bibliography

A detailed account of the role of isospin in nuclear physics may be found in

Wilkinson, D. H., ed. (1969). *Isospin in Nuclear Physics* (North-Holland, Amsterdam)

Bohr, A. and Mottelson, B. R. (1969). *Nuclear Structure* (Benjamin, New York)

For the particle physics applications see Perkins (1972) in the bibliography to chapter 11.

Problems

10.1 Using isospin arguments, explain how some of the energy levels of the nuclei $^{18}O(Z = 8)$, $^{18}F(Z = 9)$ and $^{18}Ne(Z = 10)$ are related.

10.2 Show that the operator $2T_z^2 - T_x^2 - T_y^2 = 3T_z^2 - T^2$ is a tensor operator of second degree $T_q^{(k)}$ with $k = 2$ and $q = 0$. You can start from T_+^2 and use equation (7.52).

10.3 The electric dipole operator, which governs certain transitions (E.1) in nuclei is given by a sum $\sum_p \boldsymbol{r}_p$ over the position vectors $\boldsymbol{r}_p$ of the protons p. Show that this may be written in the form $\sum_i (\frac{1}{2} + t_z(i))\boldsymbol{r}_i$, summing over all nucleons i. Hence show that E.1 transitions between any two states with $T = 0$ are forbidden. (Use the fact that the operator $\sum_i \boldsymbol{r}_i$ is the centre-of-mass vector of the entire nucleus and so, being independent of all internal coordinates, cannot cause any transitions.)

10.4 Assuming conservation of isospin in the reaction $^{16}O + {}^{2}H \rightarrow {}^{18}F$, what would be the isospin of the resulting states of ^{18}F? Assume that the ^{16}O target is in its ground state and that the ^{2}H has $T = 0$.

10.5 Deduce the ratio of cross sections for the two reactions $^{16}O + {}^{3}He \rightarrow {}^{18}F + p$ and $^{16}O + {}^{3}He \rightarrow {}^{18}Ne + n$, where the states of ^{18}F and ^{18}Ne both have $T = 1$.

11

The Group SU_3 with Applications to Elementary Particles

The construction of very high energy particle accelerators in the past twenty-five years has led to the discovery of an increasing number of new particles. By 'new' we mean particles other than those like the electron, proton and neutron which are known to be constituents of the stable atoms of which matter is composed. One may even regard the neutron as the first of the new particles since it is not, by itself, stable. It will β-decay with a mean life of about 15 minutes into a proton, electron and what is known as an antineutrino. The neutron achieves stability only when combined with protons in nuclei.

The neutron was discovered as early as 1932 following collisions produced by a beam of α-particles from a naturally radioactive source. In such a beam the α-particles have an energy of about 5 MeV. When the energy of a beam of protons exceeds about 300 MeV there is sufficient energy available in the collision with a proton in the target for a π-meson to be created through the mass–energy relation $E = Mc^2$. At higher energies of about 1000 MeV ($= 1$ GeV) the K-mesons are created, in pairs, and at still higher energies of about 4 GeV the antiproton is created together with another proton. Having produced π-mesons in proton–proton collisions one may then form a beam of these π-mesons and study the results of their collisions with protons in a second target. In the same way, beams of K-mesons are produced and their collisions studied. One sees therefore that there is great scope for producing a variety of new particles in all these different kinds of collisions. At first this produced a confused mass of results but gradually it has been possible to see order emerging from chaos and the present position is that the new particles may be very beautifully classified by using the isospin and a larger group of unitary transformations in three dimensions, which we call SU_3. The extension from the isospin group SU_2 to the new group SU_3 incorporates a new physical quantum number called the 'hypercharge' which in many respects plays a role like the electric charge.

In the first three sections of this chapter we give a brief description and listing of the properties of some of the more stable elementary particles including a discussion of the physical reasons for introducing the hypercharge—see also the first two references in the bibliography. We then spend three sections in developing the mathematical properties of the group SU_3 before returning to the physical data to see how well it fits into the SU_3 framework.

11.1 Compilation of some relevant data

The word 'hadron' was introduced in section 10.2 to describe particles which interact via the strong nuclear force. In table 11.1 we list some of the most important properties of the more stable hadrons—see also the first reference in the bibliography. The distinction between baryons and mesons, which have been listed separately in table 11.1, will be explained in section 11.3. The notation has unfortunately developed historically in a rather haphazard way. Thus p and n refer to proton and neutron, respectively, while the other particles are referred to by a Greek letter, for example Λ denotes the lambda particle. The superscript denotes the electric charge in units of e, with Δ^{++} having charge $2e$. The asterisk on Σ^* and Ξ^* is used to distinguish them from particles with the same notation but lighter mass. It is customary to use the same letter for particles with the same values for T and Y. The masses M are given in the second column by recording the value of Mc^2 in MeV. The third column gives the intrinsic spin of the particle and the fourth column gives the charge. For completeness we also list the intrinsic parity of the particle in column three, for example $\frac{1}{2}^+$ indicates a spin of $\frac{1}{2}\hbar$ and positive parity. (The precise meaning of intrinsic parity, which has nothing to do with the motion of the particle but refers only to its internal structure, will be discussed in subsection 15.7.3 of volume 2.)

We have already seen in chapter 10 how the neutron and proton may be regarded as members of a $T = \frac{1}{2}$ isospin doublet. From the table we see that all particles fall into isospin multiplets, the $(2T+1)$ members of a T-multiplet having similar masses, within a few MeV, and with consecutive values for their

Table 11.1 Properties of the lighter hadrons

(a) BARYONS

Particle	*Mass* (MeV)	*Spin*	*Charge*	*Isospin* T	M_T	*Hyper-charge* Y	*Mean life* τ (sec)	*Principal decay modes*
p	938.28	$\frac{1}{2}^+$	1	$\frac{1}{2}$	$\frac{1}{2}$	1	∞	Stable
n	939.57	$\frac{1}{2}^+$	0	$\frac{1}{2}$	$-\frac{1}{2}$	1	$1.0(10)^3$	$p+e^-+\bar{\nu}$
Λ	1115.6	$\frac{1}{2}^+$	0	0	0	0	$2.5(10)^{-10}$	$p+\pi^-$ (65 %), $n+\pi^0$ (35 %)
Σ^+	1189.4	$\frac{1}{2}^+$	1	1	1	0	$0.8(10)^{-10}$	$p+\pi^0$ (53 %), $n+\pi^+$ (47 %)
Σ^0	1192.5	$\frac{1}{2}^+$	0	1	0	0	$<(10)^{-14}$	$\Lambda+\gamma$
Σ^-	1197.4	$\frac{1}{2}^+$	-1	1	-1	0	$1.7(10)^{-10}$	$n+\pi^-$
Ξ^0	1315	$\frac{1}{2}^+$	0	$\frac{1}{2}$	$\frac{1}{2}$	-1	$2.9(10)^{-10}$	$\Lambda+\pi^0$
Ξ^-	1321	$\frac{1}{2}^+$	-1	$\frac{1}{2}$	$-\frac{1}{2}$	-1	$1.7(10)^{-10}$	$\Lambda+\pi^-$
Δ^{++}	1232[†]	$\frac{3}{2}^+$	2	$\frac{3}{2}$	$\frac{3}{2}$	1	$5(10)^{-24}$	$p+\pi^+$
Δ^+		$\frac{3}{2}^+$	1	$\frac{3}{2}$	$\frac{1}{2}$	1		$p+\pi^0$, $n+\pi^+$
Δ^0		$\frac{3}{2}^+$	0	$\frac{3}{2}$	$-\frac{1}{2}$	1		$p+\pi^-$, $n+\pi^0$
Δ^-		$\frac{3}{2}^+$	-1	$\frac{3}{2}$	$-\frac{3}{2}$	1		$n+\pi^-$
Σ^{+*}	1385[†]	$\frac{3}{2}^+$	1	1	1	0	$1(10)^{-23}$	$\Lambda+\pi$(88 %), $\Sigma+\pi$(12 %)
Σ^{0*}		$\frac{3}{2}^+$	0	1	0	0		
Σ^{-*}		$\frac{3}{2}^+$	-1	1	-1	0		
Ξ^{0*}	1532	$\frac{3}{2}^+$	0	$\frac{1}{2}$	$\frac{1}{2}$	-1	$(10)^{-22}$	$\Xi+\pi$
Ξ^{-*}	1535	$\frac{3}{2}^+$	-1	$\frac{1}{2}$	$-\frac{1}{2}$	-1		
Ω^-	1672	$\frac{3}{2}^+$	-1	0	0	-2	$1.1(10)^{-10}$	$\Xi+\pi$, $\Lambda+K^-$

† The experimental masses and lifetimes for these multiplets have large errors and we give only approximate average values.

Table 11.1 (*continued*)
(b) MESONS

Particle	*Mass* (MeV)	*Spin*	*Charge*	*Isospin* T	M_T	*Hyper-charge* Y	*Mean life* τ (sec)	*Principal decay modes*
$\pi^{\pm}$	139.6	0^-	± 1	1	± 1	0	$2.6(10)^{-8}$	$\mu + \bar{\nu}$
π^0	135.0	0^-	0	1	0	0	$0.9(10)^{-16}$	$\gamma + \gamma$
$K^{\pm}$	493.7	0^-	± 1	$\frac{1}{2}$	$\pm\frac{1}{2}$	± 1	$1.2(10)^{-8}$	$\mu + \nu(64\%)$, $2\pi(21\%)$, $3\pi(7\%)$
K^0	497.7	0^-	0	$\frac{1}{2}$	$-\frac{1}{2}$	1	‡ $0.9(10)^{-10}$	$\pi^+ + \pi^-(68\%)$, $\pi^0 + \pi^0(32\%)$
$\bar{K}^0$	497.7	0^-	0	$\frac{1}{2}$	$\frac{1}{2}$	-1	‡ $5.3(10)^{-8}$	$3\pi(38\%)$, $\pi e\nu(35\%)$, $\pi\mu\nu(27\%)$
η^0	549	0^-	0	0	0	0	$3(10)^{-19}$	$2\gamma(42\%)$, $3\pi(51\%)$, $2\pi\gamma(6\%)$

‡ The K^0 and $\bar{K}^0$ particles decay by both these modes, see subsection 16.3.6 of volume 2.

charges. Thus for example the Ξ^0 and Ξ^- particles are taken to be members of a $T=\frac{1}{2}$ doublet, while Σ^+, Σ° and Σ^- have $T=1$. These isospin assignments have been confirmed by observing selection rules of the kind described in subsection 10.2.1. Notice, however, that the relation $Q = M_T + \frac{1}{2}$ between charge Q and isospin projection M_T for the nucleons (n and p) is not satisfied for the other baryons. In section 11.2 we shall explain the introduction of a new quantum number Y, called the hypercharge and listed in column seven of the table. It is then seen that the more general relation $Q = M_T + \frac{1}{2}Y$ is satisfied by all the hadrons.

In column eight of the table we give the mean life τ of the particle, being the constant which appears in the exponential decay law $\exp(-t/\tau)$ for the particle. Thus a smaller value for τ indicates a less stable particle which decays more quickly. The principal decay mode is shown in the last column with the percentages given when more than one mode occurs with comparable probability. Here the symbol γ denotes a photon (γ-decay), while ν denotes a neutrino and μ a muon.

The insertion of a bar above a symbol, as in $\overline{K}$ or $\bar{\nu}$, denotes the antiparticle of K or ν. From field theory, and also from observation, there is an antiparticle corresponding to every particle although the antiparticle is often difficult to produce. The antiparticle has the same mass and spin as the particle but opposite charge. Some neutral particles like the π^0 are identical with their antiparticles. We need not say any more about antiparticles here but return to the subject in chapter 16 of volume 2.

Some of the heavier particles shown in table 11.1 may decay via the strong interactions in very short times of about $(10)^{-23}$ sec. It is debatable whether such a short existence really deserves the name particle. In fact, particles which decay in this way are sometimes referred to as resonances. Generally, in quantum mechanics, a resonance in a collision process is associated with an excited state of the combined system which may be unstable. Many of the particles listed in the table may be regarded as excited states of others, just as in chapter 10 we have already regarded the neutron and proton as two states of the nucleon.

11.2 The hypercharge

Like most new concepts, the idea of hypercharge was introduced tentatively at first to describe certain unexplained phenomena and gradually confidence grew as more data was found to be explained by the concept. In the first place there was the puzzle that whereas π-mesons were produced singly in proton–proton collisions, the K-mesons were produced only in pairs. Secondly, the K-mesons decayed in times of the order 10^{-10} sec. Where strong nuclear forces are concerned and typical energies are of the order of hundreds of MeV the decay times are expected, from the uncertainty relation $\Delta E\,\Delta t \approx \hbar$, to be of the order $(10)^{-23}$ sec using $\hbar = 6.6(10)^{-22}$ MeV sec. This suggested that some selection rule was preventing the decay of the K-meson into lighter particles. It was therefore postulated that all the strongly interacting particles should be labelled by a quantity Y, called the hypercharge, and that the strong interaction should conserve hypercharge. In other words in any production or decay process governed by the strong interaction the total hypercharge of initial and final states should be the same. Thus one associates $Y = 0$ with the π-mesons to allow single production. It is found that the K-mesons, which are produced in pairs, are of four kinds K^+, K^0, $\overline{K}^0$ and K^- and that by associating $Y = +1$ with the first two of these and $Y = -1$ with the last two, the conservation of Y is ensured. The pair produced is always found to contain one with $Y = +1$ and one with $Y = -1$. There are many cross-checks on the Y-assignments to different particles from the variety of collisions which have been studied and the resulting values of Y are given in the seventh column of table 11.1. Like the charge, the hypercharges of particle and antiparticle are equal and opposite. The fact that a K^0-meson with $Y = +1$, may decay into two π^0-mesons with $Y = 0$ violates conservation of hypercharge. However, as remarked above, the decay is slower by a factor of $(10)^{12}$ than expected from the strong interactions. This is just the factor to be expected if the decay takes place via the weak nuclear interaction which is responsible for β-decay, since from other data one knows the ratio of weak to strong interactions. One is therefore led to the conclusion that the strong interaction conserves hypercharge while the weak interaction does not. All other data support this conclusion.

The word 'strangeness' is sometimes used as an equivalent description for hypercharge. The strangeness is denoted by a number S related to Y by $S = Y$ for mesons and $S = Y - 1$ for baryons. (In general $S = Y - B$, where B is the baryon number defined in section 11.3.) The advantage of the use of S is that the more familiar particles like the π-meson and nucleon have $S = 0$, while the newer particles (like the K-meson) which have 'strange properties', will have non-zero values for S. For classification purposes, the hypercharge is more convenient, as we shall see in section 11.7.

11.3 Baryon number

Since the proton is much heavier than the π- and K-mesons one might ask why it is completely stable and does not decay. The only answer is to make another postulate, the conservation of 'Baryon number'. To define the word 'baryon' we recall from section 10.2 that the word 'hadron' was introduced to describe any strongly interacting particle. The strongly interacting particles are now divided into mesons which have integer spin and baryons which have half-integer spin. These two classes, mesons and baryons, are also distinguished by their mass, the baryons being the heavier, as seen from the table.

The baryon number B is defined by associating the number $B = +1$ with each baryon, $B = 0$ for each meson and $B = -1$ for each antibaryon. It is an experimental fact that the baryon number is conserved absolutely in all known reactions, even those governed by weak interactions. In this sense it has the same status as the electric charge which also is conserved without exception. The decay of a proton into mesons would involve a violation of the postulate of baryon number conservation and therefore does not take place. For completeness we remark that the electron, muon and neutrinos, which do not interact via the strong nuclear force, are called leptons. They all have a spin of $\frac{1}{2}$ and $B = 0$. The electron, μ^- and neutrino have a 'lepton number' $L = 1$, while the positron, μ^+ and antineutrino have $L = -1$.

11.4 The group SU_3

Having seen how the particles fall into isospin multiplets T and having assigned values of hypercharge Y we now leave the physics to describe the mathematical properties of the group SU_3. It will then be shown in section 11.7 how these combinations of T and Y values fall naturally into larger multiplets corresponding to irreducible representations of SU_3.

The group U_3 is defined as the set of unitary 3×3 matrices U and is a natural extension of the group U_2 described in section 10.1. It is also a special case of the group U_N described in chapter 18 of volume 2 for general values of N but we shall deduce its properties from first principles rather than quote the general results. The unitary condition imposes nine constraints on the nine complex numbers of the 3×3 matrix, leaving nine real parameters. The removal of a phase $\exp(i\phi)$ such that the remaining matrix has $\det \mathrm{U} = 1$ restricts the operations to the subgroup SU_3 which therefore contains eight real parameters. The corresponding eight infinitesimal operators must again be skew-Hermitian and have zero trace. As a natural extension of equations (10.1) the eight infinitesimal operators may be taken as

$$\mathrm{X}_1 = -\tfrac{1}{2}\begin{pmatrix} 0 & i & 0 \\ i & 0 & 0 \\ 0 & 0 & 0 \end{pmatrix}, \mathrm{X}_2 = -\tfrac{1}{2}\begin{pmatrix} 0 & 1 & 0 \\ -1 & 0 & 0 \\ 0 & 0 & 0 \end{pmatrix}, \mathrm{X}_3 = -\tfrac{1}{2}\begin{pmatrix} i & 0 & 0 \\ 0 & -i & 0 \\ 0 & 0 & 0 \end{pmatrix}$$

$$\mathrm{X}_4 = -\tfrac{1}{2}\begin{pmatrix} 0 & 0 & i \\ 0 & 0 & 0 \\ i & 0 & 0 \end{pmatrix}, \quad \mathrm{X}_5 = -\tfrac{1}{2}\begin{pmatrix} 0 & 0 & 1 \\ 0 & 0 & 0 \\ -1 & 0 & 0 \end{pmatrix}, \tag{11.1}$$

$$\mathrm{X}_6 = -\tfrac{1}{2}\begin{pmatrix} 0 & 0 & 0 \\ 0 & 0 & i \\ 0 & i & 0 \end{pmatrix}, \quad \mathrm{X}_7 = -\tfrac{1}{2}\begin{pmatrix} 0 & 0 & 0 \\ 0 & 0 & 1 \\ 0 & -1 & 0 \end{pmatrix},$$

$$\mathrm{X}_8 = - \begin{pmatrix} i & 0 & 0 \\ 0 & i & 0 \\ 0 & 0 & -2i \end{pmatrix}$$

The first three are simply the SU_2 operators on the two-dimensional subspace formed by ignoring the third basis vector. The next two pairs are analogues of X_1 and X_2 but refer to the subgroup SU_2 obtained by ignoring the second or first basis vector. The two analogues of X_3 are not linearly independent of X_3. (If they were, we would have nine operators instead of eight.) Instead we take the diagonal matrix X_8 which is skew-Hermitian and traceless and is clearly independent of X_3.

11.5 Subgroups of SU_3

An SU_2 subgroup of SU_3 may be constructed by taking any two-dimensional subspace of the three-dimensional space in which SU_3 is defined. For example the infinitesimal operators X_1, X_2 and X_3 describe such a group. So also do the operators X_4, X_5 and $\frac{1}{2}(X_3+\frac{1}{2}X_8)$ and the operators X_6, X_7 and $\frac{1}{2}(-X_3+\frac{1}{2}X_8)$. In fact these subgroups may be enlarged by noticing, for example, that X_8 commutes with X_1, X_2 and X_3. This set of four operators generate the product group $SU_2 \times U_1$. If we denote the three basis vectors of SU_3 by $\boldsymbol{e}_1$, $\boldsymbol{e}_2$ and $\boldsymbol{e}_3$ then the SU_2 subgroup relates to $\boldsymbol{e}_1$ and $\boldsymbol{e}_2$. The finite group element of U_1 may be written as $U(a)=\exp(aX_8)$ so that $U(a)\boldsymbol{e}_1=\exp(-ia)\boldsymbol{e}_1$, $U(a)\boldsymbol{e}_2=\exp(-ia)\boldsymbol{e}_2$, while $U(a)\boldsymbol{e}_3=\exp(2ia)\boldsymbol{e}_3$. The group U_1 is clearly Abelian since $U(a)\,U(b)=U(a+b)$, so that its irreducible representations are one-dimensional. In fact U_1 is isomorphic with the group $\mathscr{R}_2$ discussed in section 7.3 so that the single-valued irreducible representations are labelled by an integer which, for convenience later, is denoted by $3Y$ and the representations are given by $T^{(Y)}(a)=\exp(-i3Ya)$. In the representation $T^{(Y)}$ the infinitesimal operator X_8 has the value $-3iY$. When constructing the irreducible representations of SU_3 we shall make use of the subgroup $SU_2 \times U_1$.

Another family of subgroups may be found by restricting to the real unimodular unitary 3×3 matrices. This is just the group $\mathscr{R}_3$ described in section 7.4 as may be seen by isolating the real matrices X_2, X_5 and X_7. It is soon verified that they satisfy the commutation relations for $\mathscr{R}_3$ and in fact they are identical with the $\mathscr{R}_3$ matrices given in equation (7.24) for the representation $D^{(1)}$. We shall not study the relation between SU_3 and this subgroup $\mathscr{R}_3$ but we remark that it has been used in understanding the appearance of collective motion in light nuclei (see also section 19.2 of volume 2).

11.6 Irreducible representations of SU_3

We now investigate the structure, labelling and properties of the irreducible representations of SU_3 following the same logical development as in subsection 7.4.2 for the group $\mathscr{R}_3$. We shall make considerable use of the three SU_2 subgroups of SU_3 mentioned in the preceding section and it will also be necessary to introduce raising and lowering operations as in equation (7.27) for each of these subgroups. Thus it will be found convenient to use, in place of the X_q, the eight combinations

$$T_+ = i(X_1 + iX_2) = \begin{pmatrix} 0 & 1 & 0 \\ 0 & 0 & 0 \\ 0 & 0 & 0 \end{pmatrix}, \quad T_- = i(X_1 - iX_2) = \begin{pmatrix} 0 & 0 & 0 \\ 1 & 0 & 0 \\ 0 & 0 & 0 \end{pmatrix},$$

$$U_+ = i(X_6 + iX_7) = \begin{pmatrix} 0 & 0 & 0 \\ 0 & 0 & 1 \\ 0 & 0 & 0 \end{pmatrix}, \quad U_- = i(X_6 - iX_7) = \begin{pmatrix} 0 & 0 & 0 \\ 0 & 0 & 0 \\ 0 & 1 & 0 \end{pmatrix},$$

$$(11.2)$$

$$V_+ = i(X_4 - iX_5) = \begin{pmatrix} 0 & 0 & 0 \\ 0 & 0 & 0 \\ 1 & 0 & 0 \end{pmatrix}, \quad V_- = i(X_4 + iX_5) = \begin{pmatrix} 0 & 0 & 1 \\ 0 & 0 & 0 \\ 0 & 0 & 0 \end{pmatrix},$$

$$T_z = iX_3 = \begin{pmatrix} \frac{1}{2} & 0 & 0 \\ 0 & -\frac{1}{2} & 0 \\ 0 & 0 & 0 \end{pmatrix}, \quad Y = \tfrac{1}{3}iX_8 = \begin{pmatrix} \frac{1}{3} & 0 & 0 \\ 0 & \frac{1}{3} & 0 \\ 0 & 0 & -\frac{2}{3} \end{pmatrix}$$

As a particular case of the general property (7.7) of infinitesimal operators one can verify that the commutator of any pair of these eight matrices is a linear combination of them. For example the three matrices $T_\pm$ and T_z, being the infinitesimal operators for an SU_2 subgroup, satisfy the angular-momentum commutation relations (7.28) and (7.30)

$$[T_+, T_-] = 2T_z, \quad [T_z, T_\pm] = \pm T_\pm \tag{11.3a}$$

In the same way if we define $U_z = \frac{3}{4}Y - \frac{1}{2}T_z$ and $V_z = -\frac{3}{4}Y - \frac{1}{2}T_z$ then the **U** and **V** operators satisfy identical relations since they also refer to SU_2 subgroups

$$\begin{aligned} [U_+, U_-] &= 2U_z, \; [U_z, U_\pm] = \pm U_\pm \\ [V_+, V_-] &= 2V_z, \; [V_z, V_\pm] = \pm V_\pm \end{aligned} \tag{11.3b}$$

We shall refer to these three SU_2 subgroups as T-spin, U-spin and V-spin. From the matrices (11.2) it is also clear that Y commutes with the three SU_2 operators $T_\pm$ and T_z. For future reference we note the following commutation relations which also follow from the definitions (11.2),

$$\begin{aligned}[] [T_z, U_\pm] &= \mp\tfrac{1}{2}U_\pm,\ [Y, U_\pm] = \pm U_\pm \\ [T_z, V_\pm] &= \mp\tfrac{1}{2}V_\pm,\ [Y, V_\pm] = \mp V_\pm \end{aligned} \qquad (11.3c)$$

$$\begin{aligned}[] [T_+, U_-] &= [T_-, U_+] = [T_+, V_-] = 0 \\ [T_-, V_+] &= [U_+, V_-] = [U_-, V_+] = 0 \end{aligned} \qquad (11.3d)$$

$$\begin{aligned}[] [V_-, U_-] &= T_+,\ [U_+, V_+] = T_-,\ [U_-, T_-] = V_+, \\ [T_+, U_+] &= V_-,\ [T_-, V_-] = U_+,\ [V_+, T_+] = U_- \end{aligned}$$

Remember from section 7.2 that for a Lie group the same commutation relations hold in all representations. Thus any such relation deduced from the 3×3 matrices (11.2) is valid generally.

We now use these commutation relations to deduce the labelling and structure of the irreducible representations of SU_3 in the same way that the relations (7.26) were used in subsection 7.4.2 to deduce the irreducible representations $D^{(j)}$ of $\mathscr{R}_3$. Before setting out it is worth recalling the steps used for $\mathscr{R}_3$. We first chose a basis in which one of the operators, J_z, was diagonal, having noted that no more than one of the three operators J_q could be diagonalised at the same time. We then wrote the remaining operators as raising and lowering operators in the eigenvalues m of J_z. Finally we deduced that if j denotes the greatest value of m in the representation then there are $(2j+1)$ basis vectors with $m = j, j-1, \ldots, -j$ and the values of j could only be $0, \frac{1}{2}, 1, \frac{3}{2}, \ldots$. The irreducible representations could therefore be labelled by this number j and denoted by $D^{(j)}$. The different basis vectors could be visualised by a set of $(2j+1)$ points at unit intervals along the m-axis and symmetrically spaced about the origin. Figure 11.1 illustrates the pattern for an integral value of j. This technique is again followed for SU_3 but in an extended form. It is now possible to simultaneously diagonalise two of the eight operators and consequently we need two dimensions instead of one in which to visualise the representations of SU_3. Two numbers are necessary to distinguish the irreducible representations in place of the single number j.

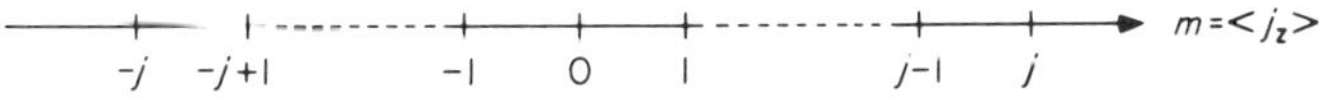

Figure 11.1

Let us for the moment denote an irreducible representation of SU_3 by a symbol D. (Later we use the more detailed notation $D^{(\lambda\mu)}$ when the labels λ and μ have been defined.) We choose a basis for the representation in which the two operators T_z and Y are diagonal. This must be possible since they commute. One cannot diagonalise more than two of the eight infinitesimal operators since this would imply the existence of more than two independent, traceless, diagonal 3×3 matrices, which is impossible. Since T_z, U_z and V_z are standard

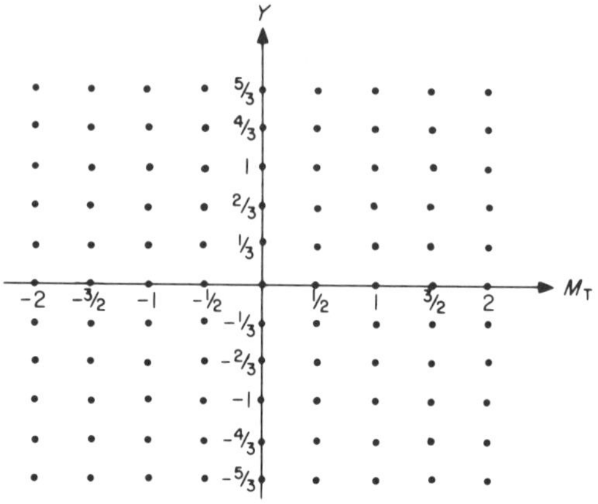

Figure 11.2

SU_2 operators we know that their eigenvalues must be 0, $\pm\frac{1}{2}$, ± 1, $\pm\frac{3}{2}$, etc. From the definitions of U_z and V_z it follows that $Y = \frac{2}{3}(U_z - V_z)$ and hence that the eigenvalues of Y must be 0, $\pm\frac{1}{3}$, $\pm\frac{2}{3}$, ± 1, $\pm\frac{4}{3}$, etc. Thus if we denote the eigenvalues of T_z and Y by M_T and Y then each basis vector will be associated with a point (M_T, Y) which lies on a rectangular grid with spacings $\frac{1}{2}$ in the M_T direction and $\frac{1}{3}$ in the Y direction, as in figure 11.2.

Having diagonalised two of the eight infinitesimal operators it is now simple to visualise the effect of the remaining six operators $T_\pm$, $U_\pm$ and $V_\pm$ in terms of movement in the (M_T, Y) plane. Since $T_\pm$ are standard raising and lowering operators with respect to M_T (see subsection 7.4.2) and since they commute with Y, the operation of $T_\pm$ on a basis vector associated with a point (M_T, Y) produces a new vector associated with the point $(M_T \pm 1, Y)$. Thus $T_\pm$ corresponds to a displacement of ± 1 in the M_T direction. In the same way, the commutation relations (11.3c) show that $U_\pm$ corresponds to displacements of $(\mp\frac{1}{2}, \pm 1)$ while $V_\pm$ corresponds to displacements of $(\mp\frac{1}{2}, \mp 1)$. These displacements are illustrated in figure 11.3.

In an irreducible representation it must be possible to generate all basis vectors from any chosen vector, using the infinitesimal operators. But from the displacements shown in figure 11.3 it is clear that Y can only be changed by integer steps and that for an odd integer change in Y, the change in M_T must be half-integral. Thus we conclude that the basis vectors in an irreducible representation lie on a more restricted grid than that of figure 11.2, namely the hexagonal-type lattice shown by crosses in figure 11.4. There are three different ways in which this lattice may be positioned on the grid of figure 11.2:

(1) With a lattice point at the origin (0, 0)
(2) With a lattice point at the point $(0, \frac{2}{3})$ (11.4)
(3) With a lattice point at the point $(0, -\frac{2}{3})$

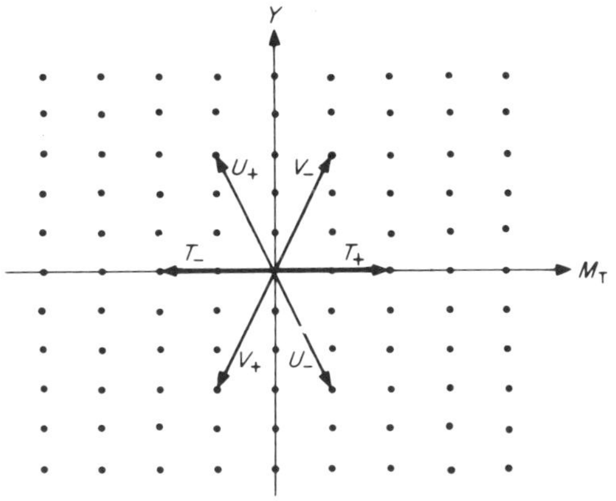

Figure 11.3

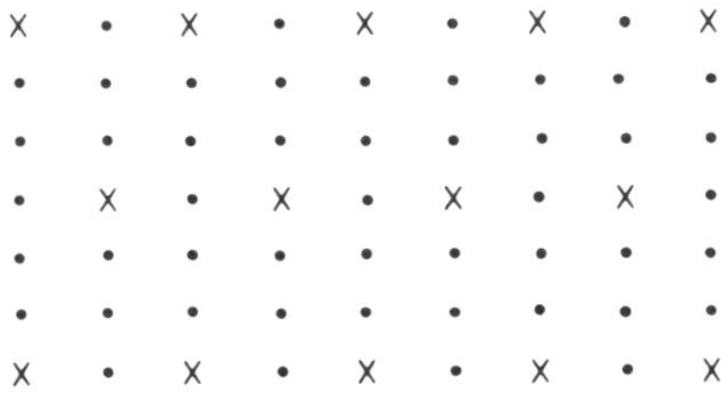

Figure 11.4

and the irreducible representations naturally fall into three corresponding types. (It is not possible to have a lattice point at $(0, \pm\frac{1}{3})$ since this would imply a value of $\pm\frac{1}{4}$ for U_z.)

The next step is to find which of the lattice points, each corresponding to a basis vector, are needed to produce an irreducible representation. Let us consider that basis vector $|\psi\rangle$ with greatest 'weight' in an irreducible representation D. By this we mean the vector with greatest value of Y, and for this Y, the greatest value of M_T. (One could alternatively define the greatest weight to have the greatest possible M_T and, for this M_T, the greatest value of Y.) This definition implies that

$$\mathrm{T}_+|\psi\rangle = \mathrm{U}_+|\psi\rangle = \mathrm{V}_-|\psi\rangle = 0 \qquad (11.5)$$

since U_+ and V_- would increase the value of Y and T_+ would increase M_T keeping Y unchanged. Hence it follows that $|\psi\rangle$ must have definite T-spin, U-spin, and V-spin with values

$$T = M_T,\ U = M_U = \tfrac{3}{4}Y - \tfrac{1}{2}M_T \text{ and } V = -M_V = \tfrac{3}{4}Y + \tfrac{1}{2}M_T$$

using the definitions given after equation (11.3a) for the operators U_z and

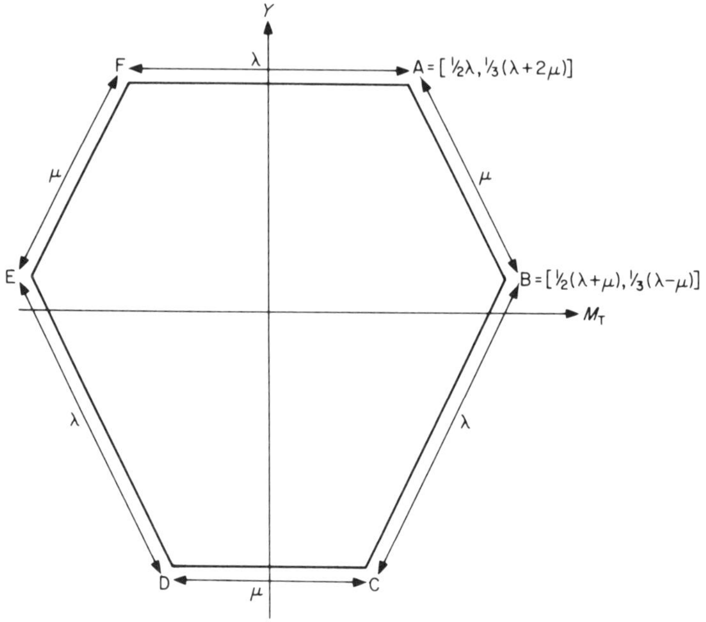

Figure 11.5

V_z. It is common practice to label this basis vector $|\psi\rangle$ by the two integers $\lambda = 2T$ and $\mu = 2U$ so that $V = \frac{1}{2}(\lambda+\mu)$ and the coordinates of the lattice point corresponding to $|\psi\rangle$ are

$$M_T = \tfrac{1}{2}\lambda,\ Y = \tfrac{1}{3}(\lambda + 2\mu) \tag{11.6}$$

From these results we may deduce that all the basis vectors of D correspond to lattice points which lie on or within the hexagon drawn in figure 11.5. The edges of the hexagon are parallel to one or other of the three lattice vectors shown in figure 11.3. The lengths λ, μ of the edges are measured in 'steps' between lattice points. (To convert these to actual lengths one would need to reduce the scale of the Y variable by a factor $\frac{1}{2}\sqrt{3}$.) Starting from the point A corresponding to the maximum weight vector $|\psi\rangle$, successive use of the operator T_- will reduce M_T by λ unit steps until the point F is reached. By definition, no lattice points may lie above the line AF. In the same way, successive operation on $|\psi\rangle$ by U_- produces the U-spin multiplet of $\mu+1$ lattice points along the line AB. Basis vectors on AB are thus given by $U_-^n|\psi\rangle$ and from this we deduce that there can be no basis vectors to the right of AB since, using equation (11.3d), $T_+U_-^n|\psi\rangle = U_-^n T_+|\psi\rangle = 0$ and also $V_-U_-^n|\psi\rangle = 0$. The coordinates of B are easily deduced, $M_T = \frac{1}{2}(\lambda+\mu)$ and $Y = \frac{1}{3}(\lambda+2\mu) - \mu = \frac{1}{3}(\lambda - \mu)$. The rest of the hexagon ABCDEF now follows from symmetry as follows. From the T-spin subgroup there is reflection symmetry about the Y-axis because of the symmetry between $\pm M_T$ in SU_2. The same symmetry in U-spin and V-spin implies a 'reflection' symmetry about the other two diagonals shown in figure 11.5. This determines the hexagon and we may then conclude that *all* basis vectors of the irreducible representation correspond to lattice points lying on or within the hexagon of figure 11.5. (The 'reflections' relate to lattice points and are true geometrical reflections only if the scale of the Y axis is reduced by $\frac{1}{2}\sqrt{3}$.)

To each lattice point on the right edge ABC there must be a T-spin multiplet running across to the left edge FED, the points of which may be reached by successive operation of T_-. In the same way, to every lattice point on AFE there must be a U-spin multiplet running across to the edge BCD and to every lattice point on BAF there is a V-spin multiplet running across to CDE. Thus starting from the point A, which is specified by the two integers λ and μ, we can generate an irreducible representation whose basis vectors correspond to the points on and within the hexagon of figure 11.5 and which we shall denote by $D^{(\lambda\mu)}$ (or sometimes simply by $(\lambda\mu)$). From the construction above, there must be a basis vector at each lattice point within the hexagon but there may actually be more than one independent basis vector at some points. It may be proved (see problem 11.6) that for each point on the perimeter of the hexagon the basis vector is unique. If the lattice points inside but adjacent to the perimeter are joined up, they form a smaller inscribed hexagon and it may be shown that each point on this hexagon corresponds to two independent basis vectors. This process continues inwards, the number of basis vectors increasing by one at each step until the hexagon becomes a triangle. From this stage inwards the number of independent basis vectors at each point remains the same. From this construction one can show (see problem 11.7) that the dimension $d(\lambda\mu)$ of the representation (i.e. the total number of independent basis vectors) is given by the expression

$$d(\lambda\mu) = \tfrac{1}{2}(\lambda+1)(\mu+1)(\lambda+\mu+2) \tag{11.7}$$

Some examples are given in figure 11.6. Notice that, when $\mu = 0$ or $\lambda = 0$, the original hexagon becomes a triangle and so there is never more than one basis vector corresponding to any lattice point.

It is worth studying the representation $D^{(11)}$ in some detail since it is the simplest example with two basis vectors at the same point and also because it is the most important representation in physical applications. It has $T = \frac{1}{2}$ multiplets at $Y = \pm 1$ and a $T = 1$ multiplet at $Y = 0$. The remaining vector at the point $(0, 0)$ must therefore have $T = 0$ since otherwise it would imply the existence of more vectors. The point $O = (0, 0)$ may be reached from the starting point $A = (\frac{1}{2}, 1)$ by a number of paths including the three shown in figure 11.7 which correspond to the following transformations:

$$\begin{array}{lll} \text{Solid line} & \text{AO}, & V_+ \,|\psi\rangle = \sqrt{2}\,|V=1\rangle \\ \text{Broken line} & \text{ABO}, & T_-U_- \,|\psi\rangle = \sqrt{2}\,|T=1\rangle \\ \text{Dot-dash line} & \text{AFO}, & U_-T_- \,|\psi\rangle = \sqrt{2}\,|U=1\rangle \end{array} \tag{11.8}$$

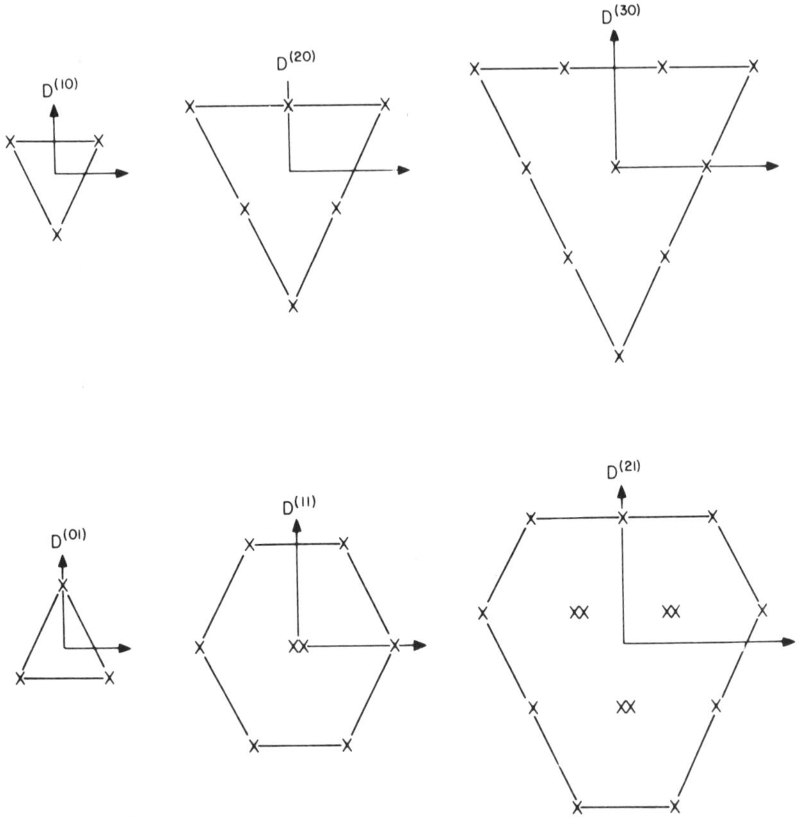

Figure 11.6

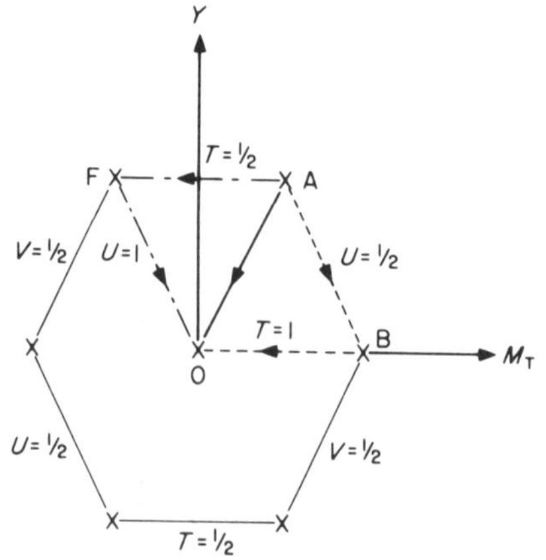

Figure 11.7

Since the state $|\psi\rangle$ has $M_V = -1$ and therefore $V = 1$ it follows that the first of these paths must simply produce a state with $V = 1$, $M_V = 0$, as indicated in equation (11.8). The factor $\sqrt{2}$ comes in the usual way through equation (7.40) from the SU_2 raising operation V_+. In the same way $\mathrm{U}_-|\psi\rangle$ is a state with $M_T = 1$ and therefore $T = 1$. Hence $\mathrm{T}_-\mathrm{U}_-|\psi\rangle$ also has $T = 1$, with $M_T = 0$. Similarly the third path produces a state with $U = 1$. We have therefore constructed three apparently different states, all corresponding to the point (0, 0) but they are not necessarily linearly independent. In fact, the commutation relation $[\mathrm{U}_-, \mathrm{T}_-] = \mathrm{V}_+$, given in equation (11.3d), shows directly from equation (11.8) the linear dependence, $|U = 1\rangle - |T = 1\rangle = |V = 1\rangle$. One therefore has a choice of which two orthogonal combinations of these three states to take as basis vectors. For many purposes it is convenient to take $|T = 1\rangle$ and $|T = 0\rangle$ which must be orthogonal. Again using the commutation relations (11.3d) we may show that the $|\mathrm{T} = 0\rangle$ state is given by $|\mathrm{T} = 0\rangle = \{|U = 1\rangle + |V = 1\rangle\}3^{-\frac{1}{2}}$. To do this it is sufficient to show that $\mathrm{T}_+(|U = 1\rangle + |V = 1\rangle) = 0$. The factor $3^{-\frac{1}{2}}$ comes about because, although the two states $|U = 1\rangle$ and $|V = 1\rangle$ are normalised, they are not orthogonal,

$$\begin{aligned}\langle V = 1|U = 1\rangle &= \tfrac{1}{2}\langle\psi|\mathrm{V}_-\mathrm{U}_-\mathrm{T}_-|\psi\rangle \\ &= \tfrac{1}{2}\langle\psi|\mathrm{U}_-\mathrm{V}_-\mathrm{T}_-|\psi\rangle + \tfrac{1}{2}\langle\psi|\mathrm{T}_+\mathrm{T}_-|\psi\rangle \\ &= \tfrac{1}{2}\langle\psi|\mathrm{U}_-\mathrm{T}_-\mathrm{V}_-|\psi\rangle - \tfrac{1}{2}\langle\psi|\mathrm{U}_-\mathrm{U}_+|\psi\rangle + \tfrac{1}{2} = \tfrac{1}{2}\end{aligned}$$

In terms of the basis vectors $|T = 1\rangle$ and $|T = 0\rangle$ we thus have

$$\begin{aligned}|U = 1\rangle &= \tfrac{1}{2}\sqrt{3}|T = 0\rangle + \tfrac{1}{2}|T = 1\rangle \\ |V = 1\rangle &= \tfrac{1}{2}\sqrt{3}|T = 0\rangle - \tfrac{1}{2}|T = 1\rangle\end{aligned}$$

and hence from orthogonality,

$$\begin{aligned}|U = 0\rangle &= \tfrac{1}{2}|T = 0\rangle - \tfrac{1}{2}\sqrt{3}|T = 1\rangle \\ |V = 0\rangle &= \tfrac{1}{2}|T = 0\rangle + \tfrac{1}{2}\sqrt{3}|T = 1\rangle\end{aligned} \tag{11.9}$$

11.6.1 Complex conjugate representations

Given a matrix representation D of any group it follows that the complex conjugate matrices D* also provide a representation of the group. To investigate D* it is simplest to consider the corresponding infinitesimal operators. Complex conjugation of the infinitesimal operators X_i in SU_3 leads, through the definitions (11.1) and (11.2), to a change in sign of T_z and Y and an interchange of raising and lowering roles for the other six operators. So far as the diagrams in the (M_T, Y) plane are concerned this amounts to a reflection in both axes and one sees from figure 11.5 that it is the same as interchanging λ and μ. Thus for example $\mathrm{D}^{(01)}$ is equivalent to the conjugate representation to $\mathrm{D}^{(10)}$ while $\mathrm{D}^{(11)}$, with $\lambda = \mu$, is self-conjugate. (For the group $\mathscr{R}_3$, or SU_2, the complex conjugation simply interchanges $+m$ and $-m$ so that $\mathrm{D}^{(j)}$ and $\mathrm{D}^{(j)*}$ are equivalent, see section 7.7.)

11.6.2 Multiplication of representations

The rule for reducing the product of two irreducible representations of $\mathscr{R}_3$ was given in equation (7.44). A corresponding procedure

$$D^{(\lambda_1\mu_1)} \times D^{(\lambda_2\mu_2)} = \dot{\sum_{\lambda\mu}} C(\lambda\mu) D^{(\lambda\mu)} \tag{11.10}$$

may be carried out for the group SU_3 but the rule for deducing the numerical coefficients $C(\lambda\mu)$ for arbitrary $(\lambda_1\mu_1)$ and $(\lambda_2\mu_2)$ is quite complicated. However for small values of the λ and μ it is simple to deduce these coefficients by making use of the subgroup $SU_2 \times U_1$ and a knowledge of the reduction of product representations for this group. The irreducible representations of $SU_2 \times U_1$ are labelled by T, Y. The product of T_1, Y_1 with T_2, Y_2 reduces to a sum of representations with $T = (T_1 + T_2)$, $(T_1 + T_2 - 1)$, . . ., $|T_1 - T_2|$ and $Y = Y_1 + Y_2$ using the familiar SU_2 (or $\mathscr{R}_3$) vector coupling rule for T and the additive rule, see section 11.5, for Y. We illustrate the method by a few examples.

Consider the product $D^{(10)} \times D^{(10)}$. Since each factor contains basis vectors given by $T = \frac{1}{2}$, $Y = \frac{1}{3}$ and $T = 0$, $Y = -\frac{2}{3}$ it follows that the product space has basis vectors labelled by the T- and Y-combinations given in the first row of table 11.2(a). (It is understood that for each T there is the usual $(2T+1)$ multiplet of basis vectors with $M_T = T, T-1, \ldots, -T$.) The notation $\frac{1}{2}, \frac{1}{2}$ indicates that two independent $T = \frac{1}{2}$ vectors may be produced by taking $T = \frac{1}{2}$ in the first factor and $T = 0$ in the second or vice versa. The highest weight of these entries is $M_T = 1$, $Y = \frac{2}{3}$ which implies that the product must contain the representation $D^{(20)}$. We use here the relations $\lambda = 2M_T$, $\mu = \frac{1}{2}(3Y - 2M_T)$ relating λ and μ to the values of M_T and Y in the vector of maximum weight, see equation (11.6). Now by subtraction of the known T, Y combinations of $D^{(20)}$, shown in the second lines of the table, we see that the remainder fits precisely the known pattern of $D^{(01)}$. Thus we conclude that

$$D^{(10)} \times D^{(10)} = D^{(20)} \dot{+} D^{(01)}$$

In the same way, table 11.2(b) illustrates the reduction

$$D^{(10)} \times D^{(01)} = D^{(11)} \dot{+} D^{(00)}$$

and, by the same procedure,

$$D^{(11)} \times D^{(11)} = D^{(22)} \dot{+} D^{(03)} \dot{+} D^{(30)} \dot{+} 2D^{(11)} \dot{+} D^{(00)} \tag{11.11}$$

Table 11.2(a)

	$Y = \frac{2}{3}$	$-\frac{1}{3}$	$-\frac{4}{3}$
$D^{(10)} \times D^{(10)}$	$T = 1, 0$	$\frac{1}{2}, \frac{1}{2}$	0
$D^{(20)}$	$T = 1$	$\frac{1}{2}$	0
$D^{(01)}$	$T = 0$	$\frac{1}{2}$	—

Table 11.2(b)

	$Y = 1$	0	-1
$D^{(10)} \times D^{(01)}$	$T = \frac{1}{2}$	1, 0, 0	$\frac{1}{2}$
$D^{(11)}$	$T = \frac{1}{2}$	1, 0	$\frac{1}{2}$
$D^{(00)}$	—	0	—

We have written down equation (11.11) partly because it will be used in an application later on but also because it illustrates that the group SU_3 is not 'simply reducible' in the sense defined in section 4.17. The representation $D^{(11)}$ occurs twice in the reduction. The results above may also be obtained from the general rule for products of representations of U_N given in section 18.5 of volume 2.

11.7 Classification of the hadrons into SU_3 multiplets

Having developed some familiarity with the irreducible representations of SU_3 we now return to the elementary particles of table 11.1 in section 11.1 to see how they fit into the SU_3 multiplets. Physically we associate the T-spin SU_2 group with isospin and the SU_3 operator Y with the hypercharge. Thus SU_3 is an enlargement of the isospin group, incorporating hypercharge.

The first eight baryons in table 11.1 all have spin $\frac{1}{2}$. (It is for this reason that we have listed Ξ before Δ even though they are slightly heavier.) If we now look at the values of T and Y for these particles we see that they are *precisely* those of the $D^{(11)}$ representation of SU_3 which was illustrated in figures 11.6 and 11.7. Furthermore, the remaining ten baryons in table 11.1, which all have the same spin of $\frac{3}{2}$, have the T, Y combinations of the $D^{(30)}$ representation shown in figure 11.6. The mesons listed in table 11.1 also fit into the $D^{(11)}$ representation of SU_3. Figure 11.8 shows the positioning of these particles in the appropriate SU_3 diagrams. It is common practice to refer to the representations $D^{(11)}$ and $D^{(30)}$ as the octet and decuplet, respectively, because they have dimensions eight and ten.

The precise way in which the particles arrange themselves into SU_3 multiplets suggests that the group SU_3 has some physical significance. We have already seen in chapter 10 that the strong interaction is invariant under the isospin SU_2 group which is a subgroup of SU_3. Furthermore, since the strong interaction also conserves hypercharge Y, the larger subgroup $SU_2 \times U_1$ described in section 11.5 is a symmetry group, where Y is the infinitesimal operator of U_1. The irreducible representations of this product group are labelled simply by the pair of labels T, Y appropriate to each separate group. We now take the further tentative step, suggested by the beautiful way in which particles fit into SU_3 multiplets, of supposing that the group SU_3 itself may be a symmetry group. However, there is a difficulty. The small mass differences between members of an isospin multiplet was ascribed to the electromagnetic forces which are not isospin invariant. If the symmetry of the strong interaction is extended from SU_2 to SU_3 then one would expect that the mass differences of all members of an SU_3 multiplet would be of the same order—a few MeV. But table 11.1 shows that the differences are of hundreds of MeV. We must therefore conclude that the strong interactions is not completely SU_3 invariant but that it contains a medium-strong part which

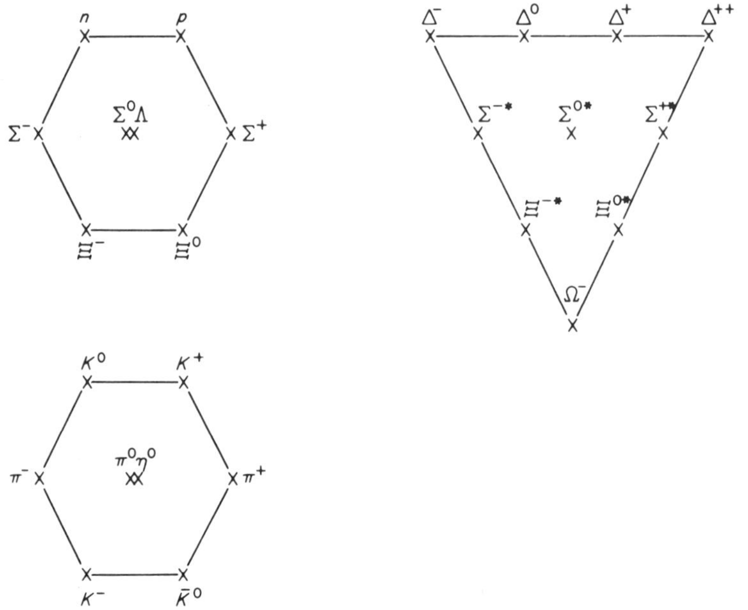

Figure 11.8

breaks the SU_3 symmetry. A study of the relative strengths of various reactions using methods analogous to those of subsection 10.2.1 leads to the same conclusion.

In the remaining sections of this chapter we go on to test the SU_3 symmetry by looking at various properties of the particles. The question of why no particles have been found which belong to representations simpler than $D^{(11)}$ or $D^{(30)}$, such as the basic $D^{(10)}$ representations, will be discussed in chapter 12. It is perhaps worth remarking that, when the group SU_3 was first taken seriously, in the early 1960s, the Ω^- had not yet been observed, although all the other nine members of the $D^{(30)}$ representation were known. The first observation of the Ω^- in 1964, with just the right properties to fit into the remaining vacancy in the $D^{(30)}$ representation, provided impressive confirmation of the SU_3 picture.

11.8 The mass-splitting formula

In subsection 10.1.2 we described the mass splitting of a nuclear isospin multiplet by considering the transformation properties of the Coulomb force which was responsible for the splitting. We argued that this was analogous to the Zeeman splitting of an angular momentum multiplet. We now apply the same general argument to describe the mass splitting between different isospin multiplets within an SU_3 multiplet. For this purpose we shall ignore the small splitting within an isospin multiplet and take the mean mass for each T. We are concerned with understanding the differences of several hundreds of MeV between, for example, the nucleon and the Σ-particles. Such splittings are due to the medium-strong force which must be an isospin and hypercharge invariant but not an SU_3 invariant. Of course we do not know what form this force has, unlike the earlier problem in subsection 10.1.2 with the Coulomb force. We must therefore make some assumption about its transformation properties under SU_3. Looking at the simple SU_3 representations drawn in figure 11.6 we see that the only basis vector with the properties $T = Y = 0$ occurs in the $D^{(11)}$ representation, at the origin. For this and other more detailed reasons we suppose that the medium-strong force which breaks SU_3 symmetry has the transformation properties of $D^{(11)}$ with $T = Y = 0$; it transforms like the Λ-particle.

With this assumption and the use of first-order perturbation theory the mass splittings $\Delta M(T, Y)$ of states T, Y belonging to the same representation $(\lambda\mu)$ are given by the expectation value

$$\Delta M(T, Y) = \langle(\lambda\mu)TY|\mathrm{H}[(11)T = Y = 0]|(\lambda\mu)TY\rangle \qquad (11.12)$$

where H denotes the medium-strong interaction. We may now use the Wigner–Eckart theorem (4.62) to relate the mass splittings.

We first note that, since the group SU_3 is not simply reducible, the sum over t in equation (4.62) will generally contain more than one term. (This differs from the group $\mathscr{R}_3$ for which the label t was not required.) In general therefore there will be more than one unknown reduced matrix element in equation (11.12). For the particular case of interest, where the operator transforms like $D^{(11)}$, it may be shown that there are, at most, two reduced matrix elements. We are particularly interested in matrix elements for states belonging to the $D^{(11)}$ and $D^{(30)}$ representations appropriate to the particles illustrated in figure 11.8. The range of t-values for these cases follows from the reductions of product representations

$$\begin{aligned} D^{(11)}\times D^{(30)} &= D^{(41)} \dotplus D^{(22)} \dotplus D^{(30)} \dotplus D^{(11)} \\ D^{(11)}\times D^{(11)} &= D^{(22)} \dotplus D^{(03)} \dotplus D^{(30)} \dotplus 2D^{(11)} \dotplus D^{(00)} \end{aligned} \qquad (11.13)$$

deduced as shown in section 11.6. Since $D^{(30)}$ appears once in the first of these equations and $D^{(11)}$ appears twice in the second we conclude that only one reduced matrix element occurs for the case $(\lambda\mu) = (30)$ in equation (11.12), while two reduced matrix elements occur for the case $(\lambda\mu) = (11)$. Closed formulae for the relevant Clebsch–Gordan coefficients are available, leading to the celebrated mass-splitting formulae

$$\langle(30)TY|\mathrm{H}[(11)T = Y = 0]|(30)TY\rangle = aY$$

and

$$\langle(11)TY|\mathrm{H}[(11)T = Y = 0]|(11)TY\rangle = bY + c[T(T+1) - \tfrac{1}{4}Y^2 - 1] \qquad (11.14)$$

where a, b and c are unknown constants arising from the reduced matrix elements of the perturbation $\mathrm{H}[(11)T=Y=0]$ together with constant numerical factors from the Clebsch–Gordan coefficients. The dependence on T and Y exhibited in equations (11.14) may be tested against the known baryon masses given in table 11.1. In deriving equations (11.14) we have neglected any electromagnetic contributions to the masses so that we shall be unable to explain the small mass differences between members of the same isospin multiplet. We therefore ignore these differences and discuss only the mean values of the mass for each isospin multiplet, which are listed below in MeV:

Decuplet (30): $Y=1, 1232$. $Y=0$, 1385. $Y=-1$, 1533. $Y=-2$, 1672.
Octet (11): $Y=1, T=\frac{1}{2}$, 939. $Y=0, T=1$, 1193. $Y=0, T=0$, 1116.
$Y=-1, T=\frac{1}{2}$, 1318.

One sees immediately the linear dependence on Y for the decuplet with a value for a of about -150 MeV. For the octet, there are only three mass differences and two unknowns b and c. From $Y=\pm 1$ we deduce $b \sim -190$ MeV and using $Y=0, T=1$ we have $c \sim 43$ MeV. The formula (11.14) then predicts the value 86 MeV for the difference between $Y=0$, $T=1$ and $Y=0$, $T=0$ in reasonable agreement with the experimental 80 MeV. Thus we conclude that the assumption of a symmetry breaking term of the type (11), $T=Y=0$ is consistent with the data to an accuracy of the order of 10 MeV.

Before leaving this topic it is instructive to see how the formulae (11.14) may be deduced using the operator equivalent method of section 7.4.7. To use the operator equivalent method we must construct two independent operators with the transformation properties $(11)T=Y=0$ and whose matrix elements are known. Two operators are required in this problem because of the non-simply reducible nature of SU_3. As with the group $\mathscr{R}_3$, we try to construct such operators from the infinitesimal operators of the group, in this case the eight operators (11.2) of SU_3. We first notice from the commutation relations (11.3) that these eight operators have a T, Y structure as follows

$$\begin{array}{ll} \mathrm{U}_+, \mathrm{V}_- & \text{have } Y=1, T=\frac{1}{2} \\ \mathrm{T}_\pm, \mathrm{T}_z & \text{have } Y=0, T=1 \\ \mathrm{Y} & \text{has } Y=0, T=0 \\ \mathrm{V}_+, \mathrm{U}_- & \text{have } Y=-1, T=\frac{1}{2} \end{array} \tag{11.15}$$

Comparing this with the structure of the $\mathrm{D}^{(11)}$ representation suggests that the eight infinitesimal operators of SU_3 transform like the representation $\mathrm{D}^{(11)}$ and a detailed comparison with the transformation properties of the basis vectors of $\mathrm{D}^{(11)}$ confirms this. (The analogous result for $\mathscr{R}_3$ is that the three operators J_q transform like the vector representation $\mathrm{D}^{(1)}$ of $\mathscr{R}_3$.) In particular, the operator Y has the required properties $(11)T=Y=0$ for the mass splitting problem and explains the first term in equation (11.14). For the decuplet this completes the problem but for the octet a second operator is required. One can now try to construct such an operator from quadratics in the infinitesimal operators. Clearly the two operators T^2 and Y^2 have the necessary $T = Y = 0$ properties but each is a mixture of (22), (11) and (00), see the second of equations (11.13). Some rather detailed algebra shows that the combination

$$\mathrm{T}^2 - \tfrac{1}{4}\mathrm{Y}^2 - \tfrac{1}{6}\mathrm{C}_2 \tag{11.16}$$

has the required (11) character where C_2 is the Casimir operator to be defined in section 11.10. Inserting the values $T(T+1)$ for T^2 and the expression (11.21) with $\lambda = \mu = 1$ for C_2 completes the derivation of equation (11.14). The operator (11.16) may also be constructed using the Cartesian notation of section 11.10 and arguing that the operators $\sum_k \mathrm{A}^j_k \mathrm{A}^k_i - \frac{1}{3}\delta_{ij}\mathrm{C}_2$ transform like A^j_i.

11.9 Electromagnetic effects

We first note from section 11.1 that the charge is given empirically by $Q = M_T + \frac{1}{2}Y$ and from the commutation relations (11.3)

$$[\mathrm{T}_z + \tfrac{1}{2}\mathrm{Y}, \mathrm{U}_\pm] = [\mathrm{T}_z + \tfrac{1}{2}\mathrm{Y}, \mathrm{U}_z] = 0 \tag{11.17}$$

Thus the charge commutes with the U-spin operators and in fact the charge Q has the same connection with U-spin that Y has with T-spin. In particular just as particles on the same horizontal line in figure 11.8 have the same value of Y, so particles on the line of a U-spin multiplet, in the direction of $\mathrm{U}_\pm$ in figure 11.3 (of section 11.6) have the same charge. This is clear from figure 11.8. It is therefore suggested that the electromagnetic interaction is a U-spin invariant and this hypothesis may be tested by comparing the electromagnetic properties of different members of an SU_3 multiplet.

The magnetic moments μ provide one such test. Since the magnetic moment operator M is linear in the charge, one supposes that it transforms like the charge, namely $\mathrm{M}[(11)U = Q = 0]$. The structure is analogous to that of the medium-strong mass splitting operator discussed in the previous section, except that T and Y have been replaced by U and Q, equivalent to a 120° rotation in the representation diagram. We therefore have the corresponding formula to (11.14).

$$\mu = \langle (11)TY | \mathrm{M}[(11)U = Q = 0] | (11)TY \rangle = \mathrm{d}Q + e\{U(U+1) - \tfrac{1}{4}Q^2 - 1\} \tag{11.18}$$

which implies that the magnetic moments of the eight particles in the baryon octet of figure 11.8 are all given in terms of only two constants. Unfortunately, not many of these moments have yet been measured but some tests of equation (11.18) can be made. For example this equation implies that $\mu(p) = \mu(\Sigma^+)$ and that $\mu(n) = 2\mu(\Lambda)$, whereas the measured values are

$$\mu(p) = 2.79, \quad \mu(\Sigma^+) = 2.5 \pm 0.7$$
$$\mu(n) = -0.95, \quad \mu(\Lambda) = -0.73 \pm 0.16$$

in units of the nuclear magneton ($e\hbar/2M_p c$).

The small mass differences between different members of each isospin multiplet are presumed to be due to electromagnetic effects. The assumption of U-spin invariance for this electromagnetic perturbation leads to relations between the mass differences and a more detailed assumption concerning the SU_3 character of the perturbation gives further relations. Such a calculation is in principle similar to that given in subsection 10.1.2 and we give no details here.

11.10 Casimir operators

For the group $\mathscr{R}_3$ we constructed in section 7.5 the Casimir operator $\mathbf{J}^2$ which commuted with all group operators and hence through Schur's lemma was a multiple $j(j+1)$ of the unit matrix in an irreducible representation $D^{(j)}$. The representation label j could be regarded as denoting the eigenvalue $j(j+1)$ of $\mathbf{J}^2$. For the group SU_3 which needs two labels λ and μ for its irreducible representations there are two independent Casimir operators, one quadratic and the other cubic in the infinitesimal operators.

The simplest way to construct them is to use an obviously Cartesian notation for the operators (11.2) which brings out the symmetry between the three dimensions. We use the symbol A_i^j to denote the 3×3 matrix with matrix elements.

$$(A_i^j)_{kl} = \delta_{ik}\delta_{jl} - \tfrac{1}{3}\delta_{ij}\delta_{kl} \tag{11.19}$$

so that for $i \neq j$ the matrix A_i^j is zero, except for the entry 1 at the intersection of the row i with the column j. When $i = j$ this is modified by the subtraction of $\frac{1}{3}\delta_{kl}$ so that A_i^j remains traceless. The three diagonal matrices are related by $A_1^1 + A_2^2 + A_3^3 = 0$ leaving eight independent matrices as in equation (11.2). The detailed correspondence is $A_1^2 = T_+$, $A_2^1 = T_-$, $A_2^3 = U_+$, $A_3^2 = U_-$, $A_3^1 = V_+$, $A_1^3 = V_-$, $A_3^3 = -Y$, $A_1^1 - A_2^2 = 2T_z$. From their definition we see that the commutation relations of the A_i^j are given very simply

$$[A_i^j, A_k^l] = \delta_{jk} A_i^l - \delta_{il} A_k^j \tag{11.20}$$

from which it follows that the operator $C_2 = \sum_{i,j} A_i^j A_j^i$ commutes with all A_k^l and may therefore be taken as a Casimir operator. In the same way, the operator $C_3 = \sum_{ijk} A_j^k A_i^j A_k^i$ also commutes with all the group operators and therefore serves as another Casimir operator. This process could be extended but all higher order operators would be expressible in terms of C_2 and C_3.

The eigenvalues of C_2 and C_3 in any irreducible representation $D^{(\lambda\mu)}$ may now be evaluated quite simply by studying their operation on the basis vector of highest weight. Any basis vector would lead to the same result and we choose the maximum weight for convenience. Using the commutation relations (11.20), we may write

$$\begin{aligned} C_2 &= \sum_i (A_i^i)^2 + \sum_{i<j} (A_i^j A_j^i + A_j^i A_i^j) \\ &= \sum_i (A_i^i)^2 + \sum_{i<j} (2A_j^i A_i^j + A_i^i - A_j^j) \end{aligned}$$

However, for the state $|\psi\rangle$ of maximum weight, $A_i^j|\psi\rangle = 0$ for $i<j$ and for the diagonal operators $A_3^3|\psi\rangle = -Y|\psi\rangle = -\frac{1}{3}(\lambda+2\mu)|\psi\rangle$ and $(A_1^1 - A_2^2)|\psi\rangle = 2M_T|\psi\rangle = \lambda|\psi\rangle$, while $A_1^1 + A_2^2 + A_3^3 = 0$. Thus finally,

$$C_2|\psi\rangle = \{\tfrac{2}{3}(\lambda^2+\mu^2+\lambda\mu) + 2(\lambda+\mu)\}|\psi\rangle \tag{11.21}$$

The cubic operator C_3 may be evaluated in a similar way. However, it is simpler to define $C_3' = \frac{1}{2}\sum_{ijk}(A_j^k A_i^j A_k^i + A_i^j A_j^k A_k^i) = C_3 + \frac{3}{2}C_2$ which in the representation $D^{(\lambda\mu)}$ has an eigenvalue given by

$$C_3'|\psi\rangle = \tfrac{1}{9}(\lambda-\mu)(2\lambda+\mu+3)(2\mu+\lambda+3)|\psi\rangle \tag{11.22}$$

The values of C_2 and C_3 may be used to define the irreducible representation of SU_3 in the same way that the value of $\mathbf{J}^2$ is used in $\mathscr{R}_3$.

Bibliography

Data on elementary particles may be found in

Particle data group (1974). *Phys. Lett.*, **50B,** 1

For an introduction to particle physics we suggest

Perkins, D. H. (1972). *Introduction to High Energy Physics* (Addison Wesley, Reading, Mass.)

More details of the use of SU_3 in particle physics may be found in

Kokkedee, J. J. J. (1969). *The Quark Model* (Benjamin, New York)

Problems

11.1 Show that the infinitesimal operators of SU_3 must have zero trace.

11.2 Given an arbitrary skew-Hermitian 3×3 matrix with elements X_{ij}, deduce the first three coefficients a_q when X is expressed $X = \sum_q a_q X_q$ as a linear combination of the eight matrices (11.1) and show that the coefficients are real.

11.3 Show directly that the irreducible representations of the group U_1 are labelled by integers.

11.4 Show that the three operators X_6, X_7 and $\frac{1}{2}(-X_3 + \frac{1}{2}X_8)$ form an SU_2 subgroup (U-spin) of SU_3 and that the Hermitian operator $Q = i(X_3 + \frac{1}{6}X_8)$ commutes with them.

11.5 Construct the weight diagram for the representation (22) of SU_3. Show that it has dimension 27 and deduce the combinations of T and Y for the different basis vectors.

11.6 Deduce the multiplicities stated in section 11.6 for the basis vectors corresponding to the lattice points of the weight diagram of an arbitrary irreducible representation of SU_3 by the following arguments: (a) From the vector $|\psi\rangle$ with greatest weight generate the vectors $(T_-)^n|\psi\rangle$ which constitute the top row of the diagram. (b) Show that the vectors in the next row can only be the T-multiplets built on the vectors $U_-|\psi\rangle$ and $V_+|\psi\rangle$. (c) Show that these two multiplets must be independent unless $\lambda = 0$ or $\mu = 0$, i.e. the diagram is triangular.

11.7 Deduce the formula (11.7) for the dimension of an SU_3 representation by noting that the number of lattice points on the outer hexagon is $3(\lambda + \mu)$ and using the known multiplicities on the interior hexagons and triangles. (You will need to deduce that the multiplicity of the triangles is $(\mu + 1)$ and that the number of lattice points on a side of the largest triangle is $(\lambda - \mu + 1)$, where we have taken $\lambda \geq \mu$.)

11.8 Use the technique given in section 11.6.2 to deduce the reductions

$$D^{(11)} \times D^{(10)} = D^{(21)} \dot{+} D^{(02)} \dot{+} D^{(10)} \text{ and}$$
$$D^{(11)} \times D^{(30)} = D^{(41)} \dot{+} D^{(22)} \dot{+} D^{(30)} \dot{+} D^{(11)}$$

of product representations. Deduce also the reduction (11.11).

11.9 If some new baryons were found to belong to the representation (22) use the results of question 11.5 to deduce their quantum numbers T, Y and Q.

11.10 Use the explicit technique of problem 7.8 to deduce the Clebsch–Gordan coefficients for the state $|(30)\, Y = 1, T = M_T = \frac{3}{2}\rangle$ which occurs in the reduction of the product $D^{(30)} \times D^{(11)}$.

11.11 Verify that the Casimir operators C_2 and C_3, defined in section 11.10, commute with the infinitesimal operators A_i^j.

12

Supermultiplets in Nuclei and Elementary Particles —the Groups SU_4 and SU_6 and Quark Models

In the discussion of atomic structure in chapter 8 the spin-independence of the interaction led to the symmetry group $\mathscr{R}_3^S$ and the quantum number S. In nuclei we have seen in chapter 10 how the charge-independence of the interaction led to the isospin quantum number T through the group SU_2. In the present chapter we first study the consequences of the assumption that the interaction between nucleons in a nucleus is independent of both spin and charge. This brings together the groups $\mathscr{R}_3^S$ and SU_2 into a larger group SU_4. Moving on to elementary particles we carry out the analogous step of bringing together the groups $\mathscr{R}_3^S$ and SU_3 within the larger group SU_6. Finally we describe some recent tentative ideas for the structure of the baryons and mesons themselves as composite particles built from even more elementary objects which have been given the rather fanciful name 'quarks'.

12.1 Supermultiplets in nuclei

A nucleus is composed of A nucleons, each of which is described by the usual position coordinates and in addition has four independent intrinsic states. These states correspond to the two possible spin states $\pm\frac{1}{2}$ and the two possible charge states, proton or neutron, and it is convenient to denote them by $p\uparrow$, $n\uparrow$, $p\downarrow$, $n\downarrow$ with the arrows $\uparrow\downarrow$ denoting the spin state $\pm\frac{1}{2}$. In exact analogy with the introduction of the isospin group SU_2 in the two-dimensional space of the states p, n, see section 10.1, we may now introduce the group SU_4 of unitary transformations in the four-dimensional space of a nucleon defined by the states $p\uparrow$, $n\uparrow$, $p\downarrow$ and $n\downarrow$. If the interaction between nucleons is not only charge-independent but also spin-independent then it follows that the group SU_4 is a symmetry group for the nucleus—see Wigner and Feenberg (1941) in the bibliography. In practice the nuclear interaction has quite a strong spin-dependence so that SU_4 is only an approximate symmetry group. Nevertheless, we give a brief description of SU_4 and its consequences, partly as an introduction to the use of SU_6 in the next section. Both groups are special cases of the group SU_N discussed in chapter 18 of volume 2.

Just as the group SU_3 has $3^2-1=8$ parameters, so the group SU_4 has $4^2-1=15$ parameters. The fifteen infinitesimal operators for a single nucleon are s_q, t_q and $\mathrm{s}_q\mathrm{t}_{q'}$, where $\mathbf{s}$ and $\mathbf{t}$ denote the usual single-particle spin and isospin operators with $q,\ q'=x,\ y,\ z$. For a system of A nucleons the infinitesimal operators are $\mathrm{S}_q=\sum_{i=1}^{A}\mathrm{s}_q(i)$, $\mathrm{T}_q=\sum_{i=1}^{A}\mathrm{t}_q(i)$ and $\mathrm{Y}_{qq'}=2\sum_{i=1}^{A}\mathrm{s}_q(i)\mathrm{t}_{q'}(i)$. (Notice that $\mathrm{Y}_{qq'}\neq 2\mathrm{S}_q\mathrm{T}_{q'}$.) The six operators S_q and T_q clearly generate the subgroup $\mathscr{R}_3^S\times SU_2$, or $SU_2^S\times SU_2^T$ because of the homomorphism between $\mathscr{R}_3$ and SU_2. The upper index S or T is used to distinguish SU_2 in spin from SU_2 in isospin. (The operator $\mathrm{Y}_{qq'}$ has no connection with the hypercharge which was denoted by Y in the previous chapter.)

The properties of the irreducible representations of SU_4 may be deduced by a natural extension of the arguments used in section 11.6 for SU_3. Three of the fifteen operators may be simultaneously diagonalised and these are usually chosen to be S_z, T_z and Y_{zz}. (We stress that, whereas the suffix z on S_z refers to a geometrical direction in ordinary space, the suffix z on T_z refers only to the abstract isospin space related to the charge degree of freedom.) Thus the SU_4 diagrams corresponding to figure 11.6 are three-dimensional and the irreducible representations are conventionally labelled by $\mathrm{D}^{(PP'P'')}$. The three numbers P, P' and P'' denote the values of S_z, T_z and Y_{zz} in the basis vector with maximum weight. In other words, P is the greatest value M_S of S_z which occurs in the representation, while P' is the greatest value M_T of T_z occurring in the set of basis vectors with $M_S=P$, and P'' is the greatest value of Y_{zz} among the basis vectors with $M_S=P$, $M_T=P'$. Because it involves much algebra we shall not attempt to deduce the properties of the general irreducible representations as we did for SU_3. (See chapter 18 of volume 2 for a general discussion of SU_N.) Instead, we shall study in detail the simple representations occurring with two or three nucleons making use of the permutation symmetry of the states and the subgroup $SU_2^S \times SU_2^T$. Since $SU_2^S \times SU_2^T$ is a subgroup of SU_4, it follows that any irreducible representation of SU_4 is expressible as the sum, in the sense of section 4.18, of irreducible representations $\mathrm{D}^{(S)} \times \mathrm{D}^{(T)}$ of $SU_2^S \times SU_2^T$. Thus any irreducible representation $\mathrm{D}^{(PP'P'')}$ of SU_4 is composed of a set of S, T combinations and is said to form a 'supermultiplet'

$$\mathrm{D}^{(PP'P'')} = \dot{\sum_{S,T}} m_{S,T} \mathrm{D}^{(S)} \times \mathrm{D}^{(T)} \tag{12.1}$$

where the $m_{S,T}$ are positive integers or zero.

The four states of a single particle constitute the space in which SU_4 is defined and hence clearly form the basis for a four-dimensional irreducible representation of SU_4 which is $\mathrm{D}^{(\frac{1}{2}\frac{1}{2}\frac{1}{2})}$. Since all four states have $S = T = \frac{1}{2}$, this representation does not reduce on restriction to the subgroup $SU_2^S \times SU_2^T$ and we have $\mathrm{D}^{(\frac{1}{2}\frac{1}{2}\frac{1}{2})} = \mathrm{D}^{(\frac{1}{2})} \times \mathrm{D}^{(\frac{1}{2})}$.

For two particles there is a total of $4^2 = 16$ states of spin and charge and this 16-dimensional space L provides a representation of SU_4. However it is not irreducible, as we may show by dividing the space into two parts $L = L_s + L_a$, where L_s contains states which are symmetric under simultaneous permutation of the spin and isospin coordinates of the two particles and L_a contains the antisymmetric states. The infinitesimal operators S_q, T_q and $\mathrm{Y}_{qq'}$ of SU_4 are symmetric under this permutation and hence cannot mix the symmetric and antisymmetric states. Thus the two subspaces L_s and L_a each provide a representation of SU_4. To find the structure of these two representations with respect to the subgroup $SU_2^S \times SU_2^T$ we use the result, deduced in subsection 8.6.4, that a state with $S = 1$ is symmetric while $S = 0$ is antisymmetric with respect to permutation of the spin coordinates. Using the corresponding result for isospin it follows that, under simultaneous permutation in both spin and isospin, the symmetry of the states is as follows:

$$L_s\colon \begin{cases} S = 1, T = 1 \\ S = 0, T - 0 \end{cases} \quad L_a\colon \begin{cases} S = 1, T = 0 \\ S - 0, T = 1 \end{cases} \tag{12.2}$$

Here, we have used the simple results that, for example, the product of two antisymmetric functions gives a symmetric function, a process which is actually the analysis of a product representation for the group $\mathscr{S}_2$. Using the known dimension $(2S+1)$ of the representation $\mathrm{D}^{(S)}$ of SU_2 one finds the dimensions 10 and 6, respectively, of L_s and L_a which checks with the known dimension 16 of L. From general arguments it can be shown that the representations generated in L_s and L_a are irreducible and following the notation introduced above they are denoted by $\mathrm{D}^{(111)}$ and $\mathrm{D}^{(100)}$, respectively, with the reductions (12.1) becoming

$$\begin{aligned} D^{(111)} &= D^{(1)} \times D^{(1)} \dotplus D^{(0)} \times D^{(0)} \\ D^{(100)} &= D^{(1)} \times D^{(0)} \dotplus D^{(0)} \times D^{(1)} \end{aligned} \tag{12.3}$$

As explicit examples of these states we have

$$\begin{aligned} |T = S = 1, M_T = M_S = 1\rangle &= p^{\uparrow}(1)\, p^{\uparrow}(2) \\ |T = 1, S = 0, M_T = 1, M_S = 0\rangle &= \{p^{\uparrow}(1)p^{\downarrow}(2) - p^{\uparrow}(2)p^{\downarrow}(1)\}/2^{\frac{1}{2}} \\ |T = 1, S = 0, M_T = M_S = 0\rangle &= \{p^{\uparrow}(1)n^{\downarrow}(2) - p^{\uparrow}(2)n^{\downarrow}(1) \\ &\qquad - p^{\downarrow}(1)n^{\uparrow}(2) + p^{\downarrow}(2)n^{\uparrow}(1)\}/2 \end{aligned} \tag{12.4}$$

For three particles we follow the same procedure but make use of some simple properties of the symmetric group $\mathscr{S}_3$ as in subsection 8.6.4. There is now a total of $4^3 = 64$ possible states and this vector space is divided into four parts, $L = L_s + L_{m_1} + L_{m_2} + L_a$, corresponding to the three irreducible representations $T^{(s)}$, $T^{(m)}$ and $T^{(a)}$ of $\mathscr{S}_3$. Since $T^{(m)}$ is two-dimensional, there are two subspaces L_{m_1} and L_{m_2} corresponding to states which transform like row 1 or row 2 of $T^{(m)}$. This division of L is an example of the use of equation (4.52) involving projection operators for $\mathscr{S}_3$. The symmetry of the SU_4 group operators with respect to permutation of particles again prevents any coupling between the four subspaces of L so that each subspace provides a representation of SU_4. Again, these four representations of SU_4 are irreducible. The two representations provided by L_{m_1} and L_{m_2} are, however, equivalent since these two spaces are transformed into each other by permutations, which commute with the SU_4 operators. To find the reduction of these representations on restriction to the subgroup $SU_2^S \times SU_2^T$ we start from the result, deduced in subsection 8.6.4, that the $S = \frac{3}{2}$ state of three spin $\frac{1}{2}$ particles is totally symmetric $T^{(s)}$, while $S = \frac{1}{2}$ has mixed symmetry $T^{(m)}$. Using the corresponding result in isospin and the known reduction of product representations in $\mathscr{S}_3$ (see subsection 8.6.4)

$$\begin{gathered} T^{(s)} \times T^{(s)} = T^{(s)}, \quad T^{(s)} \times T^{(m)} = T^{(m)}, \\ T^{(m)} \times T^{(m)} = T^{(s)} \dotplus T^{(m)} \dotplus T^{(a)} \end{gathered}$$

it follows that, under simultaneous permutations in both spin and isospin, the symmetry of the states is as follows:

$$L_s\colon \begin{cases} S = \frac{3}{2}, T = \frac{3}{2} \\ S = \frac{1}{2}, T = \frac{1}{2} \end{cases} \qquad L_m\colon \begin{cases} S = \frac{3}{2}, T = \frac{1}{2} \\ S = \frac{1}{2}, T = \frac{3}{2} \\ S = \frac{1}{2}, T = \frac{1}{2} \end{cases} \qquad L_a\colon \ S = \frac{1}{2}, T = \frac{1}{2} \tag{12.5}$$

The dimension of these three representations is seen to be 20, 20 and 4, respectively, and allowing for the duplication $L_{m_1} + L_{m_2}$ we have the check on the total dimension of L, $64 = 20 + 2(20) + 4$. These representations are denoted by $D^{(\frac{3}{2}\frac{3}{2}\frac{3}{2})}$, $D^{(\frac{3}{2}\frac{1}{2}\frac{1}{2})}$ and $D^{(\frac{1}{2}\frac{1}{2}-\frac{1}{2})}$ and the reductions (12.1) are given by

$$
\begin{aligned}
D^{(\frac{3}{2}\frac{3}{2}\frac{3}{2})} &= D^{(\frac{3}{2})} \times D^{(\frac{3}{2})} \dotplus D^{(\frac{1}{2})} \times D^{(\frac{1}{2})} \\
D^{(\frac{3}{2}\frac{1}{2}\frac{1}{2})} &= D^{(\frac{3}{2})} \times D^{(\frac{1}{2})} \dotplus D^{(\frac{1}{2})} \times D^{(\frac{3}{2})} \dotplus D^{(\frac{1}{2})} \times D^{(\frac{1}{2})} \\
D^{(\frac{1}{2}\frac{1}{2}-\frac{1}{2})} &= D^{(\frac{1}{2})} \times D^{(\frac{1}{2})}
\end{aligned} \tag{12.6}
$$

Let us now return to the physics of nuclear structure. If the interaction were spin and isospin independent then SU_4 would be a symmetry group and we should expect large degeneracies corresponding to the dimensions d of the irreducible representations $D^{(PP'P'')}$ of SU_4. Experimentally one does not see such degeneracies, which is not unexpected because of the known strong spin-dependence of the nuclear forces. Nevertheless, the SU_4 symmetry is not completely destroyed and its presence shows up most clearly in the lighter nuclei through selection rules for β-decay, where the relevant operator is proportional to $Y_{qq'}$. The systematics of nuclear binding energies also lend some support to the SU_4 picture—see Franzini and Radicati (1963) in the bibliography.

Finally we remark that the states of different permutation symmetry in (12.5) must be combined with appropriate functions of the position coordinates of the particles to give a complete wave function which is anti-symmetric and therefore satisfies the Pauli principle. Thus the representation $D^{(\frac{1}{2}\frac{1}{2}-\frac{1}{2})}$ must be combined with a function of $T^{(s)}$ symmetry in the position coordinates. This leads to a form of Hund's rule in nuclei, according to which the supermultiplets with the smallest values of $(P\ P'\ P'')$ lie lowest. The arguments for this are very similar to those given in subsection 8.6.5 for atoms and in subsection 10.1.1 for nuclear isospin.

12.2 Supermultiplets of elementary particles

We have seen in the previous section how various combinations of S and T are brought together within an irreducible representation of SU_4. We also recall from section 11.7 that the baryons were found to belong to the SU_3 octet if they had spin $S = \frac{1}{2}$ and the decuplet if they had $S = \frac{3}{2}$. One is therefore tempted first to construct the product group $SU_2^S \times SU_3$ and then to enlarge it to SU_6 in an attempt to see if the empirical associations between S and $(\lambda\mu)$, quoted above, arise naturally from irreducible representations of SU_6. There is no need here to discuss the group SU_6 in detail since we can deduce all the properties we require by following the method of section 12.1. (A general account of SU_N is nevertheless given in chapter 18 of volume 2.)

We define the six-dimensional space of SU_6 by the product space of the basic three-dimensional space $D^{(10)}$ of SU_3 and a spin space with $S = \frac{1}{2}$, so that it also provides the representation $D^{(\frac{1}{2})} \times D^{(10)}$ of the product group $SU_2^S \times SU_3$. The fact that the baryons listed in table 11.1 have spins up to $S = \frac{3}{2}$ means that we must construct at least triple products of the basic SU_6 representation. The triple products may be studied by the same procedure which led to the results (12.5). As before, the $S = \frac{3}{2}$ state is totally symmetric $T^{(s)}$, while $S = \frac{1}{2}$ has mixed symmetry $T^{(m)}$. For the SU_3 factors the discussion of multiplication of representations in section 11.6.2 tells us that

$$D^{(10)} \times D^{(10)} \times D^{(10)} = D^{(30)} \dotplus 2D^{(11)} \dotplus D^{(00)}$$

and it may be shown (see problem 12.6) that $D^{(30)}$ is totally symmetric, $D^{(11)}$ has mixed symmetry while $D^{(00)}$ is totally antisymmetric. Thus, taking spin and SU_3 together, we have the analogue to (12.5)

$$L_s\colon \begin{cases} S = \frac{3}{2} & D^{(30)} \\ S = \frac{1}{2} & D^{(11)} \end{cases}$$

$$L_m\colon \begin{cases} S = \frac{3}{2} & D^{(11)} \\ S = \frac{1}{2} & D^{(30)} \\ S = \frac{1}{2} & D^{(11)} \\ S = \frac{1}{2} & D^{(00)} \end{cases} \tag{12.7}$$

$$L_a\colon \begin{cases} S = \frac{3}{2} & D^{(00)} \\ S = \frac{1}{2} & D^{(11)} \end{cases}$$

The set of states with the same permutation symmetry again forms the basis for an irreducible representation of SU_6. Immediately we see that the symmetric states L_s have precisely the combinations of S with $(\lambda\mu)$ which are observed empirically for the baryons. This irreducible representation of SU_6 has dimension $d = 6 \times 7 \times 8/1 \times 2 \times 3 = 56$ which checks with the dimension of its $SU_2^S \times SU_3$ components $(4 \times 10) + (2 \times 8) = 56$. There is evidence that the heavier baryon resonances, with odd parity, fall into the 70-dimensional representation of SU_6 given in (12.7) for states of mixed symmetry L_m. We shall not go into detail here but simply remark that the mesons also fit precisely into a 35-dimensional irreducible representation of SU_6 which is formed in the product of the basic six-dimensional representation with its complex conjugate; see also section 12.3.

The group SU_6 not only explains the associations between S and $(\lambda\mu)$ but also leads to mass-splitting formulae and relations between electromagnetic properties if some assumption is made about the transformation properties of the relevant operators. These relations cannot conflict in any way with the SU_3 results described in sections 11.8 and 11.9 because SU_6 contains SU_3 as a subgroup. The role of SU_6 is to give further relations in addition to those arising from SU_3. We illustrate this effect by considering the magnetic moments of the baryons. The SU_3 result of section 11.9 showed that the moments in the octet could all be expressed in terms of two constants d and e through equation (11.18), while the moments in the decuplet were given simply by fQ in analogy with the first of equations (11.14), where f is a further constant. With the group SU_6 the simplest assumption for the operator is that it transforms like a member of the 35-dimensional representation (like the mesons) with the properties $S = 1$, (11) $U = Q = 0$, under the $SU_2^S \times SU_3$ subgroup. We may now use the Wigner–Eckart theorem in SU_6 and a study of the appropriate products of representations, as in equation (11.13), tells us that for matrix elements between states belonging to the 56-dimensional representation of SU_6 only one reduced matrix element is involved. Thus, instead of the magnetic moments of the octet and decuplet of baryons being given in terms of three unknown constants in SU_3, the use of SU_6 gives them all in terms of a single constant! In particular this technique leads to the result $\mu_p/\mu_n = -\frac{3}{2}$ compared with the known ratio $\mu_p/\mu_n = -1.46$, a remarkable agreement. We shall deduce this ratio using an explicit model in the next section but it is a direct consequence of the SU_6 assumptions.

12.3 The three-quark model

The introduction of SU_3 and SU_6 has been done in an extremely empirical way, depending heavily on the fitting of the particles into irreducible representations according to their properties—charge, hypercharge, isospin and spin. It was seen that the baryons fitted into representations which were most easily constructed by a triple product $D^{(10)} \times D^{(10)} \times D^{(10)}$ of the simplest non-trival representation of SU_3. Exactly the same structure persisted in SU_6, the relevant representations again being most easily pictured as arising from the triple product of the simplest representation of SU_6 corresponding to $D^{(10)}$ of SU_3 and spin $S = \frac{1}{2}$. So far, this idea of building up the baryons as a triple product has been a mathematical one but its persistence suggests the possibility that the baryons may physically be composed of three basic objects which have spin $S = \frac{1}{2}$ and have three possible intrinsic states corresponding to the representation $D^{(10)}$ of SU_3. These objects have been given the name 'quark' by Gell–Mann, a word taken, rather arbitrarily, from a poem in *Finnegan's Wake* by James Joyce. We hasten to add that there is no experimental evidence for the existence of quarks, although searches have been in progress for ten years. However, it is possible that even the biggest accelerators which have been used to date have insufficient energy to produce quarks—see Goldhaber and Smith (1975) in the bibliography. It is nevertheless of considerable interest to see what properties the quark should possess according to SU_6 and what further properties of the baryons and mesons would follow from a quark model. The present section is concerned with such questions.

From figure 11.6 of section 11.6 and the relation $Q = M_T + \frac{1}{2}Y$ of section 11.1 we see that the three quark states of $D^{(10)}$, which we denote by u, d and s, are as follows:

$$
\begin{aligned}
u&: T = \tfrac{1}{2},\ M_T = \tfrac{1}{2},\ Y = \tfrac{1}{3},\ Q = \tfrac{2}{3} \\
d&: T = \tfrac{1}{2},\ M_T = -\tfrac{1}{2},\ Y = \tfrac{1}{3},\ Q = -\tfrac{1}{3} \\
s&: T = 0,\ M_T = 0,\ Y = -\tfrac{2}{3},\ Q = -\tfrac{1}{3}
\end{aligned}
\tag{12.8}
$$

(The notation p, n, λ is sometimes used instead of u, d, s.) Since a baryon is composed of three quarks we must also have $B = \frac{1}{3}$ for all three quark states.

Including spin, denoted as in section 12.1 by an arrow, the six possible quark states are $u\uparrow, u\downarrow, d\uparrow, d\downarrow, s\uparrow, s\downarrow$. Immediately we see that the quark must have some very unusual properties; not only does it have non-integer baryon number and hypercharge but it also has a non-integer charge. However, there seems to be no real objection to such a state of affairs and we must await the test of experiment. (A more complicated model is described in section 12.4 which avoids these unusual features.) Starting from (12.8) we may now build up the baryon wave functions from quarks in the manner of equation (12.4) for SU_4.

From figure 11.8 of section 11.7 and the definitions (12.8) it is clear that the wave function for the Δ^{++} particle with $M_S = \frac{3}{2}$ must be a simple product $|\Delta^{++}, M_S = \frac{3}{2}\rangle = u\uparrow(1)\ u\uparrow(2)\ u\uparrow(3)$. (The notation Δ^{++} implies $T = \frac{3}{2}$, $Y = 1$). The wave functions for all other particles in the decuplet may then be constructed using the lowering operators T_- and U_- in T-spin and U-spin. To obtain values of M_S other than $\frac{3}{2}$ we must of course use the ordinary spin lowering operator S_-. The particles Δ^- and Ω^- which lie at the other vertices of the decuplet triangle will also have simple product wave functions when $M_S = \frac{3}{2}$, corresponding to the single quark states $d\uparrow$ and $s\uparrow$, respectively, instead of $u\uparrow$. For the octet, the situation is a little more complicated but if we denote by $|d\downarrow u\uparrow u\uparrow\rangle$ the normalised symmetric state formed from three quarks occupying states $d\downarrow$ $u\uparrow$ and $u\uparrow$ then it is clear from the values of M_S, M_T and Y that the proton state must be a linear combination of the kind

$$|p, M_S = \tfrac{1}{2}\rangle = \alpha|d\downarrow u\uparrow u\uparrow\rangle + \beta|d\uparrow u\uparrow u\downarrow\rangle \tag{12.9}$$

But it must have the property that $\mathrm{T}_+|p, M_S = \frac{1}{2}\rangle = 0$, since this distinguishes it from the Δ^+ particle, see figure 11.8. Using the fact that $\mathrm{T}_+ = \sum_{i=1}^{3} \mathrm{t}_+(i)$, where $\mathrm{t}_+|u\rangle = 0$ and $\mathrm{t}_+|\mathrm{d}\rangle = |u\rangle$, we soon deduce that $\alpha = \sqrt{\frac{2}{3}}$, $\beta = -\sqrt{\frac{1}{3}}$. Similarly for the neutron

$$|n, M_S = \tfrac{1}{2}\rangle = \sqrt{\tfrac{2}{3}}|u\downarrow d\uparrow d\uparrow\rangle - \sqrt{\tfrac{1}{3}}|u\uparrow d\uparrow d\downarrow\rangle \tag{12.10}$$

The value $-\frac{3}{2}$ for the ratio of the nucleon magnetic moments may now be calculated from equations (12.9) and (12.10) if we assume the relevant operator to be given by

$$\mu_z = \mu_0 \sum_{i=1}^{3} \mathrm{Q}(i)\, \mathrm{s}_z(i) \tag{12.11}$$

where i runs over the quarks. For the quarks themselves the moments are then $\frac{1}{3}$, $-\frac{1}{6}$ and $-\frac{1}{6}$ for u, d and s, respectively, in units of μ_0. For the two states $|d\downarrow u\uparrow u\uparrow\rangle$ and $|d\uparrow u\uparrow u\downarrow\rangle$ of three quarks the moments are given by $\frac{1}{6}+\frac{1}{3}+\frac{1}{3}=\frac{5}{6}$ and $-\frac{1}{6}+\frac{1}{3}-\frac{1}{3}=-\frac{1}{6}$, respectively, leading to the value $\frac{2}{3}(\frac{5}{6})+\frac{1}{3}(-\frac{1}{6})=\frac{1}{2}$ for μ_p/μ_0. In the same way we deduce from equation (12.10) the value $\frac{2}{3}(-\frac{2}{3})+\frac{1}{3}(\frac{1}{3}) = -\frac{1}{3}$ for μ_n/μ_0 giving finally $\mu_p/\mu_n = -\frac{3}{2}$. Although we have used the explicit form (12.11) for the magnetic moment operator, the ratios which we have deduced have a more general validity. In fact they may be deduced entirely from the SU_6 vector-coupling coefficients and depend only on the assumption that the magnetic moment operator transforms according to a definite irreducible representation of SU_6 in addition to having the transformation properties (11) $T = Y = 0$ under the group SU_3. (In the partition notation of chapter 18 this representation of SU_6 is [21111] which is obtained from the product of the basic representation [1] and its complex conjugate $[1]^* = [11111]$.) Again in units of μ_0, the magnetic moment of the Δ^{++} particle has the value $+1$ and the constants d, e and f used in equation (11.18) and towards the end of section 12.2 are given by $d = \frac{1}{3}\mu_0$, $e = -\frac{1}{3}\mu_0$, $f = \frac{1}{2}\mu_0$.

If the anti-quark is introduced and supposed to transform like the complex conjugate representation of that for the quark, then the 35-dimensional SU_6 representation of the mesons is consistent with the mesons being composed of a quark anti-quark pair. In terms of the $SU_2 \times SU_3$ subgroup this pair gives rise to the following states

$$(D^{(\frac{1}{2})} \times D^{(10)}) \times (D^{(\frac{1}{2})} \times D^{(01)}) = D^{(1)} \times (D^{(11)} \dotplus D^{(00)}) \dotplus D^{(0)} \times (D^{(11)} \dotplus D^{(00)})$$

using the equation preceding (11.11) and the familiar product for SU_2. Of these product states it can be shown that $D^{(0)} \times D^{(00)}$ is an SU_6 invariant and the remaining 35 form an irreducible representation of SU_6([2 1 1 1 1] in the notation of chapter 18). This remainder provides just those quantum numbers required by the mesons, namely an octet and a singlet with $S = 1$ and an octet with $S = 0$. The $S = 0$ octet was discussed in section 11.7; the $S = 1$ octet contains the ϕ, ρ and K^* mesons and the singlet corresponds to the ω meson.

Having presented a favourable picture of the quark model so far, we now describe some of the mysteries and apparent difficulties. The fact that quarks have not yet been seen in collisions produced by accelerators suggests that the mass of a quark exceeds 5000 MeV. Since the mass of a baryon is only 1000 MeV this implies a colossal binding energy in excess of 14 000 MeV. Compared with nuclear binding energies of the order of 10 MeV this demands an attractive force between quarks which is many times stronger than the strong nuclear force between nucleons. It has been called the superstrong force. However, there could be some reason, as yet unknown, why quarks are not produced as real physical particles, in which case their mass need not be great. In fact there are some results which support this possibility, but we shall not discuss them here—see, however, Goldhaber and Smith (1975).

With a strong attractive force one would expect the three quarks in a baryon to have a wave function which was symmetrical in their position coordinates. This would be like the three nucleons in the nucleus ^{3}He or the four nucleons in an α-particle. However, if we were to do this, then the result shown in (12.7) that the SU_6 part of the wave function is also symmetric would imply that the complete wave function is symmetric. Since the quarks have $S = \frac{1}{2}$ this then conflicts with the usual situation where half-integer particles obey Fermi–Dirac statistics, and have antisymmetric wave functions. The only escape from this dilemma in the three-quark model is to suppose that quarks obey a type of statistics intermediate between Fermi–Dirac and Bose–Einstein. Such a situation is called 'parastatistics' and is not forbidden by the general spin-statistics theorem (see subsection 16.3.4 of volume 2). One would be led to assume 'parafermi-statistics of order three' which means that the wave functions may be symmetric in not more than three particles. (In this language, the usual Fermi–Dirac statistics would be called parafermi-statistics of order one).

If one were to use simple non-relativistic quantum mechanics to describe the three attracting quarks in a baryon then one might suppose that they moved rather like the nucleons in a nucleus. The latter are known to move as if they were in a smooth attractive potential well, centred at their centre of mass, and a harmonic oscillator is a reasonable first approximation to this well. A number of simple calculations have been made for the quark model using this approach and they meet with remarkable success. For example, one would expect the ground state to be simply given by the configuration $(1s)^3$ with all three quarks in the lowest $(1s)$ orbit of the oscillator. This configuration would be assumed for all particles belonging to the 56-dimensional representation of SU_6. Excited states would most simply be formed by exciting one quark into the next $(1p)$ orbit to give the configuration $(1s)^2(1p)$ which has angular momentum $L = 1$ and odd parity. From this configuration one would expect to form states of the type $T^{(s)}$ or $T^{(m)}$ under permutation. (With two quarks in the same $(1s)$ state it is clearly impossible to form states of type $T^{(a)}$.) However the $T^{(s)}$ state is not acceptable since it can be shown to differ from the ground state $(1s)^3$ only through the motion of the centre of mass of the baryon. It does not therefore describe any new internal state of the baryon and has arisen only because of the simplicity of our 'shell model'. It follows therefore that the first excited state has an orbital wave function of type $T^{(m)}$. Assuming that the complete wave function for the baryon is again totally symmetric, $T^{(s)}$, as it was in the ground state this implies that the SU_6 part of the excited state must also be of type $T^{(m)}$. From equation (12.7) this is seen to be the 70-dimensional representation of SU_6. We therefore expect to find a set of odd parity excited states with an orbital $L = 1$ which will combine with the S given for $T^{(m)}$ in (12.7) to give total spins $J = (L+S), \ldots, |L-S|$ in the usual way. In other words we expect to find (1) a decuplet with $J = \frac{1}{2}$ and $\frac{3}{2}$; (2) an octet with $J = \frac{1}{2}, \frac{3}{2}$ and $\frac{5}{2}$; (3) an octet with $J = \frac{1}{2}$ and $\frac{3}{2}$; (4) a singlet with $J = \frac{1}{2}$ and $\frac{3}{2}$. Such a structure is found in the experimental data with masses in the range 1405 to 1710 MeV but in the interests of simplicity we did not include them in table 11.1.

12.4 The nine-quark model

Many of the difficulties of the three-quark model may be avoided if one postulates more elaborate properties for the quark. The most promising model of this kind supposes that, in addition to its spin $\frac{1}{2}$ and the SU_3 states, u, d and s, a quark has additional degrees of freedom described by three states a, b, c corresponding to another SU_3 group which we call SU_3'. (For no good reason, these three states are sometimes said to describe the 'colour' of the quark.) Thus, apart from spin, there are nine possible intrinsic states of the quark, u_a, u_b, u_c, d_a, d_b, . . ., etc. To construct a complete wave function for the baryons, built from three quarks, we must then include a factor relating to the new SU_3' degree of freedom. The natural assumption to make is that the SU_3' state is totally antisymmetric. Thus a totally antisymmetric state could be formed with the highly successful symmetric SU_6 structure described in the preceding section. The Fermi statistics for quarks would then be restored. The only totally antisymmetric SU_3' state is the singlet $D^{(00)}$ (see section 12.2) so that the SU_6 classification (12.7) of the baryons is unaltered. Had the SU_3' representation been of dimension >1 then we would have had a contradiction, since an unobserved multiplicity of baryons would have been predicted. To explain why the singlet representation of SU_3' is appropriate for the observed baryons one must postulate a term in the Hamiltonian which depends on the SU_3' variables and which has the property that $D^{(00)}$ has lowest energy. Such a term is not difficult to construct and in fact the Casimir operator C_2 described in section 11.10 has just this property. One would then expect the other possible SU_3' representations, like $D^{(11)}$, to come at much higher energy. Since we are now dealing with the superstrong forces, it is not surprising that these excited SU_3' states have not yet been seen.

A second advantage of the nine-quark model is that it explains how the three-quark system (baryons) and the quark–anti-quark system (mesons) may have bound states while the two-quark and four-quark–one anti-quark systems remain unbound and unobserved. This was always a puzzle in the three-quark model. The difference lies in the greater number of possible quark states, enabling the interaction to be chosen to be repulsive in some quark–quark states and attractive in others so that bound states occur only rarely, in the $D^{(00)}$ states described above.

A further advantage of the nine-quark model is that the charge and hypercharge of the quarks may now be integers. This is achieved by generalising the relation $Q = M_T + \frac{1}{2}Y$ of section 11.1 for the charge by writing instead $Q = M_T + \frac{1}{2}Y + X$, where X is the value of one of the SU_3' operators, analogous to Y or Q in SU_3. For the $D^{(00)}$ representation, corresponding to all the observed particles so far, we must have $X = 0$ so that the relation is unaltered. However, for other SU_3' representations, $X \neq 0$ and for the quarks in particular one can arrange for X to take the values $\frac{1}{3}$, $\frac{1}{3}$ and $-\frac{2}{3}$. Combined with the values $\frac{2}{3}$, $-\frac{1}{3}$ and $-\frac{1}{3}$ for $M_T + \frac{1}{2}Y$ in the quark states u, d, s, as given in equation (12.8), this leads to the following values for the charges of the nine quark states, 1, 1, 0, 0, 0, -1, 0, 0, -1.

We stress that much of this section is speculative since all known particles correspond to the $D^{(00)}$representation of SU_3'. However, the circumstantial evidence in support of the nine-quark model is building up from a number of directions, including the possibility of unifying the theories of weak and electromagnetic interactions—see Goldhaber and Smith (1975) in the bibliography. The role of symmetry in physics is well illustrated here. It helps to construct and describe a number of possible scenarios for the real world. We must await the results of experiment to tell us which, if any, is in agreement with the facts.

12.5 Charm

A simple extension of the quark model has been suggested to explain the observed strangeness selection rule in some weak interaction processes. One supposes that there is a fourth quark, denoted by c, which is an SU_3 singlet with $T = Y = 0$ but that it differs from the usual three quarks u, d and s in a new quantum number called 'charm' (again, there is no good reason for this name). If C is used to denote the charm, then the new quark has charm $C = 1$ while u, d and s have $C = 0$. The new quark must have the same charge $\frac{2}{3}e$ as the quark u so that the relation between charge and isospin projection, see section 11.1, must be generalised to $Q = M_T + \frac{1}{2}Y + \frac{2}{3}C$. Once again, this new concept of charm is a theoretical speculation but if it is valid one expects a richer variety of hadrons to emerge when quarks are combined. They may be classified into multiplets of the SU_4 group obtained by adding the fourth quark to SU_3. In addition to the hadrons described earlier, there will be some with non-zero charm. If the new quark is heavier than the others, then these particles would be heavier also and this could explain why they have not yet been seen. However, it is being suggested that some unexpected new particles which have just been observed (early 1975), may be described as excitations of either the charm or the colour degree of freedom—Goldhaber and Smith (1975).

Addendum (mid-1978).

Although no quarks have yet been observed (see Jones, L. W. (1977). *Rev. Mod. Phys.*, **49**, 717) recent experimental and theoretical work continues to support the quark model of hadrons, including the additional three degrees of freedom referred to as colour. The device mentioned in section 12.4 by which quarks achieve integer charge has not found favour over the fractional charges given in equation (12.8) but this possibility is not yet excluded. In analogy with electrodynamics the interaction of quarks is now called 'chromodynamics' and the corresponding field particle, analogous to the photon, is called the 'gluon'. These possible constituents of the hadrons are referred to generally as 'partons'. The word 'flavour' has been introduced to describe the four quark states u, d, s and c so that one has a product space between flavour and colour and a supposed symmetry group SU_4 (flavour) $\times$ SU_3 (colour). However, the story is far from complete since the properties of one of the newest mesons, the upsilon or U-particle, suggest the need for a fifth and possibly a sixth flavour for the quarks. These two new proposed quarks have been called 'top' and 'bottom' quarks.

Second addendum (late 1983)

The ideas described in the 1978 addendum have all gained ground since that time. The continuing failure to observe free quarks is not now seen as an objection to the quark model since the theory allows quarks to be 'confined'. This is brought about by an attractive force which increases with the distance between quarks and so prevents the escape of a single quark from a baryon or meson. There is now some evidence for the *b*-quark (bottom or beauty) and a preliminary report supporting the existence of a *t*-quark (top or truth). Like the *s*- and *c*-quarks they have zero isospin and give rise to a succession of symmetry groups starting with the isospin SU_2 with just *u*- and *d*-quarks and progressing through SU_3 with *u*, *d* and *s* to SU_6 with *u*, *d*, *s*, *c*, *b* and *t* although the amount of symmetry breaking increases with the number of 'flavours'.

Impressive advances have been made towards a unification of the theories of the weak and electromagnetic interactions (see Bailin, D. (1982), *Weak Interactions* (second edition), Adam Hilger, Bristol). The left-helicity electron and the electron neutrino are regarded as a doublet in an SU_2 group called 'weak isospin' while the right-helicity electron is a singlet in the same group. A 'weak hypercharge' is introduced, as in section 11.7, to give an $SU_2 \times U_1$ group which then contains the electromagnetic current operator and provides a framework for a unified theory of weak and electromagnetic forces. This $SU_2 \times U_1$ group is to be seen as a generalisation of the usual gauge transformation of the electromagnetic field (see section 16.3.5 of volume 2, which is described simply by the group U_1). An even more ambitious programme is to combine this $SU_2 \times U_1$ group with the colour SU_3 group which we called SU'_3 in section 12.4. In quantum chromodynamics (QCD), this SU_3 is the gauge group for the strong interactions so that a field theory based on the product $SU_3 \times SU_2 \times U_1$ gives hope of a 'grand unified theory' (GUT) which encompasses strong, weak and electromagnetic forces. These gauge groups are distinct from the flavour groups of SU_2 and SU_3 discussed in the main body of this chapter in connection with isospin and hypercharge. The two types of group co-exist without conflict because they act in different spaces.

Bibliography

For a discussion of supermultiplets in nuclei we suggest the early paper

Wigner, E. P. and Feenberg, E. (1941). *Rep. Prog. Phys.*, **8,** 274 and the more recent analysis by

Franzini, P. and Radicati, L. A. (1963). *Phys. Rev. Lett.*, **6,** 322

The use of SU_6 in particle physics is discussed in Perkins (1972) and Kokkedee (1969) in the bibliography to chapter 11.

There is an excellent review on hypothetical particles by

Goldhaber, A. S. and Smith, J. (1975). *Rep. Prog. Phys.*, **38,** 731

which discusses, among other things, the existence of quarks and has an excellent bibliography on quark models and the new concepts of colour and charm.

Problems

12.1 Write down all fifteen Hermitian traceless 4×4 matrices in the basis $p^{\uparrow}, n^{\uparrow}, p^{\downarrow}, n^{\downarrow}$, as a generalisation of equation (11.1), and express them as linear combinations of s_q, t_q and $Y_{qq'} = s_q t_{q'}$.

12.2 Use the properties of the spin-matrices s_q and t_q to show that $[S_x, Y_{xq}] = 0$, $[S_x, Y_{yq}] = iY_{zq}$, $[Y_{xy}, Y_{yz}] = 0$ and $[Y_{xy}, Y_{xz}] = it_x$.

12.3 By writing down the appropriate products of three single-particle states, verify that the values of $(P\ P'\ P'')$ for the three SU_4 representations given in equation (12.5) are as given, namely $(\frac{3}{2}\frac{3}{2}\frac{3}{2})$, $(\frac{3}{2}\frac{1}{2}\frac{1}{2})$ and $(\frac{1}{2}\frac{1}{2}-\frac{1}{2})$.

12.4 Use the arguments of subsection 8.6.5 and 10.1.1 and section 12.1 to show that, for a nucleus with three valence nucleons in p-orbits, the lowest group of $T = \frac{1}{2}$ states are expected to have (in LS-coupling) $S = \frac{1}{2}$ with $L = 1$ or 3, while the lowest states with $T = \frac{3}{2}$ have $S = \frac{1}{2}$ with $L = 1$ or 2.

12.5 Use the method given in section 11.6 to show that $D^{(10)} \times D^{(20)} = D^{(30)} \dot{+} D^{(11)}$ and hence deduce that $D^{(10)} \times D^{(10)} \times D^{(10)} = D^{(30)} \dot{+} 2D^{(11)} \dot{+} D^{(00)}$.

12.6 In the reduction in the previous question show that $D^{(30)}$ is totally symmetric, $D^{(00)}$ is totally antisymmetric and $D^{(11)}$ has mixed symmetry under permutation of the three factors.

12.7 Use the methods explained in section 12.3 to construct the quark wave functions for the Σ^+-particle and for the Λ-particle. (Confine your attention to the state with $M_S = \frac{1}{2}$.)

Appendix 1: Character Tables for the Irreducible Representations of the Point Groups

The classes are labelled using the notation of section 9.1 by a typical element preceded by the number of elements in the class. Occasionally, a rotation or reflection is given a suffix to indicate the axis of the rotation or the plane of the reflection. The z-axis is always taken along the principal axis for rotations.

The irreducible representations are labelled by the chemical (Muliken) notation as follows. One-dimensional representations are labelled by A or B. A is used if the character of the smallest rotation (proper or improper) about the principal axis is $+1$, B is used if it is -1. Two, three and four-dimensional representations are labelled by E, T and U, respectively. (Pairs of complex conjugate representations are also denoted by E since corresponding states are usually degenerate, see section 5.10.) A superscript $+$ or $-$ is used to indicate the parity if the inversion is a group element.

Table A.1 contains the characters for the eleven proper point groups appearing in the first row of table 9.2 of section 9.6. The characters for the isomorphic groups in rows three and four of table 9.2 are, of course, the same so that each individual table is provided with a set of class and irreducible representation labels appropriate for each isomorphic group. The character tables for improper point groups which contain the inversion can be obtained using the relationships of table 9.2 and are not therefore listed separately unless they are isomorphic with one of the proper point groups. The direct product relationships are, however, shown in the table.

The character tables for the double-valued representations of the point groups are listed in table A.2 for the eleven double groups corresponding to the first row of table 9.2. Tables for the remaining groups follow from the relationships in that table. In the tables for the groups $\bar{C}_n$ we have listed only the characters for the unbarred elements, the character for $\bar{G}_a$ being $-\chi(G_a)$. In the remaining tables we have listed all the classes. In fact, two headings have been given to the classes, the first being the one usually adopted in the literature. The second heading has been added to illustrate the way that particular elements fall into the classes. For instance, the group $\bar{D}_4$, discussed in section 9.7, has two classes $2C_4$ and $2\bar{C}_4$ but we note that, although the element C_4 is in the first of these, the element C_4^3 is not in the same class as it was in the single group because the identity is not now a rotation through 2π. In fact, the element C_4 is in the same class as $C_4^{-1} = \bar{E}C_4^3 = \bar{C}_4^3$. It must also be remembered that the double groups $\bar{\mathscr{G}}$ also have single-valued representations with characters given in the tables for the ordinary group $\mathscr{G}$ for the elements G_a and with $\chi(\bar{G}_a) = \chi(G_a)$.

Table A.1

C_1	E
A	1

C_2			E	C_2
	S_1		E	σ
		S_2	E	I
A	A_1	A^+	1	1
B	A_2	A^-	1	-1

The groups S_1 and S_2 are frequently denoted C_s and C_i, respectively.
$C_{2h} \equiv C_2 \times S_2$

Table A.1 (continued)

C_3	E	C_3	C_3^2
A	1	1	1
E	1	ε	ε^2
	1	ε^2	ε

$\varepsilon = \exp(2\pi i/3)$
$S_6 = C_3 \times S_2$ (also isomorphic with C_6)
$C_{3h} = C_3 \times S_1$ (also isomorphic with C_6)

C_4		E	C_4	C_2	C_4^3
	S_4	E	S_4	C_2	S_4^3
A	A	1	1	1	1
B	B	1	-1	1	-1
E	E	1	i	-1	$-i$
		1	$-i$	-1	i

$C_{4h} = C_4 \times S_2$

C_6			E	C_6	C_3	C_2	C_3^2	C_6^5
	C_{3h}		E	S_3	C_3	σ_h	C_3^2	S_3^5
		S_6	E	S_6	C_3	I	C_3^2	S_6^5
A	A_1	A	1	1	1	1	1	1
B	A_2	B	1	-1	1	-1	1	-1
E_1	E_1	E_1	1	ε	ε^2	1	ε	ε^2
			1	ε^2	ε	1	ε^2	ε
E_2	E_2	E_2	1	$-\varepsilon^2$	ε	-1	ε^2	$-\varepsilon$
			1	$-\varepsilon$	ε^2	-1	ε	$-\varepsilon^2$

$\varepsilon = \exp(2\pi i/3)$
$C_{6h} = C_6 \times S_2$

Table A.1 (continued)

D_2	C_{2v}	C_{2h}	E	C_{2z}	C_{2y}	C_{2x}
			E	C_2	σ_y	σ_x
			E	C_2	σ_h	I
A	A_1	A^+	1	1	1	1
B_3	B_1	B^+	1	−1	−1	1
B_1	A_2	A^-	1	1	−1	−1
B_2	B_2	B^-	1	−1	1	−1

$D_{2h} = D_2 \times S_2$

D_3	C_{3v}	E	$2C_3$	$3C_2$
		E	$2C_3$	$3\sigma_v$
A_1	A_1	1	1	1
A_2	A_2	1	1	−1
E	E	2	−1	0

$D_{3d} = D_3 \times S_2$ (also isomorphic with D_6)
$D_{3h} = D_3 \times S_1$ (also isomorphic with D_6)

D_4	C_{4v}	D_{2d}	E	C_4^2	$2C_4$	$2C_2$	$2C_2'$
			E	C_2	$2C_4$	$2\sigma_v$	$2\sigma_d$
			E	C_2	$2S_4$	$2C_2'$	$2\sigma_d$
A_1	A_1	A_1	1	1	1	1	1
A_2	A_2	A_2	1	1	1	−1	−1
B_1	B_1	B_1	1	1	−1	1	−1
B_2	B_2	B_2	1	1	−1	−1	1
E	E	E	2	−2	0	0	0

$D_{4h} = D_4 \times S_2$

Table A.1 (continued)

D_6				E	C_2	$2C_3$	$2C_6$	$3C'_2$	$3C''_2$
	C_{6v}			E	C_2	$2C_3$	$2C_6$	$3\sigma_v$	$3\sigma'_v$
		D_{3h}		E	σ_h	$2C_3$	$2S_3$	$3C_2$	$3\sigma'_v$
			D_{3d}	E	I	$2C_3$	$2S_6$	$3C_2$	$3\sigma_v$
A_1	A_1	A_1	A_1^+	1	1	1	1	1	1
A_2	A_2	A_2	A_2^+	1	1	1	1	−1	−1
B_1	B_1	A_3	A_1^-	1	−1	1	−1	1	−1
B_2	B_2	A_4	A_2^-	1	−1	1	−1	−1	1
E_1	E_1	E_1	E^-	2	−2	−1	1	0	0
E_2	E_2	E_2	E^+	2	2	−1	−1	0	0

$D_{6h} = D_6 \times S_2$

T	E	$3C_2$	$4C_3$	$4C_3^2$
A	1	1	1	1
E {	1	1	ε	ε^2
	1	1	ε^2	ε
T	3	−1	0	0

$T_h = T \times S_2$

O		E	$8C_3$	$3C_2$	$6C_2$	$6C_4$
	T_d	E	$8C_3$	$3C_2$	$6\sigma_d$	$6S_4$
A_1	A_1	1	1	1	1	1
A_2	A_2	1	1	1	−1	−1
E	E	2	−1	2	0	0
T_1	T_1	3	0	−1	−1	1
T_2	T_2	3	0	−1	1	−1

$O_h = O \times S_2$

Table A.2

$\bar{C}_1$	E
$\bar{A}$	1

Table A.2 (continued)

$\overline{C}_2$	E	C_2
$\overline{A}_1$	1	$-i$
$\overline{A}_2$	1	i

$\overline{C}_3$	E	C_3	C_3^2
$\overline{E}$	1	γ	γ^2
	1	γ^{-1}	γ^{-2}
$\overline{B}$	1	-1	1

$\gamma = \exp(\pi i/3)$

$\overline{C}_4$	E	C_4	C_2	C_4^3
$\overline{E}_1$	1	δ	i	δ^3
	1	δ^{-1}	$-i$	δ^{-3}
$\overline{E}_2$	1	δ^3	$-i$	δ
	1	δ^{-3}	i	δ^{-1}

$\delta = \exp(\pi i/4)$

$\overline{C}_6$	E	C_6	C_3	C_2	C_3^2	C_6^5
$\overline{E}_1$	1	ω	ω^2	i	ω^4	ω^5
	1	ω^{-1}	ω^{-2}	$-i$	ω^{-4}	ω^{-5}
$\overline{E}_2$	1	i	-1	$-i$	1	i
	1	$-i$	-1	i	1	$-i$
$\overline{E}_2$	1	ω^5	$-\omega^4$	i	$-\omega^2$	ω
	1	ω^{-5}	$-\omega^{-4}$	$-i$	$-\omega^{-2}$	ω^{-1}

$\omega = \exp(\pi i/6)$

$\overline{D}_2$	E	$\overline{E}$	$2C_{2z}$	$2C_{2y}$	$2C_{2x}$
	E	$\overline{E}$	$C_{2z}, \overline{C}_{2z}$	$C_{2y}, \overline{C}_{2y}$	$C_{2x}, \overline{C}_{2x}$
$\overline{E}$	2	-2	0	0	0

Table A.2 (continued)

$\bar{D}_3$	E	$\bar{E}$	$2C_3$	$2\bar{C}_3$	$3C_2$	$3\bar{C}_2$
	E	$\bar{E}$	$C_3, \bar{C}_3^2$	$\bar{C}_3, C_3^2$	$3C_2$	$3\bar{C}_2$
$\bar{E}_1$	1	-1	-1	1	i	$-i$
	1	-1	-1	1	$-i$	i
$\bar{E}_2$	2	-2	1	-1	0	0

$\bar{D}_4$	E	$\bar{E}$	$2C_4^2$	$2C_4$	$2\bar{C}_4$	$4C_2$	$4C'_2$
	E	$\bar{E}$	$C_4^2, \bar{C}_4^2$	$C_4, \bar{C}_4^3$	$\bar{C}_4, C_4^3$	$2C_2, 2\bar{C}_2$	$2C'_2, 2\bar{C}'_2$
$\bar{E}_1$	2	-2	0	$\sqrt{2}$	$-\sqrt{2}$	0	0
$\bar{E}_2$	2	-2	0	$-\sqrt{2}$	$\sqrt{2}$	0	0

$\bar{D}_6$	E	$\bar{E}$	$2C_2$	$2C_3$	$2\bar{C}_3$	$2C_6$	$2\bar{C}_6$	$6C'_2$	$6C''_2$
	E	$\bar{E}$	$C_2, \bar{C}_2$	$C_3, \bar{C}_3^2$	$\bar{C}_3, C_3^2$	$C_6, \bar{C}_6^5$	$\bar{C}_6, C_6^5$	$3C'_2, 3\bar{C}'_2$	$3C''_2, 3\bar{C}''_2$
$\bar{E}_1$	2	-2	0	1	-1	$\sqrt{3}$	$-\sqrt{3}$	0	0
$\bar{E}_2$	2	-2	0	1	-1	$-\sqrt{3}$	$\sqrt{3}$	0	0
$\bar{E}_3$	2	-2	0	-2	2	0	0	0	0

$\bar{T}$	E	$\bar{E}$	$6C_2$	$4C_3$	$4\bar{C}_3$	$4C_3^2$	$4\bar{C}_3^2$
	E	$\bar{E}$	$3C_2, 3\bar{C}_2$	$4C_3$	$4\bar{C}_3$	$4C_3^2$	$4\bar{C}_3^2$
$\bar{E}$	2	-2	0	1	-1	-1	1
$\bar{U}$	2	-2	0	ε	$-\varepsilon$	$-\varepsilon^2$	ε^2
	2	-2	0	ε^2	$-\varepsilon^2$	$-\varepsilon$	ε

$\varepsilon = \exp(2\pi i/3)$

Table A.2 (continued)

$\bar{O}$	E	$\bar{E}$	$8C_3$	$8\bar{C}_3$	$6C_4^2$	$3C_4$	$3\bar{C}_4$	$12C_2$
	E	$\bar{E}$	$4C_3, 4\bar{C}_3^2$	$4\bar{C}_3, 4C_3^2$	$3C_4^2, 3\bar{C}_4^2$	$3C_4, 3\bar{C}_4^3$	$3\bar{C}_4, 3C_4^3$	$6C_2, 6\bar{C}_2$
$\bar{E}_1$	2	−2	1	−1	0	$\sqrt{2}$	$-\sqrt{2}$	0
$\bar{E}_2$	2	−2	1	−1	0	$-\sqrt{2}$	$\sqrt{2}$	0
$\bar{U}$	4	−4	−1	1	0	0	0	0

International notation

In chapter 9 we pointed out that, apart from the Schoenflies notation which we used in the text, there is also an international notation for point groups. In this system, an n-fold rotation axis C_n is labelled n and mirror planes denoted by m. If there is a mirror plane (σ_h) perpendicular to an n-fold axis C_n we write $\frac{n}{m}$ while, if the mirror plane contains the n-fold axis C_n, we write nm. The general improper element is taken to be a product of the inversion with a rotation. It is called a roto-inversion and denoted by $\bar{n} = IC_n$. The mirror rotation, $S_n = C_n\sigma_h$ used in chapter 9 may be written in terms of the $\bar{n}$ by using the relation $\sigma_h = C_2 I$. For example,

$$\bar{4} = IC_4 = \sigma_h C_2 C_4 = C_4^{-1}\sigma_h = S_4^{-1}$$

The full international symbol for a group is a list of the non-equivalent generators but, in practice, a shorter symbol is normally used. In table A.3 we list both the full symbol and the shorter symbol when these are different.

Table A.3

Schoenflies notation	*Full international notation*	*Shorter symbol (where different)*
C_n	n	
S_1 or C_{1h} or C_s	m	
S_2 or C_i	$\bar{1}$	
S_4	$\bar{4}$	
S_6	$\bar{3}$	

Table A.3 (continued)

Schoenflies notation	*Full international notation*	*Shorter symbol (where different)*
C_{2h}	$\frac{2}{m}$	$2/m$
C_{3h}	$\bar{6}$	
C_{4h}	$\frac{4}{m}$	$4/m$
C_{6h}	$\frac{6}{m}$	$6/m$
C_{2v}	$2mm$	
C_{3v}	$3m$	
C_{4v}	$4mm$	
C_{6v}	$6mm$	
D_2 or V	222	
D_3	32	
D_4	422	
D_6	622	
D_{2h}	$\frac{2}{m}\frac{2}{m}\frac{2}{m}$	mmm
D_{3h}	$\bar{6}m2$	
D_{4h}	$\frac{4}{m}\frac{2}{m}\frac{2}{m}$	$4/mmm$
D_{6h}	$\frac{6}{m}\frac{2}{m}\frac{2}{m}$	$6/mmm$
D_{2d}	$\bar{4}2m$	
D_{3d}	$\bar{3}\frac{2}{m}$	$\bar{3}m$
T	23	
T_h	$\frac{2}{m}\bar{3}$	$m3$
T_d	$\bar{4}3m$	
O	432	
O_h	$\frac{4}{m}\bar{3}\frac{2}{m}$	$m3m$

Appendix 2: Solutions to Problems in Volume 1

Chapter 2

2.3

	E	C_4	C_4^2	C_4^3	C_2	C_2'	C_2''	C_2'''
E	E	C_4	C_4^2	C_4^3	C_2	C_2'	C_2''	C_2'''
C_4	C_4	C_4^2	C_4^3	E	C_2''	C_2'''	C_2'	C_2
C_4^2	C_4^2	C_4^3	E	C_4	C_2'	C_2	C_2'''	C_2''
C_4^3	C_4^3	E	C_4	C_4^2	C_2'''	C_2''	C_2	C_2'
C_2	C_2	C_2'''	C_2'	C_2''	E	C_4^2	C_4^3	C_4
C_2'	C_2'	C_2''	C_2	C_2'''	C_4^2	E	C_4	C_4^3
C_2''	C_2''	C_2	C_2'''	C_2'	C_4	C_4^3	E	C_4^2
C_2'''	C_2'''	C_2'	C_2''	C_2	C_4^3	C_4	C_4^2	E

Classes: E; C_4^2; $\{C_4, C_4^3\}$; $\{C_2, C_2'\}$; $\{C_2'', C_2'''\}$. (C_2 and C_2' denote rotations about axes perpendicular to the edges, while the axes of C_2'' and C_2''' are diagonals of the square.)

2.5 E; $\{P_{12}, P_{13}, P_{23}\}$; $\{\binom{123}{231}, \binom{123}{312}\}$.

2.6

	E	R	σ	$I = R\sigma$
E	E	R	σ	I
R	R	E	I	σ
σ	σ	I	E	R
I	I	σ	R	E

The group is called C_{2h} and may be written as direct products $\{E, R\} \times \{E, \sigma\}$ or $\{E, I\} \times \{E, R\}$ or $\{E, I\} \times \{E, \sigma\}$.

2.7 There are six classes E; C_2; $\{C_3, C_3^2\}$; $\{C_6, C_6^5\}$; $\{C_2', C_2'', C_2'''\}$; $\{C_2'''', C_2''''', C_2''''''\}$. (Here $C_2 = C_6^3$ and the primed axes lie in the *xy*-plane.)

Chapter 3

3.2 $(\frac{3}{2}, \frac{3}{2}, 0) + (-\frac{1}{2}, \frac{1}{2}, 3)$.

3.3 $f_2(x) = (45/8)^{\frac{1}{2}}(x^2 - \frac{1}{3})$, $f_3(x) = (175/8)^{\frac{1}{2}}(x^3 - \frac{3}{5}x)$.

3.4 $T_{01} = 0$, $T_{11} = \frac{3}{5}$, $T_{02} = (4/45)^{\frac{1}{2}}$.

3.5 Periodic, $f(-1) = \exp(i\alpha)\, f(1)$.

3.6. $\boldsymbol{e}_1' = R\boldsymbol{e}_1 = \boldsymbol{e}_2$, $\boldsymbol{e}_2' = -\boldsymbol{e}_1$, $\boldsymbol{e}_3' = \boldsymbol{e}_3$. $T(R)x = \bar{x} = y$, $T(R)y = -x$, $T(R)x^2 = y^2$, $T(R)xy = -yx$. For the 45° rotation, $\boldsymbol{e}_1' = (\boldsymbol{e}_1 + \boldsymbol{e}_2)/2^{\frac{1}{2}}$, $\boldsymbol{e}_2' = (-\boldsymbol{e}_1 + \boldsymbol{e}_2)/2^{\frac{1}{2}}$, $\boldsymbol{e}_3' = \boldsymbol{e}_3$, $\bar{x} = (x + y)/2^{\frac{1}{2}}$, $\bar{y} = (-x + y)/2^{\frac{1}{2}}$.

3.7 $f_1 = (5/4)^{\frac{1}{2}}x^2$, $f_2 = (405/224)^{\frac{1}{2}}(y^2 - \frac{5}{9}x^2)$, $f_3 = \frac{3}{2}xy$. $T_{11} = 5/9$, $T_{22} = -5/9$, $T_{33} = -1$, $T_{12} = T_{21} = (56)^{\frac{1}{2}}/9$, $T_{13} = T_{31} = T_{23} = T_{32} = 0$. Eigenvalues are 1 and -1 (twice) with eigenvectors $(x^2 + y^2)$, $(x^2 - y^2)$ and xy.

Chapter 4

4.2

$$T(C_4) = \begin{pmatrix} 0 & -1 & 0 \\ 1 & 0 & 0 \\ 0 & 0 & 1 \end{pmatrix}, \quad T(C_2) = \begin{pmatrix} 1 & 0 & 0 \\ 0 & -1 & 0 \\ 0 & 0 & -1 \end{pmatrix},$$

$$T(C_2') = \begin{pmatrix} -1 & 0 & 0 \\ 0 & 1 & 0 \\ 0 & 0 & -1 \end{pmatrix}, \quad T(C_2'') = \begin{pmatrix} 0 & -1 & 0 \\ -1 & 0 & 0 \\ 0 & 0 & -1 \end{pmatrix},$$

$$T(C_2''') = \begin{pmatrix} 0 & 1 & 0 \\ 1 & 0 & 0 \\ 0 & 0 & -1 \end{pmatrix}, \quad T(C_4^2) = \begin{pmatrix} -1 & 0 & 0 \\ 0 & -1 & 0 \\ 0 & 0 & 1 \end{pmatrix},$$

$$T(C_4^3) = \begin{pmatrix} 0 & 1 & 0 \\ -1 & 0 & 0 \\ 0 & 0 & 1 \end{pmatrix}, \quad T(E) = \begin{pmatrix} 1 & 0 & 0 \\ 0 & 1 & 0 \\ 0 & 0 & 1 \end{pmatrix}.$$

4.4 yz and xz.
4.5 $\chi = (3, 1, -1, -1, -1)$. Reduces to $A_2 \dot{+} E$.
4.6 See appendix 1.
4.9 $\chi = (6, 0, 2)$.
4.10 See appendix 1.
4.11 $x^2 + y^2$, z, $x^2 - y^2$, xy, (x and y) or (xz and yz).
4.12 $\chi = (4, 0, 4, 0, 0) = A_1 \dot{+} A_2 \dot{+} B_1 \dot{+} B_2$.
4.13 $(x_1 x_2 + y_1 y_2)$, $(x_1 y_2 - y_1 x_2)$, $(x_1 x_2 - y_1 y_2)$, $(x_1 y_2 + y_1 x_2)$.
4.14 $A_1 \to A$, $A_2 \to A$, $B_1 \to B$, $B_2 \to B$, $E \to E$.
4.15 y^3.

Chapter 5

5.2 Some two-fold degeneracies. Allowed transitions are $A_1 \to E$ or A_2, $A_2 \to A_1$ or E, $B_1 \to B_2$ or E, $B_2 \to B_1$ or E, $E \to E$, A_1, A_2, B_1, B_2.
5.5 x and y form an E-doublet and z an A_2-singlet.

Chapter 6

6.1 The modes A^-, B_1^+, B_2^- and B_3^- appear once, while A^+ appears twice. No degeneracies. See figure S.1.

Figure S.1

6.2 The modes A_1 and A_4 appear once and E_2 appears twice. Each of the E_2 modes is doubly degenerate.

6.3 Of the fundamental modes, B_2^- and B_3^- are excited by infrared and A^+ and B_1^+ by Raman transitions.

Chapter 7

7.1 The functions $(x+iy)^2$, $(x-iy)^2$, (x^2+y^2), z^2, $(x+iy)z$, $(x-iy)z$ transform according to $\mathrm{T}^{(m)}$ with $m = 2, -2, 0, 0, 1, -1$, respectively.

7.2 No identity.

7.6 $\mathrm{D}^{(2)} = 2\mathrm{T}^{(3)} \dotplus \mathrm{T}^{(1)}$.

7.7 Both coefficients have the value $(2)^{-\frac{1}{2}}$.

7.8 $C(121{,}000) = -(2/5)^{\frac{1}{2}}, C(121{,}1-10) = (3/10)^{\frac{1}{2}}$,
$C(121, -1, 1, 0) = (3/10)^{\frac{1}{2}}$.

7.14 Invariant $\sum_m (-1)^m Y_m^{(2)} Y_{-m}^{(2)}$.

Chapter 8

8.1 Multipolarities are 2^k, where $k = 2, 3, 4, 5$ or 6.

8.2 Minimum $j = 1$.

8.3 $C(2\tfrac{1}{2}\tfrac{3}{2}, 2-\tfrac{1}{2}\tfrac{3}{2}) = (4/5)^{\frac{1}{2}}$, $C(2\tfrac{1}{2}\tfrac{3}{2}, 1\tfrac{1}{2}\tfrac{3}{2}) = -(1/5)^{\frac{1}{2}}$.

Chapter 9

9.2 (a) D_3, (b) D_4.

9.3 (a) $A_1 \to A_1$, $T_1 \to A_2 \dotplus E$, $T_2 \to A_1 \dotplus E$, $E \to E$,
(b) $A_1 \to A_1$, $A_2 \to B_1$, $E \to A_1 \dotplus B_1$, $T_1 \to A_2 \dotplus E$, $T_2 \to B_2 \dotplus E$.

9.6 A_1; $x^2+y^2+z^2$. E: x^2-y^2 and $3z^2-r^2$. T_2; xy, yz and zx.

9.7 $\mathrm{D}^{(3)} = A_2 \dotplus T_1 \dotplus T_2$. The f-state splits into two triplets and a singlet.

9.8 (a) Yes. (b) No.

9.9 $A_1 \to B_1, B_2, B_3$. $B_1 \to A_1, B_2, B_3$. $B_2 \to A_1, B_1, B_3$. $B_3 \to A_1, B_1, B_2$. These are the allowed transitions for electric dipole if the parity also changes and for magnetic dipole if it does not.

9.11 (a) Group is T_d. Modes are A_1, E(doublet) and T_2(triplet).
(b) Group is T_d. One singlet A_1, one doublet E and two triplets T_2.
(c) Group is D_{2h}. All twelve modes are singlets: three A^+, one A^-, two B_2^+, one B_3^+, two B_1^-, one B_2^- and two B_3^-.

9.12 Splits into a triplet T_2 and a doublet E.

9.13 $j = 5/2$ split into a quartet U' and a doublet E' with $j = 3/2$ remaining as a quartet U'.

9.14 (a) Allowed transitions are $A_1 \to A_2, E$. $A_2 \to A_1, E$. $E \to A_1, A_2, E$.
(b) $A_1 \to A_2, E$. $A_2 \to A_1, E$. $B_1 \to B_2, E$. $B_2 \to B_1, E$. $E \to A_1, A_2, B_1, B_2, E$.

Chapter 10

10.1 States in ^{18}O and ^{18}Ne with $T \geq 1$ have analogues in ^{18}F, which in addition has states with $T = 0$.

10.4 $T = 0$.

10.5 $\frac{1}{2}$.

Chapter 11

11.2 $a_2 = -2$ (real part of X_{12}), $a_1 = -2$ (imaginary part of X_{12}), $a_3 = i(X_{11} - X_{22})$.

11.5 See figure S.2. $Y = \pm 2$ with $T = 1$, $Y = \pm 1$ with $T = \frac{3}{2}$ or $\frac{1}{2}$ and $Y = 0$ with $T = 2$, 1 or 0.

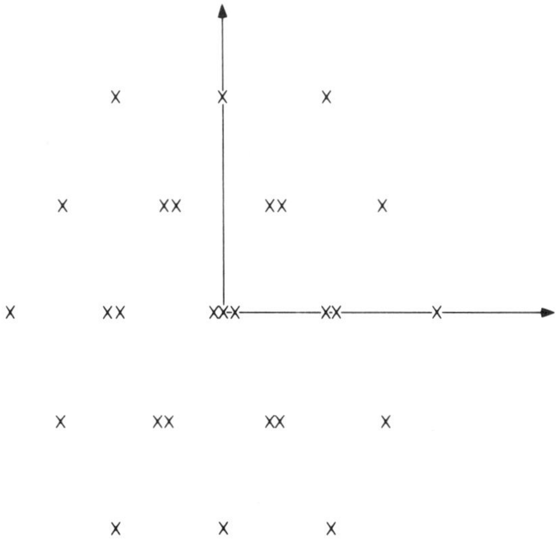

Figure S.2

11.9 Put $Q = M_T + \frac{1}{2}Y$ in the solution of problem 11.5

11.10 $(3/8)^{\frac{1}{2}}|(30)1\,\tfrac{3}{2}\tfrac{3}{2}, (11)010\rangle - (2/8)^{\frac{1}{2}}|(30)1\,\tfrac{3}{2}\tfrac{1}{2}, (11)011\rangle -$
$(1/8)^{\frac{1}{2}}|(30)1\,\tfrac{3}{2}\tfrac{3}{2}, (11)000\rangle + (2/8)^{\frac{1}{2}}|(30)011, (11)1\,\tfrac{1}{2}\tfrac{1}{2}\rangle$
(There is freedom in the choice of phases.)

Chapter 12

12.1 For example,

$$2\mathrm{t}_0 = \begin{pmatrix} 1 & 0 & 0 & 0 \\ 0 & -1 & 0 & 0 \\ 0 & 0 & 1 & 0 \\ 0 & 0 & 0 & -1 \end{pmatrix}, \quad 2\mathrm{y}_{00} = \begin{pmatrix} 1 & 0 & 0 & 0 \\ 0 & -1 & 0 & 0 \\ 0 & 0 & -1 & 0 \\ 0 & 0 & 0 & 1 \end{pmatrix},$$

$$2(\tfrac{1}{2}+\mathrm{s}_z)\mathrm{t}_x = \begin{pmatrix} 0 & 1 & 0 & 0 \\ 1 & 0 & 0 & 0 \\ 0 & 0 & 0 & 0 \\ 0 & 0 & 0 & 0 \end{pmatrix}, \quad 2(\tfrac{1}{2}+\mathrm{s}_z)\mathrm{t}_y = \begin{pmatrix} 0 & -i & 0 & 0 \\ i & 0 & 0 & 0 \\ 0 & 0 & 0 & 0 \\ 0 & 0 & 0 & 0 \end{pmatrix}$$

12.7 $|\Sigma^+\rangle = (2/3)^{\frac{1}{2}}|u\uparrow\ u\uparrow\ s\downarrow\ \rangle - (1/3)^{\frac{1}{2}}|u\uparrow\ u\downarrow\ s\uparrow\rangle$
$|\Lambda\rangle = (1/2)^{\frac{1}{2}}|u\uparrow\ s\uparrow\ d\downarrow\ \rangle - (1/2)^{\frac{1}{2}}|d\uparrow\ s\uparrow\ u\downarrow\ \rangle$

Index to Volumes 1 and 2*

* Pages 1–280 refer to volume 1 and pages 281–557 refer to volume 2